Studies in Computational Intelligence

Volume 1207

The series "Studies in Computational Intelligence" (SCI) publishes new developments and advances in the various areas of computational intelligence—quickly and with a high quality. The intent is to cover the theory, applications, and design methods of computational intelligence, as embedded in the fields of engineering, computer science, physics and life sciences, as well as the methodologies behind them. The series contains monographs, lecture notes and edited volumes in computational intelligence spanning the areas of neural networks, connectionist systems, genetic algorithms, evolutionary computation, artificial intelligence, cellular automata, self-organizing systems, soft computing, fuzzy systems, and hybrid intelligent systems. Of particular value to both the contributors and the readership are the short publication timeframe and the world-wide distribution, which enable both wide and rapid dissemination of research output.

Indexed by SCOPUS, DBLP, WTI AG (Switzerland), zbMATH, SCImago.

All books published in the series are submitted for consideration in Web of Science.

Jitendra Bhatia · Sudeep Tanwar ·
Joel J. P. C. Rodrigues · Malaram Kumhar
Editors

Deep Learning Based Solutions for Vehicular Adhoc Networks

Springer

Editors
Jitendra Bhatia
Computer Science and Engineering
Institute of Technology, Nirma University
Ahmedabad, Gujarat, India

Sudeep Tanwar
Computer Science and Engineering
Institute of Technology, Nirma University
Ahmedabad, Gujarat, India

Joel J. P. C. Rodrigues
Federal University of Piauí (UFPI)
Brazil and Instituto de Telecomunicações
Piauí, Portugal

Malaram Kumhar
Computer Science and Engineering
Institute of Technology, Nirma University
Ahmedabad, Gujarat, India

ISSN 1860-949X ISSN 1860-9503 (electronic)
Studies in Computational Intelligence
ISBN 978-981-96-5189-4 ISBN 978-981-96-5190-0 (eBook)
https://doi.org/10.1007/978-981-96-5190-0

This Springer imprint is published by the registered company Springer Nature Singapore Pte Ltd.
The registered company address is: 152 Beach Road, #21-01/04 Gateway East, Singapore 189721, Singapore

Preface

Vehicular ad hoc networks (VANETs) are networks formed by vehicles communicating with each other and roadside infrastructure forming a complex network. These vehicles are equipped with various types of sensors that generate data in huge quantities that need to be processed for efficient communication between nodes. In recent years, VANETs have emerged as promising technology for intelligent transportation systems (ITS) to improve road safety, traffic efficiency, and overall driving experience. The book *Deep Learning Based Solutions for Vehicular Adhoc Networks* begins with discussions on how the transportation system has been converted into Intelligent Transportation System (ITS). The use of VANETs is increasing in the development of ITS to enhance road safety, traffic efficiency, and driver comfort. However, the dynamic nature of vehicular environments and the high mobility of vehicles pose significant challenges to designing and implementing VANETs and ensuring reliable and efficient communication.

Deep learning, a subset of machine learning, has the potential to revolutionize VANETs to enable various applications such as traffic management, collision avoidance, and infotainment. DL has demonstrated great potential in addressing various challenges involved in VANETs by leveraging its ability to learn from vast data and make accurate predictions. In this book, we review the state-of-the-art DL-based approaches for various applications in VANETs, including routing, congestion control, autonomous driving, and security. In addition, this book provides a comprehensive analysis of these approaches' advantages and limitations and discusses their future research directions. Our study in this book shows that DL-based techniques can significantly improve the performance and reliability of VANETs. Still, in-depth research is required to address the challenges of deploying these methods in real-world scenarios. Finally, the book discussed the potential of DL-based VANETs in supporting other emerging technologies, such as autonomous driving and smart cities. It explored the simulation/emulation tools for practical exposure to the vehicular ad-hoc network.

The book is organized into five sections. The first section is focused on the background and preliminaries in Vehicular Ad Hoc Networks, which includes two chapters. The second section illustrates the Applicability of Deep Learning in Vehicular

Ad Hoc Networks with well-structured four chapters. The third section discusses about the Traffic Modeling in Vehicular Ad Hoc Networks which includes four chapters. The fourth section covers the Implementation and Deployment of Deep Learning in Vehicular Ad Hoc Networks with three chapters. Finally, the last section focused on the Future Research Directions in Deep Learning for Vehicular Ad Hoc Networks with three chapters.

Introduction to Vehicular Ad Hoc Networks

First chapter "Emerging Trends in Vehicular Communication and Routing for Intelligent Transportation Systems" explores the complexities of this evolving landscape and introduces innovative techniques to improve dynamic traffic routing, cooperative accident prevention, and adaptive traffic signal control. This chapter studies the integration of advanced technologies such as Dedicated Short-Range Communications (DSRC), 5G, VANETs, Light Fidelity (Li-Fi) and Blockchain in a simulated city infrastructure. Dynamic Traffic Routing (DTR) is a proposed method to regularly optimize real-time traffic shapes according to the changing circumstances, whereas Cooperative Accident Avoidance (CAA) enables vehicle communication and coordination to minimize collision risks. Also, ATSC adapts signals in real-time to traffic changes and optimally reduces stoppages on intersections. The study concludes that these innovations could make urban travel safer, more efficient, and more adaptable to evolving traffic demands.

Second chapter "Architecture and Protocols for Data Transmission in VANET" focuses on limitations through deep learning-enhanced solutions, using neural networks, convolutional models, and reinforcement learning. These approaches maximize intrusion detection, increase QoS resource allocation and enhance data dissemination techniques like broadcasting and geocasting. Even in dynamic circumstances, the suggested approaches guarantee dependable and secure communication by providing predictive and adaptable capabilities. The outcome demonstrates that deep learning significantly enhances VANET performance, leading to better traffic management, improved safety, and efficient coordination of autonomous vehicles.

Applicability of Deep Learning in Vehicular Ad Hoc Networks

Third chapter "Integration of Deep Learning Algorithms into VANET" explores the key components and challenges of VANETs, which includes but not limited to security concerns, protocols for communication, use of deep learning and machine learning for analyzing traffic flow, etc. This chapter also addresses the deployment requirements of the VANET architectures with special attention being given to the energy efficiency, resource utilization and randomness of the VANET environment. Since vehicles swiftly change their locations, which results in interactions with different topologies and network dynamics. This chapter proposes a novel framework for VANETs that addresses intricacies of the VANET environment while leveraging the potentials of communication technology and deep learning to address the underlying challenges.

Fourth chapter, "Impact of Deep Learning on VANET Security: An Analysis", presents a detailed analysis of deep learning frameworks such as DeepVCM and DeepADV for intrusion detection and anomaly detection, respectively, demonstrating their superiority over conventional methods. Additionally, few DL-based solutions for secure communication and trust management in VANETs including a framework for Sybil attack detection and trust computation using deep neural networks are discussed in this chapter. The comparative study of mitigation strategies highlights the effectiveness of DL in automating feature extraction and improving detection accuracy. This chapter presents multiple methods of notable breakthroughs in VANET security by examining these DL procedures, ensuring a more reliable and safe vehicular communication environment.

Fifth chapter, "An Intrusion System for Vehicular Ad Hoc Networks Using Deep Learning", investigates the application of deep learning-based recurrent neural networks (RNNs) for enhanced auto encoder networks and intrusion detection in automobiles. To effectively comprehend the vehicle's boundary behaviour and identify invasive behaviour, two algorithms are merged. The correctness and efficacy of the proposed model were verified through an analysis of actual vehicle data. The experiment's results show that integrating the two technologies can successfully and dependably identify aberrant border behaviour.

Sixth chapter, "DLAR: Deep Learning-Based Adaptive Routing Strategies for Vehicular Ad-Hoc Networks (VANETs)", presents an intelligent deep route learning technique for estimating highway traffic density. The proposed work does a comparison study using different performance measurement factors, such as coverage ratio, packet delivery ratio (PDR), and control overhead. The proposed methods compared with existing methods and found that our suggested route increases the aggregate packet delivery ratio (PDR) by 23% over the current method.

Intelligent Traffic Modeling in Vehicular Ad Hoc Networks

Seventh chapter, "Deep Learning for Traffic Prediction in VANET", discusses the role of deep learning technologies in traffic prediction in VANETs. It gives importance to different methodologies, architectures, and practical implementations. This chapter refers to strategies on data preparation-data cleaning, normalization, and augmentation. Data preparation with its key demands-normalization and augmentation makes the input data correct for the deep learning model application; thus, correctness also relies on relevance. This chapter presents a comprehensive review of deep learning algorithms for traffic prediction in VANETs which can provide much-needed insight to academics, practitioners, and stakeholders interested in intelligent transportation systems.

Eighth chapter, "Traffic Prediction and Modeling in Vehicular Ad Hoc Networks", provides a study of various modeling techniques in VANETs like static statistical methods, machine learning techniques, deep learning techniques, and hybrid models. The proposed chapter evaluates the strengths and limitations of these methods in predicting various traffic parameters, such as traffic volume, congestion levels, and travel times.

Ninth chapter, "Intelligent Data Dissemination in Vehicular Networks: Leveraging Reinforcement Learning", explores the potential of Reinforcement Learning (RL) to address these challenges by enabling adaptive routing protocols that dynamically adjust to network conditions. It provides a comprehensive overview of the fundamentals of RL and examines how these concepts can be applied to develop RL-based routing strategies in VANETs. Through detailed analysis and discussion, the chapter demonstrates the ability of RL to enhance the scalability, QoS, and overall performance of data dissemination in VANETs, paving the way for more robust and efficient vehicular communications in future ITS deployments.

Tenth chapter, "Smart Traffic Forecasting in Urban Junctions: A Deep Learning and VANET Approach", presents a method to predict congestion at intersections using advanced deep learning techniques. Using data from Vehicle Ad-Hoc Networks (VANET), this study focuses on understanding how various factors shape traffic patterns. These include weather, time of day, and special events. A key feature of the approach is its integration with VANET systems. This enables real-time solutions like adjusting traffic lights, offering alternative routes, and sending alerts to drivers. The work concludes by identifying areas where research could expand to refine smart traffic management further.

Implementation and Deployment of Deep Learning in Vehicular Ad Hoc Networks

Eleventh chapter, "ANN Optimized Fractal Microstrip Patch Antenna Design for Vehicular Ad-Hoc Networks", explores a novel fractal geometry-based microstrip patch antenna design using Artificial neural networks (ANNs). The existing and traditional deep learning approaches, like Support Vector Machine (SVM) and Genetic Algorithm (GA) in the patch antenna design process, are reviewed in this chapter, and their efficiency, accuracy, and computational effort are highlighted. The ability of ANN to effectively handle the nonlinear relationships in the antenna design process and its performance metrics in comparison with traditional techniques are highlighted in the chapter. The proposed method establishes a foundation for advanced research in utilizing ML & DL for antenna design, paving the way for more efficient and advanced vehicular communication.

Twelfth chapter, "Deep Learning Driven Detection and Classification of Attacks in Vehicular Communication Systems", provides an efficient IDS strategy for binary classification based on deep learning. The best model is determined using the keras tuner library, and the implementation results suggest that proposed strategy achieves the accuracy of 99% in classification of messages in CAN bus network. The results outperform the classification accuracy with respect to the supervised machine learning-based approaches.

Thirteenth chapter, "Case Studies and Real-World Deployment Examples of VANETs (Vehicular Ad-Hoc Networks)", focuses on the current progression in VANET technology and its extensive analysis and study of real-world deployment examples which showcase their transformative impact on modern-day transportation systems. The chapter starts with an introduction to VANET constitutive, followed by a communication of the revolutionary implications for road safety, need of traffic flow management, environmental monitoring with respect to VANET, and user its experience. By presenting detailed case studies, this chapter aims to lighten up new applications of VANET technology and their execution across various arena. These case studies provide new insights into improved road safety through cutting-edge systems for real-time incident sensing, emergency vehicle notification, and any road hazard alerts. This chapter fuses up the trends and future of emergency response applications and advanced traffic flow monitoring and regulation techniques, including state-of-the-art traffic signal control and dynamic routing approaches, in building a more efficient and sustainable transportation infrastructure.

Future Research Directions in Deep Learning for Vehicular Ad Hoc Networks

Fourteenth chapter, "Deep Learning Algorithms and Methods for VANET-Based Congestion Management: Implementation Concerns and Limitations", discusses the main problems and difficulties in implementing Intelligent Transportation Systems (ITS). These include the need to minimize delays in communication between automobiles and Roadside Units (RSUs), assuring a continuous flow of traffic, and improving road safety. It explores the possibilities of using machine learning techniques to tackle these difficulties. In the end, it discusses the future direction and challenges and analyzed a case study that demonstrated a situation related to VANET.

Fifteenth chapter, "Navigating the Future of Transportation with the Support of Large Language Models and AI", discusses how transportation is evolving swiftly and how major issues require innovative solutions. It discusses how LLMs and AI might improve transportation for people, efficiency, and the environment. It emphasizes the importance of enhanced user experiences, self-driving cars, real-time traffic control, and data-driven transportation innovations. It examines the technological concepts that enable this shift. It emphasizes how large language models simplify human-car communication. It also discusses the merits and downsides of this transformation, emphasizing data privacy, security, and infrastructure strengthening.

Sixteenth chapter, "Exploring the Prospectives of Vehicle Ad-Hoc Networks (VANETs): Applications, Challenges and Future Directions", inspects the assorted applications of VANETs, encompassing the smart transportation systems where it inhibits intelligence in it and along with management of traffic where it responds for any emergency conditions. This chapter delivers a wide-ranging investigation of VANET design and implementation, highlighting the encounters, solicitations, and future prospectives. This chapter sightsees evolving progresses and potential guidelines for the developing VANET technologies. Additionally, innovative communication standards and diverse architectures available in VANETs are deliberated under future explorations.

Ahmedabad, India Dr. Jitendra Bhatia
Ahmedabad, India Prof. Malaram Kumhar
Ahmedabad, India Prof. (Dr.) Sudeep Tanwar
Piauí, Portugal Prof. (Dr.) Joel J. P. C. Rodrigues

Contents

About the Editors

Dr. Jitendra Bhatia received the B.Tech degree from Hemchandra Acharya North Gujarat University, in 2004, the M.Tech degree (Computer Science and Engineering) in 2012 and Ph.D. (Specialization in Vehicular Ad Hoc Networks) in 2019 from Institute of Technology, Nirma University, Ahmedabad Gujarat India. He is currently associated with Computer Science and Engineering Department at Institute of Technology, Nirma University, as Associate Professor. He has over 17 years of teaching experience at undergraduate and postgraduate level. He is Recipient of the "Best Idea2Innovate Award" in 2019 by Brihasti (Claris) foundation. His main research area of interest is Protocol Design for Intelligent Transportation System, Software Defined Networking, Vehicular Networks and Cloud Computing. He has been Session Chair and Keynote Speaker in various international conferences. He has published/presented a number of papers at reputed International and National level Journals and Conferences. He has authored or co-authored more than 40 technical research papers published in leading journals and conferences from the IEEE, Elsevier, Springer, and Wiley. Some of his research findings are published in top-cited journals, such as IEEE IoT, Elsevier Computer Communication, Springer: Peer to Peer Network Applications, and IJCS Wiley.

Dr. Sudeep Tanwar (Senior Member, IEEE) is working as Professor at the Nirma University, India. He is also Visiting Professor with WSG University, Bydgoszczy, Poland, Jan Wyzykowski University, Poland, and the University of Pitesti, Romania. He is Senior Member of IEEE and also Vice Chair of IEEE Computer Society Gujarat Section. He received B.Tech in 2002 from Kurukshetra University, India, M.Tech (Honor's) in 2009 from Guru Gobind Singh Indraprastha University, Delhi, India and Ph.D. in 2016 with specialization in Wireless Sensor Network. He has authored 05 books and edited 25 books, more than 400 technical articles, including top cited journals and conferences, such as IEEE TNSE, IEEE TVT, IEEE TII, IEEE TGCN, IEEE TCSC, IEEE IoTJ, IEEE NETWORKS, IEEE WCM, ICC, IWCMC, GLOBECOM, CITS, and INFOCOM. He initiated the research field of blockchain technology adoption in various verticals in the year 2017. His H-index as per Google Scholar and

Scopus is 73 and 60, respectively. His research interests include blockchain technology, D2D communication, Deep Learning/machine Learning, wireless sensor networks, fog computing, smart grid, and the IoT. He is Member of the Technical Committee on Tactile Internet of IEEE Communication Society. Recently, he has been awarded cash prize of Rs, 50,000 for publishing papers with 5+ Impact factor and publication of books with Springer, IET & CRC under the scheme of "Faculty Awards and Incentives" of Nirma University for the year 2019–2020. He has been awarded the Best Research Paper Awards from IEEE SCIoT 2022, IEEE IWCMC-2021, IEEE ICCCA-2021, IEEE GLOBECOM 2018, IEEE ICC 2019, and Springer ICRIC-2019. He has won Dr. K. W. Wong Annual Best Paper Prize (with 750 USD) for 2021 sponsored by Elsevier (publishers of JISA). He has served many international conferences as Member of the Organizing Committee, such as Publication Chair for FTNCT-2020, ICCIC 2020, and WiMob2019, and General Chair for IC4S 2019, 2020, 2021, 2022, ICCSDF 2020, FTNCT 2021. He is also serving the editorial boards of COMCOM-Elsevier, IJCS-Wiley.

Joel J. P. C. Rodrigues [S'01, M'06, SM'06, F'20] is Professor at the Federal University of Piauí, Brazil; and Senior Researcher at the Instituto de Telecomunicações, Portugal. He has been Professor at the National Institute of Telecommunications, Brazil; University of Beira Interior (UBI), Portugal, and Visiting Professor at the University of Fortaleza (UNIFOR), Brazil. He received the Academic Title of Aggregated Professor in computer science and engineering from UBI, the Habilitation in computer science and engineering from the University of Haute Alsace, France, a Ph.D. degree in computer science and engineering and an M.Sc. degree from the UBI, and a five-year B.Sc. degree (licentiate) in computer science and engineering from the University of Coimbra, Portugal. His main research interests include IoT and sensor networks, e-health technologies, vehicular communications, and mobile and ubiquitous computing. Prof. Rodrigues is Leader of the Next Generation Networks and Applications (NetGNA) research group (CNPq), IEEE Distinguished Lecturer, Member Representative of the IEEE Communications Society on the IEEE Biometrics Council, and President of the scientific council at ParkUrbis—Covilhã Science and Technology Park. He was Director for Conference Development—IEEE ComSoc Board of Governors, Technical Activities Committee Chair of the IEEE ComSoc Latin America Region Board, Past-Chair of the IEEE ComSoc Technical Committee (TC) on eHealth and the TC on Communications Software, Steering Committee Member of the IEEE Life Sciences Technical Community and Publications Co-Chair. He is Editor-in-Chief of the International Journal of E-Health and Medical Communications and Editorial Board Member of several high-reputed journals (mainly, from IEEE). He has been General Chair and TPC Chair of many international conferences, including IEEE ICC, IEEE GLOBECOM, IEEE HEALTHCOM, and IEEE LatinCom. He had been awarded several Outstanding Leadership and Outstanding Service Awards by IEEE Communications Society and several best papers awards. Prof. Rodrigues is Member of the Internet Society, Senior Member ACM, and Fellow of IEEE.

Prof. Malaram Kumhar is working as Assistant Professor in Computer Science and Engineering Department. He has more than 18 years of teaching experience. He received his BE degree in Computer Engineering from Rajasthan University, Jaipur, and M.Tech in Computer Science and Engineering from Nirma University, Ahmedabad. He received Gold Medal for securing first position in M.Tech in the year 2014. His research interests include Internet of Medical Things (IoMT), Software Defined Networking and Fog computing. He has published/presented a number of papers at reputed International and National level Journals and Conferences. He has authored or co-authored more than 20 technical research papers published in leading journals and conferences from the IEEE, Elsevier, Springer, and Wiley.

Emerging Trends in Vehicular Communication and Routing for Intelligent Transportation Systems

S. Shiek Aalam, M. Sivakumar, S. Anjali Devi, Putti Venkata Siva Teja, and Nagendra Singh

Abstract Emerging technology in urban transportation is revolutionizing how automobiles interact within smart mobility systems. This chapter explores the complexities of this evolving landscape and introduces innovative techniques to improve dynamic traffic routing, cooperative accident prevention, and adaptive traffic signal control. This chapter studies the integration of advanced technologies such as Dedicated Short-Range Communications (DSRC), 5G, Vehicular Ad Hoc Networks (VANETs), Light Fidelity (Li-Fi) and Blockchain in a simulated city infrastructure. Dynamic Traffic Routing (DTR) is a proposed method to regularly optimize real-time traffic shapes according to the changing circumstances, whereas Cooperative Accident Avoidance (CAA) enables vehicle communication and coordination to minimize collision risks. Also, ATSC adapts signals in real-time to traffic changes and optimally reduces stoppages on intersections. The study concludes that these innovations could make urban travel safer, more efficient, and more adaptable to evolving traffic demands. The proposed method shows outperformance with 92% increase in Dynamic Traffic and also improved with other parameters evaluation.

Keywords Adaptive traffic signal control · Blockchain · Cooperative collision avoidance · Dynamic traffic routing · Future transportation · Intelligent communication · Li-fi · Smart cities · Smart mobility

---

S. Shiek Aalam · M. Sivakumar
Department of Computer Science and Engineering, Saveetha School of Engineering, Saveetha Institute of Medical and Technical Sciences, Saveetha University, Chennai, Tamil Nadu, India

S. Anjali Devi
Department of Computer Science and Engineering, Koneru Lakshmaiah Education Foundation, Vaddeswaram, Guntur, Andhra Pradesh, India

P. V. S. Teja
Department of Mechanical Engineering, Dhanekula Institute of Engineering and Technology, Vijayawada, Andhra Pradesh, India

N. Singh (✉)
Department of Electrical and Electronics Engineering, Trinity College of Engineering and Technology, Karimnagar, Telangana, India
e-mail: nsingh007@rediffmail.com

J. Bhatia et al. (eds.), *Deep Learning Based Solutions for Vehicular Adhoc Networks*, Studies in Computational Intelligence 1207,
https://doi.org/10.1007/978-981-96-5190-0_1

1 Introduction

Smart transportation systems' vehicle-to-vehicle interaction has transformed road use. Technology is set to advance, and cutting-edge transportation and information networks might revolutionize how we think about mobility. A detailed look into automobile communication's future reveals how smart transportation systems might make life better, harder, and more intriguing. In the twenty-first century, normal transportation has had to be rethought due to more people coming to cities and living in them. Adding cutting-edge communication technology appears like a bright light in a world with worsening traffic, environmental concerns, and road safety issues [1]. Communications between vehicles (V2V) and between vehicles and infrastructure usher in a new era in which automobiles are smart, aware entities that improve transportation efficiency.

The IoT, which connects vehicles, infrastructure, and people, is what powers this transformative journey. Cars and their surroundings may be able to collaborate on a flexible transportation system by sharing information electronically. This research examines smart transportation vehicle communication. It examines the technology, its use, and its consequences on urban planning and social ties [2]. We must comprehend automotive communication technologies before beginning our inquiry. Connected gadgets make smart transportation systems operate. This base includes 5G and DSRC. Cars can perceive and respond by combining sensors, motors, and communication [3].

V2V communication lets autos share speed, position, and threats. Real-time-communicating municipal vehicles can anchor networks. But the vehicle-to-infrastructure link allows automobiles to communicate with smart junctions, stop signs, and traffic signals. All of them improve traffic, remove congestion, and speed things up. Future automobile communication depends on road safety [4].

Connected automobiles reduce collision risk like a moving safety net. Nearby cars are notified immediately of sudden stops or obstacles. It allows everyone to behave securely. With time, this makes transportation more dependable and automobiles safer. Vehicle-to-vehicle interaction benefits more than safety. It aids in traffic control and environmental protection. Smart transportation solutions enhance traffic flow and save time and energy by using real-time data. Smart traffic signals may adjust their cycles depending on real-time data from linked automobiles, reducing delays and pollution [5]. This has made city transit more sustainable and environmentally friendly. It's difficult to achieve future automotive communication goals. Cars sharing data on a massive network raise privacy and safety concerns. Strong security must protect sensitive data from unauthorized access.

The interoperability of communication standards is another major issue that requires industry cooperation to build universal models. We must navigate a complicated network of new and old technology to get there. It's apparent that vehicle communication will go beyond transportation. Culture and how individuals, groups, and the city interact evolve. People with mobility issues will benefit most from linked and self-driving automobiles. Smart transportation technologies might make transportation more equitable and open [6]. Future automobile interaction in smart transportation networks will offer new ideas, issues, and opportunities. This research examines how intricate our environment is by looking at technology that advances us and difficulties that challenge our inventiveness [7].

In transportation, there are indications that distinct change is imminent. Shifts are being made in the ways people stay, learn, build and even engage in activities in communities through the use of smart systems and vehicle communication. With regard to this, a brief look at invariant neural networks (INNs) is made in this post to depict pictures from a number of viewpoints efficiently and effectively. The purpose is to learn what a person needs to know in order to use INNs to create photographs that are the best they could be. Another potential goal is to achieve an understanding of citizens, INN means. To do this, we shall explore the mathematical and conceptual particulars of neural network invariance. In fact, the goal of the present investigation is to understand how useful it is to introduce invariance properties into the organizational structure of neural networks with a view of using this capability to produce imagery that is resistant to noise. By using this theory and research [8], authors want to have a certain insight into what INNs are and how they work.

As synthesis of picture demands the speed, there is a problem of lessening data and, therefore, it is obligatory to investigate INNs. We are looking forward to identify how INNs learn in lower dimensions than the actual features and so reduce the need to gather and analyze huge datasets. One of the goals of the present study is to enhance the usability of picture synthesis especially for applications that take place in near real time with prolonged resource content. It is at the top of the wish list that there should be the ability of robust INNs to produce great images even if the data that are being used to feed them have some serious drawbacks. In reality they will be challenged with all kinds of raw data during the real testing process—data that can be noisy—have irrelevant values or other problems [9]. If INNs are to be used in areas where the quality of data is likely to be low then one needs to know the drawbacks of INNs and then find ways of getting better data quality. However, much research is still needed to understand how INNs perform between datasets, and regarding data unseen by the model. We would like to know how versatile the models of different INNs are by testing how well they perform when faced with scenarios different from that of their learning set. From how far they are able to generalize, one can get a feeling of the revolutionary possibilities that they offer for picture synthesis in a range of applications.

Further, the research examines possible uses of INNs in medical science, computer image, and creation of media [10]. Studying these applications, we expect to identify cases where considerations of INNs could lead to their increasing efficiency and reliability. We would like See IoT networks as valuable to all types of ventures. This purpose is not in conflict with it. In this study, we will explore the current trends for INN design as well as the various architectural strategies used. Once we have accomplished this objective, we will be able to assess the performance of INN and isolate the difficulties. We aim to contribute to the development of this field and achieve this by following up to date research and providing accurate estimations for its growth. Finally, based on the research, recommendations for implementing INN are made [11].

The documentation of these best practices, issues, and eventual address is carried out with the purely noble goal of nudging practitioners, engineers, and academics across disciplines to embrace INNs where applicable. Finally, we would like to invite a large number of people to utilize INNs for the generation of pictures to be made in a short time and with a high level of reliability for the purpose of practical application with reference to the gap between hypothesis and experiment.

1.1 Research Gap

The research gap earmarked herein is the static nature of current approaches towards the management of live flow of traffic, accidents and signals. Still, modern efforts do not always meet expectations in terms of efficiency and flexibility of the methods used. This work seeks to overcome these shortcomings through an inclusive and holistic approach incorporating the DSRC, 5G and blockchain technologies.

1.2 Motivation

The research focuses on the increasing problem of congestion in the urban areas as well as increased accident frequency. Most modern systems fail to factor in the option of real time adjustments hence some of the major drawbacks. To address these problems, this work employs the following technologies: Dedicated Short-Range Communications, 5G, Vehicular Ad Hoc Networks, Light Fidelity, blockchain solutions. The purpose is to come up with an effective and efficient traffic control and prevention of accidents by improving on the availability and the control of traffic networks in urban environments.

1.3 Research Contribution

This work seeks to propose techniques that can be used to enrich smart transport solutions inclusive of Dynamic Traffic Routing, Cooperative Accident Avoidance, and Adaptive Traffic Signal Control. By integrating sophisticated solutions like Dedicated Short-Range Communications, 5G, Vehicular Ad Hoc Networks, Li-Fi and even blockchain solutions, the proposed methods are proven to be capable of real-time traffic flow optimization, enhanced vehicle-interaction to avoid or at least mitigate accidents, and adaptive traffic signal control. These solutions can meet current problems in urban mobility, which has a more efficient, flexible, and safer transportation system. Application of these technologies is expected to redefine smart travel and traffic systems within cities.

2 Related Works

Given the goal of making sense of automotive communication for intelligent transportation networks, it is worthwhile to first gain an appreciation of the connected that formed this paradigm shift. To advance ITS, the literature presents a variety of communication protocols and applications as working systems. Communication by vehicles and processes and standards related to it are emerging. Early programmes enabled transmission from one vehicle to the other by using Dedicated Short Range Transmission [12]. Among them, the technology built upon the IEEE 802.11p and enabling automobiles to communicate with other automobiles was called DSRC, which received the most research attention. It is for this reason that Vehicle-to-Vehicle Communication systems have shown that autos can make a vehicle communicate life-saving accident alarms.

The advances in communications standards have inclined towards cellular technology especially 5G network technology. The possibility of the 5G networks reinforcing and expanding car broadcasting is still under an investigation. For realizing smart transportation and experiencing its advantages in low latency and fast data transmission, real time application is mandatory [13]. Thus, the need to introduce new topics into the discussion is apparent, for example to employ different communication patterns in intelligent transportation systems. Actual applications as well as field experiments have also helped in the assessment of the principles of vehicular communication. The Connected Vehicle Pilot Deployment Programme of the United States Department of Transportation has proved that widespread vehicle to vehicle and vehicle to infrastructure connectivity has its advantages and disadvantages. these programmes examine the reliability, dependability, impact on society and efficiency of incorporating vehicle communication into transport systems.

Such observations are increasing due to the evident interest given to the effects of vehicle technology on at once privacy and safety. Linked automobiles pose high risks to privacy and security for data. Among the methods investigated by the researchers in an attempt to protect automobile communication systems, there were cryptography, secure key management and intrusion detection. Those challenges need to be solved to regain people's confidence and develop ITS. Due to multiple field studies, there is no gap distinguishing artificial intelligence from car digital messaging. Machine learning was implemented in order to enhance the abilities of intelligent transportation systems to predict and find. Presumably, the use of AI with the help of predictive analytics might improve response time to emergencies or safety and policy on the roads [14].

Worse, people want much improved self-driving mobility solutions as AI and IT are integrated increasingly. In the future, the communication between vehicles in intelligent transportation system will involve several integrated functions. These tasks include the establishment of communication standards within the system, the application of such protocols, threat analysis procedures and coordination of specialists in the various disciplines. The best way to understand the problems and prospects of smart transportation is to apply theoretical knowledge to actual practice. These important pieces are orientated for the future, in which vehicle-to-vehicle and vehicle-to-infrastructure communication will actualize radical shifts in mobility within cities.

Table 1 compares six vehicle communication methods for speed, latency, reliability, scalability, security, and privacy [15]. QuantumSync employs quantum communication, and SwiftConnect uses 5G to demonstrate future smart transportation network approaches.

3 Approach for Traffic Management

Intelligent Cooperative Traffic Management (ICTM) is a novel approach to smart transportation systems with a vehicle-to-vehicle interface. It aims to enhance traffic flow, safety, and economics by connecting automobiles and infrastructure. DTR means "dynamic traffic routing."

The suggested intelligent cooperative traffic management system relies heavily on dynamic traffic routing. DTR allows automobiles to communicate in real time to discover the optimum journey routes based on a cost function. Equation 1 shows the cost function. Travel time and traffic are considered. The pieces enable a larger picture. Depending on their preferences, users may wish to reduce journey time or avoid popular routes. DTR analyzes a continuous stream of data from connected automobiles to find the optimal methods [16]. By responding to changing conditions, this program makes the transportation system more adaptable and responsive. DTR lets automobiles exchange real-time location, speed, and route preferences. Traffic will flow smoothly, and trips will be shorter.

$$\text{Budget} = \text{Travel Time} + \text{Cost} \tag{1}$$

DTR continually moves automobiles using a cost-based algorithm, determining the optimum routes depending on traffic [17]. Equation shows that the cost function considers trip time and traffic. Choose weightings to make sophisticated adjustments in the parameters.

Table 1 Comparative performance analysis of vehicular communication methods in smart transportation systems

Method name	Communication standard	Throughput (Mbps)	Latency (ms)	Reliability (%)	Scalability	Security measures	Privacy protection
QuantumSync [5]	Quantum communication	High	Low	99.9	High	Quantum encryption	Quantum anonymity
SwiftConnect [7]	5G	Very high	Low	98.5	High	Blockchain	Differential privacy
IntelliMesh [8]	Mesh networking	Moderate	Low-moderate	97.2	Moderate	Dynamic key exchange	Anonymous routing
OptiWave [7]	Li-Fi	High	Low	98.0	Moderate	Li-Fi security	Private light channels
NebulaLink [10]	Satellite communication	Moderate	Moderate	96.8	Very high	Satellite encryption	Secure orbital routing
QuantumGrid [9]	Grid-based communication	Very high	Low	99.5	High	Quantum blockchain	Quantum location cloaking
SmartWave [10]	RFID communication	Moderate	Low-moderate	95.7	High	RFID authentication	Secure tag broadcasting

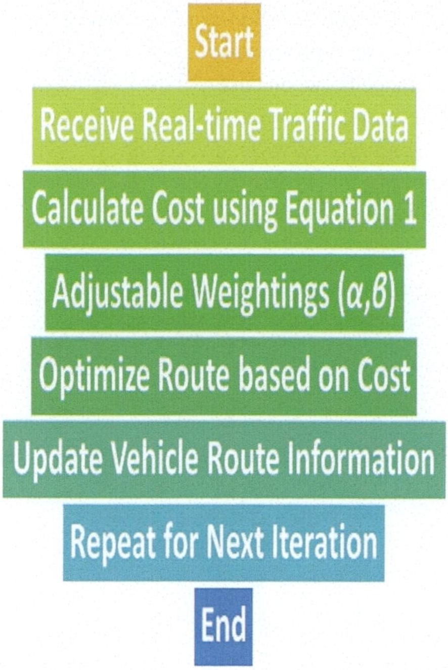

Fig. 1 Optimizing routes in real-time

This allows the DTR to swiftly adapt to changing traffic circumstances by gathering data in real time, using a dynamic equation to calculate route costs, and choosing the optimum routes depending on user preferences as shown in Fig. 1.

3.1 CCA: Preventing Collisions Together

Intelligent Cooperative Traffic Management protects users via cooperative collision avoidance. Equation 2 shows how CCA calculates accident risk from car-to-car data. This data creates the Safety Index. The Safety Index considers both the legal minimum and the actual distance between automobiles. CCA prevents collisions by examining and adjusting this index [18]. The primary idea underlying CCA is that automobiles can cooperate to promote safety. CCA predicts accidents by analyzing spatial relationships between automobiles' location, speed, and path data. To keep self-driving cars safe, CCA delivers alarms and offers course adjustments when the Safety Index approaches a particular threshold. Joint collision avoidance considerably improves transportation safety.

$$\text{Minimum Safe Distance} = \text{Current Vehicle Separation}/\text{Safety Index} \tag{2}$$

CCA use V2V communication to determine a Safety Index, described by Eq. 2. The system promotes collaborative collision avoidance by continually evaluating the distance between cars and establishing a minimum safe distance, hence reducing the likelihood of collisions.

Figure 2 provides the using information about cars' locations and speeds in real time, CCA may estimate the likelihood of collisions and take preventative measures by issuing warnings and making recommendations.

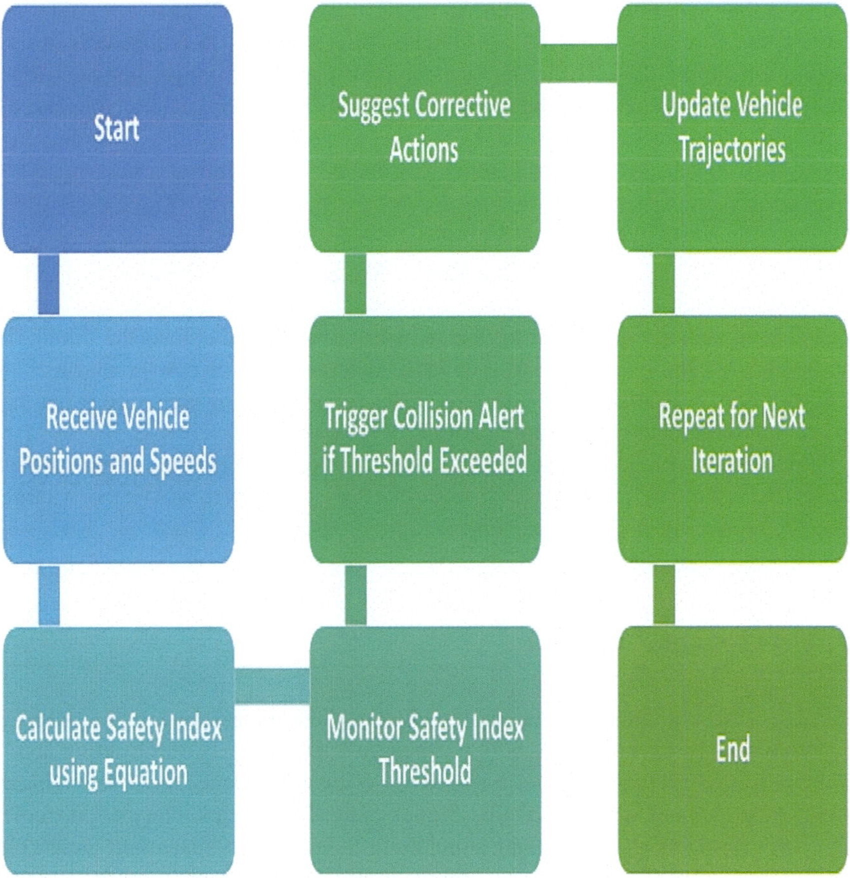

Fig. 2 Proactive safety through vehicle collaboration

3.2 *Adaptive Traffic Signal Control (ATSC)*

In order to optimize traffic signal timings in a dynamic and adaptable fashion, the proposed Intelligent Cooperative Traffic Management system relies heavily on Adaptive Traffic Signal Control. The method, summarized in Equation, accounts for both current and past traffic volumes and trends. By enabling the algorithm to prioritize current circumstances above past behavior, the parameter enables flexibility in selecting when traffic lights will turn green. ATSC solves problems caused by unpredictable occurrences and shifting traffic patterns. ATSC analyzes real-time traffic data and makes real-time adjustments to signal times to reduce congestion and improve traffic flow. The system can anticipate and adjust to predictable traffic patterns because of the inclusion of past patterns [19–22]. This adaptive method to traffic signal management presents ATSC as a dynamic orchestrator, coordinating junctions to enhance overall traffic efficiency and decrease delays for passengers.

$$\text{New Green Time} = \text{Current Traffic Volume minus} \\ \text{One Year of Average Traffic Volume} \quad (3)$$

As traffic volumes and patterns change in real time, ATSC adjusts the timing of the lights accordingly [23–25]. The dynamic determination of new green times is defined by Eq. 3, where permits a degree of discretion in giving weight to either the present circumstances or the traffic patterns of the past.

Figure 3 shows the account past traffic patterns and employing an adaptable parameter, ATSC optimizes traffic signals in real time. As a result, there will be less waiting time and more efficient traffic movement at crossings.

4 Experimental Setup

The experimental setup involves a virtual city environment integrating DSRC, 5G, VANETs, Li-Fi, and blockchain technologies. Dynamic Traffic Routing algorithms adjust routes in real-time, Cooperative Accident Avoidance facilitates vehicle communication, and Adaptive Traffic Signal Control optimizes intersection traffic [26]. These systems are tested using simulations in order to compare their effects on the traffic in terms of flow, safety and productivity in an urban environment.

The data set utilized in this work involves simulated exercises in traffic within a virtual city. Such ingredients are real-time traffic data, vehicle communication logs, past records of accidents, and signal phase recall. To assess the DT agent three proposed methods of Dynamic Traffic Routing, Cooperative Collision Avoidance, and Adaptive Traffic Signal Control are implemented and tested with 5G, DSRC, VANETs, Li-fi, and blockchain technologies in different traffic scenarios.

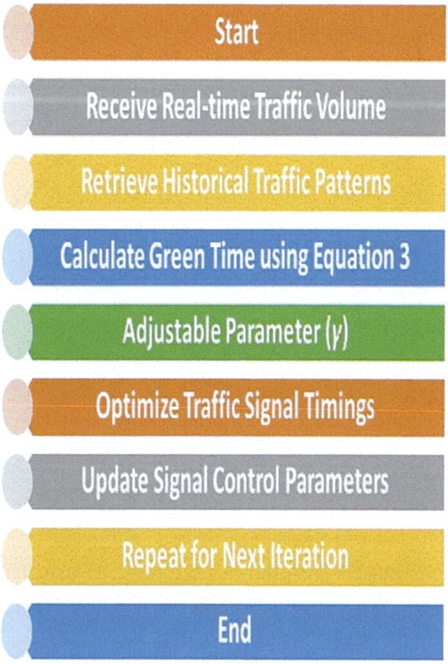

Fig. 3 Dynamic intersection management

This chapter describes experiments that were done to test smart transportation systems and methods of automobile interaction in a virtual city. It was intended to set DTR, CCA, and ATSC against blockchain, 5G, and auto-wireless networks [27, 28]. There was also a change of traffic flows, unpredictable traffic density, and special cases associated with city traffic. Feasibility, accessibility, reliability, and versatility were used as the criteria to measure the practical applicability of each method. Security, privacy and survivability to assault were also evaluated in the same simulation. Based on the tests performed, it was established that the proposed DTR algorithm was better at shifting traffic channels than existing methods. As it was suggested by the author, the proposed device could immediately response to the change in traffic conditions and increase traffic performance [29–31].

CCA was also safe; they prevented accidents by connecting with all their cars. ATSC's adaptive traffic light management reduced wait times and boosted crossing flow, which is intriguing. The recommended remedy adapts to changing traffic circumstances and goes beyond current approaches to improving travel. DSRC, 5G connectivity, VANET, Li-Fi, and blockchain performed similarly but worse in some cases. The experiment demonstrated the need for a comprehensive design for future vehicle communication systems, including dynamic routes, collision avoidance, and signal control. Studies revealed that the recommended strategies work and provided us with a better notion of how they may transform smart transportation networks. The findings show future city transportation may be safer, more flexible, and better [32–34]. This might lead to future cities using intelligent vehicle communication.

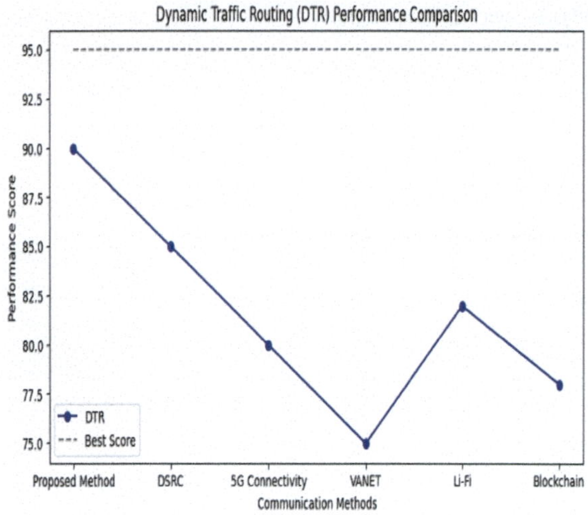

Fig. 4 Performance analysis of dynamic traffic routing mechanism

Figure 4 compares Dynamic Traffic Routing with other transmission methods to illustrate their effectiveness. The proposed DTR method, with its dynamic route design, outperforms the alternatives in managing real-time traffic. The comparison highlights how DTR optimizes traffic flow and adapts to changing conditions more efficiently than other methods, showcasing its superior capabilities in real-time traffic management and overall effectiveness in enhancing urban transportation systems.

Cooperative Collision Avoidance shown in Fig. 5 involves a constant observation of possible accidents in other to improve safety. This approach makes it possible for the vehicle to work in unison in real time and avoid an accident. By way of achieving this cooperative interaction, the strategy enhances safety measures on the road very much. The study brings out that there's strong validity in using the proposed method in minimizing the chances of collision thus increasing safety in urban transport systems and enhancing driver trust in adoption of these technologies.

In Fig. 6, we provided an overview of how ATSC can efficiently manage real-time traffic light. Operational flexibility makes Traffic Signals and Communication System (ATSC) produce minimum waiting times for vehicles at an intersection hence enhancing the interconnectivity of road systems. In contrast to conventional traffic signal systems that work with prescheduled timings, flexibility in design of ATSC allows traffic signals to be adjusted according its database results. In enhancing the overall attributed performance of the transport system, this adaptive capability reduces delays and therefore improves the mobility of the urban travel. The results emphasize the effectiveness of ATSC in traffic management for reducing congestion and improving the flow and safety rates of transportation in the urban areas.

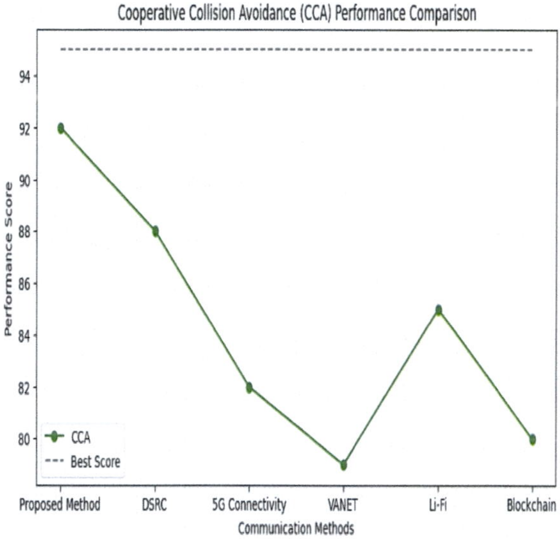

Fig. 5 Performance analysis of cooperative collision avoidance (CCA) mechanism

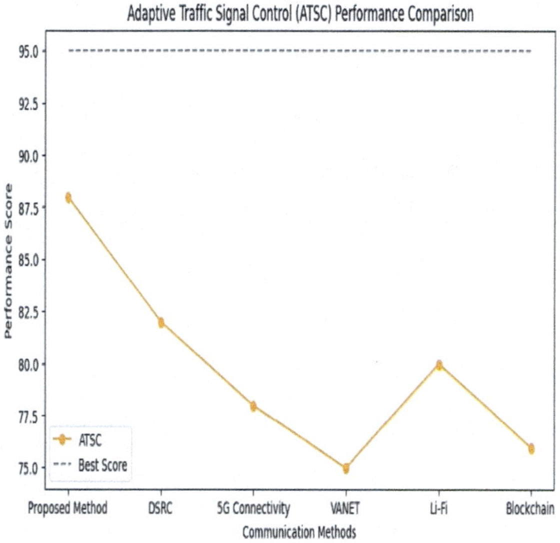

Fig. 6 Performance comparison of adaptive traffic signal control

5 Conclusion

In conclusion, ITS utilize the advanced vehicle-to-vehicle concepts and en relation to user demands for supporting the modern ITS. This work analysed such techniques like Dynamic Traffic Routing, Cooperative Collision Avoidance, and Adaptive Traffic Signal Control which have prospects of transforming urban transport systems. Substantial experimentation has substantiated these concepts in terms of their capability to better control junctions, facilitate the flow of traffic and avoid dangerous incidents through allowing multiple vehicles to collectively manage their avoidance and overall coordination. Real-time practices show that to improve this or that aspect of the work, it is necessary to develop a holistic approach requiring the use of effective technologies in communication, flexibility, and safety. The findings point out that future transportation systems can only be improved especially in the strength, safety and efficiency by using vehicle interaction. All these are of great necessity as urban areas expand and traffic challenges assume difficult dimensions. The study demonstrates that intelligent vehicle communication can address real-world traffic challenges, paving the way for a new era of smart transportation. These advancements promise to foster connected, adaptable, and sustainable urban environments as populations continue to expand, ultimately leading to more efficient and safer transportation networks.

References

1. Aguirre E et al (2017) Design and implementation of context aware applications with wireless sensor network support in urban train transportation environments. IEEE Sens J 17(1):169–178
2. Shukla S, Roy V, Prakash A (2020) Wavelet based empirical approach to mitigate the effect of motion Artifacts from EEG signal. In: 2020 IEEE 9th International conference on communication systems and network technologies (CSNT), pp 323–326. https://doi.org/10.1109/CSNT48778.2020.9115761
3. Kunkulagunta M (2024) Cloud computing applications for ERP implementation. Int J Comput Eng Technol (IJCET) 15(2):165–175. https://iaeme.com/Home/article_id/IJCET_15_02_020
4. Roy V (2023) An effective FOG computing based distributed forecasting of cyber-attacks in Internet of Things. J Cybersecurity Inf Manag 12(2):8–17
5. Raghunath V (2024) User support solution implementation for SAP ERP utilizing artificial intelligence to drive automated. Int J Artif Intell Res Dev. https://iaeme-library.com/index.php/IJAIRD/article/view/IJAIRD_02_01_012
6. Menouar H et al (2017) UAV-enabled intelligent transportation systems for the smart city: applications and challenges. IEEE Commun Mag 55(3):22–28
7. Singh N, Singh HP, Mishra A, Khare A, Swarnkar M, Almas SK (2024) Blockchain cloud computing: comparative study on DDoS, MITM and SQL injection attack. In: 2024 IEEE International conference on big data & machine learning (ICBDML), Bhopal, India, 2024, pp 73–78. https://doi.org/10.1109/ICBDML60909.2024.10577412
8. Khan AA et al (2023) MaReSPS for energy efficient spectral precoding technique in large scale MIMO-OFDM. Phys Commun 58:102057, ISSN 1874-4907. https://doi.org/10.1016/j.phycom.2023.102057
9. Raghunath V (2024) SAP cloud services on decoding cybersecurity and data privacy controls. Eur J Adv Eng Technol 11(9):95–102. https://ejaet.com/archive/volume-11-issue-9-2024/
10. Ramu K et al (2024) Deep learning-infused hybrid security model for energy optimization and enhanced security in wireless sensor networks. SN Comput Sci 5:848. https://doi.org/10.1007/s42979-024-03193-6
11. Roy V et al (2024) Network physical address based encryption technique using digital logic. Int J Sci Technol Res 9(4):3119–3122
12. Singh N, Chakrabarti T, Chakrabarti P, Panchenko V, Budnikov D, Yudaev I, Bolshev V (2023) Analysis of heuristic optimization technique solutions for combined heat-power economic load dispatch. Appl Sci 13(18):10380. https://doi.org/10.3390/app131810380

13. Kunkulagunta M (2024) Role of machine learning, data mining, and analytics. Int J Comput Eng Technol 15(2):154–164. https://iaeme.com/Home/article_id/IJCET_15_02_019
14. Zhu L et al (2014) Communication-based train control (CBTC) systems with cooperative relaying: design and performance analysis. IEEE Trans Veh Technol 63(5):2162–2172
15. Parashar V et al (2022) Aggregation-based dynamic channel bonding to maximise the performance of wireless local area networks (WLAN). Wirel Commun Mob Comput 2022:1–11, Article ID 4464447. [Online]. Available: https://doi.org/10.1155/2022/4464447
16. Shukla S, Pal S, and Singh S (2025) Managing economic load dispatch through multi-swarm statistical particle swarm optimization. 2025 2nd International Conference on Computational Intelligence, Communication Technology and Networking (CICTN), Ghaziabad, India, 2025, pp 979–983. https://doi.org/10.1109/CICTN64563.2025.10932346
17. Roy V (2021) An improved image encryption consuming fusion transmutation and edge operator. J Cybersecurity Inf Manag 8(1):42–52
18. Choi J et al (2016) Millimeter-wave vehicular communication to support massive automotive sensing. IEEE Commun Mag 54(12):160–167
19. Aravind AR, Prajapati M, Arunkumar E, Kumar R, Kumar H, Rao SPVS (2024) A way of optimization of wireless sensor network using TSCH. In: 4th International conference on advance computing and innovative technologies in engineering, ICACITE 2024, pp 326–330. https://doi.org/10.1109/ICACITE60783.2024.10616840
20. Bavkar D et al (2023) Deep hybrid model with trained weights for multimodal sarcasm detection. Lect Notes Netw Syst, pp 179–194. [Online]. Available: https://doi.org/10.1007/978-981-99-5166-6_13
21. Raghunath V (2024) Investigation on cloud security frameworks, problems and proposed solutions. Eur J Adv Eng Technol 11(9):103–110. https://ejaet.com/archive/volume-11-issue-9-2024/
22. Singh N, Chandel P, Shukla S, Soni A, Sen D, Khan S (2024) Design of microstrip antenna for multipurpose wireless communication, pp 12–18, eBook ISBN9781003422440. https://doi.org/10.1201/9781003422440
23. Kunkulagunta M (2024) Studying of exploring based on impacts of artificial, intelligence & machine learning on enterprise resource planning. Int J Artif Intell Mach Learn (IJAIML). 3(1):102–112. https://iaeme.com/Home/article_id/IJAIML_03_01_010
24. Roy V et al (2024) Detection of sleep APNEA through heart rate signal using convolutional neural network. Int J Pharm Res 12(4):4829–4836
25. Singh N, Prajapati GK (2018) Wireless sensor network security problems-a review. Int J Emerg Technol Adv Eng J 8(1), ISSN 2250-2459
26. Paramasivam P, Gowthaman N, Srivastava VM (2023) Analytical modeling of [001] orientation in silicon trigate rectangular nanowire using a tight-binding model. SILICON 16(6):2743–2756. https://doi.org/10.1007/s12633-024-02864-6
27. Roy V et al (2013) Image denoising by data adaptive and non-data adaptive transform domain denoising method using EEG signal. In: Kumar V, Bhatele M (eds) Proceedings of All India seminar on biomedical engineering 2012 (AISOBE 2012), Lecture Notes in Bioengineering. Springer, India, 2013. https://doi.org/10.1007/978-81-322-0970-6_2
28. Singh N, Khare A, Krishna S R, Lall S, Thakur C and Mishra A (2024) A study of signal processing techniques in power quality disturbance detection. 2024 IEEE 2nd International Conference on Innovations in High Speed Communication and Signal Processing (IHCSP), Bhopal, India, 2024, pp 1–4. https://doi.org/10.1109/IHCSP63227.2024.10960026
29. Rohini G, Gnana Kousalya C, Singh N, Patil VR (2023) Autonomous forecasting of traffic in cellular networks based on long-short term memory recurrent neural network. Cybernetics Syst 1–13 (2023). https://doi.org/10.1080/01969722.2023.2166245
30. Kumar P et al (2022) Machine learning enabled techniques for protecting wireless sensor networks by estimating attack prevalence and device deployment strategy for 5G networks. Wirel Commun Mob Comput 2022:1–15, Article ID 5713092. https://doi.org/10.1155/2022/5713092

31. Raghunath V (2024) A comprehensive review on security and privacy properties in cloud-based business and scientific workflows. Eur J Adv Eng Technol 11(9):87–94. https://ejaet.com/archive/volume-11-issue-9-2024/
32. Khatri S et al (2021) Machine learning models and techniques for VANET based traffic management: implementation issues and challenges. Peer-to-Peer Netw Appl 14:1778–1805. https://doi.org/10.1007/s12083-020-00993-4
33. Singh N, Sharma AK, Sarkar I, Prabhu S, Chadaga K (2024) IoT-based greenhouse technologies for enhanced crop production: a comprehensive study of monitoring, control, and communication techniques. Syst Sci Control Eng 12(1). https://doi.org/10.1080/21642583.2024.2306825
34. Gupta SS, Pandey TK, Raju VP, Shrivastava R, Pandey R, Nigam A, Roy V (2024) Diabetes estimation through data mining using optimization, clustering, and secure cloud storage strategies. SN Comput Sci 5(781). https://doi.org/10.1007/s42979-024-03158-9

Architecture and Protocols for Data Transmission in VANET

Archana S. Ubale, Harshada Magar, Rameez Shamalik, and Vineeta Philip

Abstract Vehicular Ad-Hoc Networks (VANETs) facilitate vehicle-to-vehicle and vehicle-to-infrastructure communication in likely environments. Poor network congestion management, vulnerability to security threats, and inadequate Quality of Service under high-demand conditions are some limitations of traditional VANET protocols. In existing intrusion detection systems rely on static rules, making them ineffective against emerging threats, while traditional data dissemination techniques struggle to distribute critical information efficiently. This chapter focuses on limitations through deep learning-enhanced solutions, using neural networks, convolutional models, and reinforcement learning. These approaches maximize intrusion detection, increases QoS resource allocation and enhance data dissemination techniques like broadcasting and geocasting. Even in dynamic circumstances, the suggested approaches guarantee dependable and secure communication by providing predictive and adaptable capabilities. The outcome demonstrates that deep learning significantly enhances VANET performance, leading to better traffic management, improved safety, and efficient coordination of autonomous vehicles.

Keywords Architecture of VANET · Protocols · Quality of Service and data transmission protocols · Medium Access Control

A. S. Ubale (✉)
AISSMS College of Engineering, Pune, India
e-mail: asubale@aissmscoe.com

H. Magar · V. Philip
AISSMS Institute of Information Technology, Pune, India
e-mail: harshada.magar@aissmsioit.org

V. Philip
e-mail: vineeta.philip@aissmsioit.org

R. Shamalik
Bharati Vidyapeeth's College of Engineering for Women, Pune, India
e-mail: rameez.shamalik@bharatividyapeeth.edu

J. Bhatia et al. (eds.), *Deep Learning Based Solutions for Vehicular Adhoc Networks*, Studies in Computational Intelligence 1207,
https://doi.org/10.1007/978-981-96-5190-0_2

1 Introduction

Vehicular Ad-Hoc Networks are pivotal for the evolution of Intelligent Transportation Systems, enabling efficient and secure data transmission among vehicles and infrastructure. VANET network represent the Ad-hoc network which is used for mobile vehicle nodes in various environment such as highway or an urban area. VANET is an intelligent traffic Transportation system. Which was initially represented by the Dedicated Short Range Communication standard (802.11p). They are incorporate LTE-V2X PC5 interface Mode 4 [1]. In the ITS applications, data is transmitted from vehicle to infrastructure is reliable and efficient. The important techniques which is used in a ITS to achieve goals of VANET which is benefits for autonomous vehicle, increasing the road safety and reduce the congestion of traffic due to this will be able to provides the convenience and comfortable entertainment services for passengers while travelling [2].

In the VANET architecture, outline the roles of On-Board Units Roadside Units (RSUs), and centralized servers in the Traffic Management Centers. It further divided communication modes into Vehicle-to-Vehicle, Vehicle-to-Infrastructure and Vehicle-to-Everything, each category has special functions in improving traffic safety, efficiency, and infotainment services [3, 4].

Medium Access Control (MAC) protocols is a central to the functioning of VANETs. Designated Short Range Communications/Wireless Access in Vehicular Environments (DSRC/WAVE) and IEEE 802.11p are two main MAC protocols [5].

IEEE 802.11p is discussed for its versatility to different vehicle circumstances, whereas DSRC/WAVE is emphasized for its durability in high-speed environments. In the routing protocols focusing the unique challenges posed by VANETs dynamic topology. Greedy Perimeter Stateless Routing and Position-based routing protocols and Geographic Source Routing are differentiating with topology-based protocols like Destination-Sequenced Distance-Vector Routing and Ad hoc On- Demand Distance Vector. GPSR is important for its efficiency in settings urban with probably -represent geographical boundaries, conversely AODV is advice for its use in route establishment, minimizing overhead in dynamic networks [6].

Cluster-based routing and broadcast protocols, comprise Urban Multi-Hop Broadcast (UMB) and Distributed Vehicular Broadcast are investigated for their efficiency in data dissemination. Cluster-based protocols in which are Clustering Oriented Inter-Vehicle Communication demonstrate benefits for scalability, while in broadcast protocols, UMB are useful in reducing broadcast storms [7].

In VANETs Security protocols are important to ensure data integrity and preserve from malicious attacks. The use of Public Key Infrastructure and Secure Efficient Ad hoc Distance Vector is used for establishing a secure communication channels, representing common security threats and mitigation strategies. A comparison shows the PKI's effectiveness approach to authentication and SEAD's efficiency in resource-limited surroundings [8].

The Quality of Service protocols, which are Vehicular Quality of Service MAC (VEQMAC) and Quality of Service Optimized Link State Routing (QOLSR), are examined based on several data kinds to satisfy varied needs for latency, reliability and throughput. Changing traffic conditions VEQMAC is known while QOLSR is evaluated for its best-fit routing choices based on link quality in Adaptive QoS management.

Inter-Vehicle Geocast (IVG) and Data Aggregation and Broadcast (DAB) protocols role are important in data dissemination protocols. This type of protocols is very useful for the efficient distribution of important information across the network with IVG being effective for focused geographical dissemination and DAB is important in reducing data redundancy.

As per the research directions and future trends taking into account the effects of developing technologies such as 5G the IOT and autonomous vehicles on VANETs. It draws attention to recent research issues and promising directions for innovation, highlighting how crucial it is for this sector to keep evolving [5].

1.1 Taxonomy of VANET Protocols and Enhancements

This section provides a structured overview of Vehicular Ad-Hoc Networks protocols and their advancements. The taxonomy categorizes VANET protocols into several key areas, highlighting both traditional methods and their deep learning-enhanced counterparts.

1.1.1 Medium Access Control Protocols

- Contention based MAC protocols.
- Deep learning enhancements in MAC protocols.

1.1.2 Routing Protocols

- Deep learning-based routing techniques.

1.1.3 Security

- Secure routing protocols with deep learning.
- Secure routing protocols with deep learning.

1.1.4 Traditional QoS Protocols

- Vehicular enhanced Quality of Service medium access control.
- Deep learning-based QoS enhancements.

1.1.5 Data Dissemination Protocols

- Traditional data dissemination techniques.
- Broadcasting and data aggregation.

2 VANET Architecture

In this section, we will discuss the architecture of Vehicular Ad Hoc Networks. We will address the primary elements of the VANET architecture, such as the different kinds of nodes, network topologies, communication protocols and a hierarchical structure that makes network administration and data transmission more effective. Additionally, we will cover the functions of various network elements and how they cooperate to deliver trustworthy vehicular communication.

2.1 Components of VANETs

As per the [9] VANET architecture which consists of several main components On-Board Units, Roadside Units and centralized server as shown in Fig. 1

2.1.1 On-Board Units

One of the components of VANET is a board unit. In particular, Ad Hoc Networks are the main computing devices embedded in vehicles that facilitate vehicle-to-everything communication. OBUs supports vehicle-to-infrastructure, vehicle-to-vehicle and vehicle-to-network interactions, which provide real-time data interchange crucial for safety and traffic management. They leverage deep learning for tasks like real-time traffic prediction and collision avoidance by analyzing data from multiple sensors, including GPS, LIDAR and cameras. These units support important functions such as improving driver safety with lane-keeping assistance and adaptive cruise control support. With secure and low-latency communication technologies like DSRC and C-V2X, OBUs contribute to the development of autonomous driving and smart traffic systems, while also integrating energy-efficient measures for modern electric and hybrid vehicles [3].

2.1.2 Roadside Units

Figure 1 shows a basic road intersection with multiple vehicles, two RSUs, and a base station. Fast wired connections are used to link the base station and the

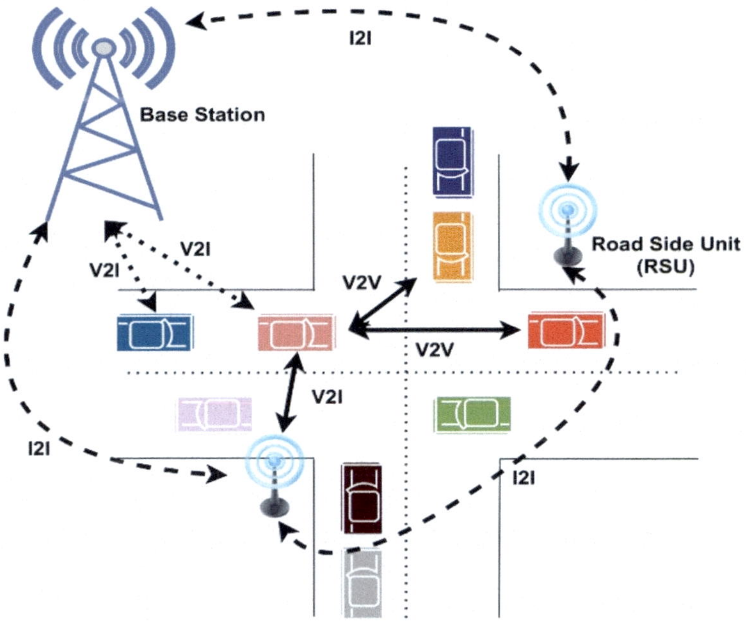

Fig. 1 VANET architecture

RSUs. In situations where a vehicle has to share data with another vehicle, it can do so in three ways. Initially, any vehicle within the range of transmission can be directly communicated with by the vehicles using V2V communication. The only option available to a vehicle using V2V communication to send a message to another vehicle outside its transmission range is multi-hop data dissemination. However, in the majority of safety–critical applications, the latency involved can be intolerable.

Second, automobiles can send cellular data straight to the base station, which can then relay it to the correct recipient. When compared to multi-hop forwarding, this forwarding method may result in a lower latency. But more traffic will be added to the already crowded terrestrial network, particularly in cities where there is a significant concentration of mobile nodes. We shall demonstrate in the following sections that due to the base station's restricted bandwidth, it is unable to meet the demands of every vehicle on the road. Utilizing the RSU in the data transmission procedure is the third option. The packets can be sent by the cars to the closest RSU. Using the fast wired connection between the RSUs, the RSU will locate the current responsible RSU of the intended receiver or destination node and utilize it to transfer the packet with the least amount of latency. As a result, RSU is a crucial network element in the architecture of future VANETs, and many deployment methodologies for RSU have been suggested thus far [10].

2.1.3 Base Station

In a traffic management system, the base station functions as a central hub that collects data from roadside units along the infrastructure and On-Board Units in vehicles. RSUs acquire information about the surroundings and traffic flow, whereas OBUs collect data on vehicle status, including location, speed, and direction. V2X and Cellular communication are the wireless networks which are used to send this data to base station. At base station data from many sources is combined into a single stream and kept in a central repository. The data goes through preprocessing to standardize, clean and filter it. The features are extracted to identify pertinent patterns. This data is then used to sophisticated deep learning algorithms, which allow the system to carry out operations like anomaly detection, traffic forecast and behavioral analysis. Use of real-time and historical data, models are taught to recognize possible risks and adjust to new traffic patterns and continuously learn, resulting in precise projections. The generated insights of traffic are shown to traffic managers as interactive visual dashboards and maps they are also conveyed to drivers in the form of alerts and recommendations. In real-time updates enhancing ongoing data analysis, this system creates a feedback loop that optimizes traffic flow, offers responsive and boosts safety, dynamic solutions to changing traffic conditions.

2.2 *Communication Network Topology in VANETs*

In this section, we will see the communication network topology in Vehicular Ad Hoc Networks. We will study the different types of network topologies used in VANETs, their characteristics and how they affect vehicular communication's effectiveness and performance. It is important to comprehend these topologies in order to develop and deploy VANET systems that work properly.

2.2.1 Vehicle-to-Vehicle Communication

In order to enhance traffic efficiency, passenger comfort and safety, vehicles transmit impeding traffic information using VANET, which is considered an important component of the intelligent transportation system. Automating message propagation via vehicle-to-vehicle and/or vehicle-to-roadside with potential roadside-to-roadside connections is crucial for VANETs. Since messages go from a vehicle to a roadside and back to the destination cars in vehicle-to-road communications, delays are a key concern. In contrast, V2V communication relies on nearby participating cars, so while latency may not be a problem, connectivity and security are serious issues. Applications linked to safety (such road merging, emergency braking, collision warning, etc.) require a small latency, which V2V communications may provide. Additionally, if vehicles communicate with one another, they can avoid colliding. Wirelessly transmitting recurring status notifications to one another [9]. To prevent any potential crashes, autonomous cruising could be implemented depending on the interactions between nearby vehicles.

2.2.2 Vehicle-to-Infrastructure Communication

V2I is a wireless Interaction between automobiles and roadside infrastructure to relay information on traffic signals, road conditions, and navigation aids. Deep learning can improve the prediction and adaptation capabilities of this communication. V2I communication is usually two directional made possible by a system of firmware, software, and hardware. Infrastructure elements like traffic lights, lane markers, and road signs can wirelessly transmit information to the vehicle and the other way around.

3 Medium Access Control Protocols

In this section we discussing about Medium Access Control protocols used in Vehicle Ad Hoc Networks. For the node access control to the communication channel, stay away from collisions and insure equitable use of network resources MAC protocols are important. We will go over the various MAC protocol their workings and how they are essential to improving the performance and durability of VANETs.

3.1 An Overview of MAC Protocols in VANETs

Medium access control protocols which make ensuring that channel resources are shared fairly and effectively among various vehicles for VANETs which can deliver packet data on time and greatly enhance traffic safety [11]. Every vehicle can interchange their information to intimate the other vehicles about the live status of the traffic flow or a hazardous circumstance such an accident. To attain these in a trustworthy and effective manner Medium Access Control protocol with minimum transmission collisions is required. Superior speed nodes, lack of infrastructure, variations in topology and their QoS requirements makes it challenging to design a MAC protocol in vehicular networks. There are many MAC protocols proposed for VANETs to make certain that every vehicle could transmit safety signals without collisions by reducing the end-to-end delay and packet loss ratio [12].

3.2 Contention Based MAC Protocols

Contention-based MAC protocols are important due to the high dynamic nature of vehicular environments, where rapidly changes in network topology and the required for real-time communication are common.

3.2.1 DSRC/WAVE

Vehicular Ad-hoc Networks ensure seamless connection between vehicles and roadside infrastructure by making use of wireless communication technologies such as Dedicated Short Range Communications and Wireless Access in Vehicular Environments. Previously VANETs have controlled communication channel access with contention-based Medium Access Control protocols like Carrier Sense Multiple Access with Collision Avoidance (CSMA/CA). However highly dynamic and congested environments usually of vehicle networks these protocols faces many difficulties such as higher mobility, interference, scalability and the necessity for strong security measures [13]. Deep learning approaches have shown to be an important method to overcome these problems and raise the effectiveness and dependability of VANETs. DL models permits real-time decision-making carry out complex data analysis, create adaptive communication techniques and offering improvements for those of conventional MAC protocols [14].

3.2.2 IEEE 802.11p

The IEEE 802.11p protocol is a derived of IEEE 802.11 e that highlight on the execution of the MAC layer within the background of Wireless Access in Vehicular Environment standard. It includes Enhanced Distributed Channel Access in tandem with Carrier Sense Multiple Access with Collision Avoidance mechanisms for access the channel. The prioritization of messages in EDCA is achieved through the utilization of Access Category. Consequently, messages categorized under AC receive the highest priority while those under ACO are considered lowest in priority. The adjustment of Contention Window and Arbitration Inter-Frame Space is determined based on the assigned AC. Messages with higher priority experience shorter waiting times for channel access in comparison to those with lower priority [11].

3.3 Deep Learning Enhancements in MAC Protocols

Deep Learning in Vehicular Ad-Hoc Networks can enhance Medium Access Control protocols in many ways to address the particular challenges posed by dynamic network topologies, high mobility and the required for real-time communication.

3.3.1 Adaptive MAC Protocols for Dynamic Environments

Although the current IEEE 802.11p/P1609.4 specified WAVE Medium Access Control protocol takes little into account future high density traffic, the wireless access in vehicular environments system might have more capacity because to its high throughput performance. In order to accomplish this, brand-new multi-channel MAC protocol that can divide Service Channel Interval and the Control Channel Interval using a dynamic time boundary rather than a fixed border. Our plan divides the CCH Interval into three distinct parts: Service Announcement Phrase, Peer-to-peer Reservation Phrase and Beacon Phrase. In the meantime, a distributed algorithm that takes into account multiple dynamic parameters determines how long low-priority PRP will last based on the actual traffic load during each Synchronization Interval [15].

3.3.2 Collision Prediction and Avoidance

VANET is dealing with a congestion issue on the network. According to the industry-standard 802.11p, the Medium Access Control protocol performs poorly, particularly under varied traffic scenarios. When channel load exceeds network capacity, congestion typically arises. This may obstruct message transmission, increase latency and lead to more collisions. So effective congestion avoidance MAC solution for VANET to help with the situation. Reducing messaging overhead and congestion of network. Deep reinforcement learning technique, makes use of a number of MAC parameters to prevent or control network congestion. Periodically, it modifies the MAC parameters by monitoring the manner in which each vehicle's neighbors communicate [16].

4 Routing Protocols

Routing protocols which are used in Vehicular Ad Hoc Networks. This protocols are important for finding the optimal paths for data transmission between different nodes in a VANET network. We highlight in this section various types of routing protocols, its mechanisms and its significance in dependable and effective communication in VANETs.

4.1 Challenges in VANET Routing

The distinct features of vehicular ad-hoc networks such as their high-speed mobility, dynamic node density, and frequent topology changes, provide a number of routing issues. The formation and application of effectively routing protocols are made more challenging by these components. In such challenging circumstances, conventional routing systems repeatedly fail to maintain dependable communication links, increase latency, increase packet loss and decrease network performance. In Modern VANET routing protocols include techniques to oversee the different communication requirement of vehicles, handle connectivity between them and accept the fast changes in topology to get around these issues [17]. Deep learning-based Potential remedies are beginning to emerge as a means of overcoming these problems since they present more intelligent and adaptable routing techniques that can manage VANETs' dynamic nature more effectively.

4.1.1 High Mobility and Dynamic Topology

High mobility and changeable topology are two major issues in VANET routing. In vehicular area networks, vehicles go at different speeds and in different directions, which results in regular topological changes and communication links transient. Due to the reorganize quickly, networks fracture and high mobility, reorganize quickly, making it challenging for conventional routing techniques to manage. Vehicles slipping beyond the reach of communication or into Routing paths may soon become invalid due to new locations. Requiring ongoing route finding and procedures for maintenance. Because protocols continuously update routing tables and discover new routes, these regular topology changes may result in high routing overhead [18].

The solution for the problems caused by dynamic topology and higher mobility in VANETs is deep learning. VANETs can predict changes in topology and the movements of vehicles, allowing for more proactive and dependable routing decisions using models for machine learning. Selecting a route in routing protocols are anticipated to stay stable for higher periods of time by using models and predictions of vehicle trajectories depends on real-time and historical data, for instance, created by recurrent neural networks and long short-term memory networks. Reduction in routing overhead and lessens the requirement for frequent route discoveries deep reinforcement learning can also be used to dynamically adjust routing techniques according to the actual state of the network and anticipated changes [19]. The as a total performance of VANETs is increased by these deep learning-based techniques, which offer stronger and effective routing remedies in the face of high mobility and changing topology.

4.2 Deep Learning-Based Routing Techniques

Strategies of deep learning-based routing in Vehicular Ad-hoc Networks enable improvement in performance and adaptability in dynamic and complicated situations, which is a very important improvement over standard routing methods. These methods can look over increase volumes of network and vehicle data by using different deep learning models, which makes routing decisions more effective and intelligent. Four main types of deep learning-based routing approaches that are investigated in this section are Position-Based Routing, Topology-Based Routing, Broadcast Routing and Cluster-Based Routing. Each one of the techniques is improved by deep learning methodologies.

4.2.1 Deep Learning Based Position-Based Routing

Deep learning-based routing strategies in Vehicular Ad-hoc Networks enable improved performance and adaptability in complicated and dynamic situations, which is a main improvement over standard routing methods. These methods can used to examine higher volumes of network and vehicle data by utilizing deep learning models, which makes additional decisions about route effective and intelligent. Broadcast Routing, Cluster-Based Routing, Topology-Based Routing and Position-Based Routing are the four main types of deep learning-based routing approaches that are analyzed in this section [14].

4.2.2 Topology-Based Routing Using Neural Networks

Determination of routing, topology-based routing uses information about network connectivity. Because deep learning offers refined techniques to anticipate different changes in network topology and adjust routing path appropriately, it can increase topology-based routing. The dynamics and structure of vehicular networks can be examined and learned from using neural networks, especially Graph Neural Networks, which permits more efficient routing path maintenance and selection. By simulating the connections and interactions between nodes in a VANET, these networks are able to expect possible link failures and make active modifications to routing techniques [20]. Topology-based routing can reply quickly to changes in network topology by utilizing deep learning, which reduces the chance of communication interruptions and boosts overall network dependability.

4.2.3 Cluster-Based Routing Enhanced by Machine Learning

In cluster-based routing vehicles are grouped into clusters, with communication between and within clusters being neglected by cluster chiefs. This type of strategy can be improved by machine learning, especially deep learning, which can maximize the cluster creation, maintenance and head selection procedures. More effective and stable clusters can be created by using deep reinforcement learning to dynamically modify cluster parameters based on the state of the network and vehicle behavior. To find the finest clustering techniques that increase network performance while minimizing energy consumption and communication overhead, DRL models can learn from the functioning data of the network [21]. Because of its flexibility, cluster-based routing works well in higher dynamic VANET situations, where more conventional clustering techniques would find it difficult to stay stable and productive.

4.2.4 Broadcast Routing with Predictive Models

By looking forward to the range of broadcasts and ideal time, deep learning can dramatically increase the efficiency of broadcast routing, which distributes messages to each node inside a network segment. Recurrent Neural Networks, a particular kind of predictive model, can determine the optimal times to broadcast messages by analyzing communication patterns and past traffic. This reduces network congestion and redundant transmissions. The routing protocol can adjust broadcast techniques to the expected and actual network conditions due to these models capacity to forecast traffic densities and arrival times of vehicle [20]. Unpredictable VANET environments broadcast routing can achieve lower latency and higher delivery success rates by using predictive deep learning models.

5 Security Protocols

5.1 Importance of Security in VANETs

VANETs, which are gradually integrated with cutting-edge technology to enable safely, efficient communication and dependable between infrastructure and vehicles, place a high priority on security. In this section explores some common attack routes and risks the significance of security in VANETs and using the deep learning algorithms to improve the secure communication and the vehicle authentication. Because VANETs offer real-time, safety–critical applications like emergency response coordination, traffic light control and collision avoidance, network security is essential. Because of their higher degree of open communication and mobility. VANETs are vulnerable to a number of security risks that might put data integrity in vehicle safety and danger. Stronger security measures aid in safeguarding confidential data, stopping harmful activity and preserving dependable communication that can cause impede traffic [22]. In this regarding, improves the overall security framework of VANETs requires incorporating cutting-edge security protocols and utilizing creative techniques.

5.1.1 Common Threats and Attack Vectors

VANETs are vulnerable to a few known dangers and attack methods, such as spoofing, eavesdropping, Sybil attacks and denial of service assaults. Eavesdropping or unauthorized communication interception can result in privacy concerns. When a hacker poses as a trustworthy infrastructure or automobile in order to give false information or obstruct communication spoofing attacks happen. DoS attacks are created to flood the network with too much traffic, which will either bring down the entire network or impede communication. In order to obtain unfair benefits and influence network protocols, Sybil attacks entail the formation of several wrong identities [23]. Very strong security procedures that can be recognize and effectively address attacks which are required to protect VANET transmission from these dangers.

5.2 *Deep Learning for Vehicle Authentication and Secure Communication*

The promising solutions for secure communication and enhancing vehicle authentication in VANETs is a Deep learning. By leveraging advanced data analysis capabilities, deep learning models can provide more advanced and techniques for security that adapts that are essential in addressing the dynamic and complex security challenges of VANETs.

5.2.1 Enhancing Public Key Infrastructure with AI

The main objective of verifying and securing communications within VANETs between automobiles, Fundamental security technique is Public Key Infrastructure. PKI's efficacy is further increased by incorporating artificial intelligence, particularly deep learning algorithms, which automate procedures like key distribution, anomaly detection and certificate management. Deep learning models are able to find out the deviations and examine certificate usage trends that can point to possible security risks like forged certificates or stolen keys. Revocation processes can be made more effective with the use of AI-enhanced PKI solutions. Revocation processes and certificate renewal, which ensuring that expired or compromised certificates are promptly revoked and preserving the VANET's integrity and reliability [24].

5.2.2 Behavioral Analysis for Security

To find and address security concerns, watches using deep learning and behavioral analysis and examines how all the vehicles behave within the network. The use of machine learning models, VANETs are able to find out unusual trends in the way that cars behave, such as dubious communication attempts or strange driving patterns that could indicate malevolent activity [25]. Recent neural networks and anomaly detection algorithms are the examples of deep learning approaches that may be able to creation of baseline behavior profiles by using past data identify deviations in real-time. This type of proactive strategy, possible risks can be identified early and countermeasures can be implemented before significant damage is caused. Behavior analysis offers a dynamic, context-aware method of mitigating and detecting security issues, which increases the overall security posture of VANETs.

5.3 *Secure Routing Protocols with Deep Learning*

For enhancing the security in current routing protocols like the Secure Efficient Distance Vector (SEAD) protocol, deep learning is radically altering safe-routing algorithms in VANETs by offering cutting-edge methods. This section investigates the different application of deep learning in intrusion detection systems of AI techniques to enhance SEAD and solve more general security concerns.

5.3.1 Improving SEAD Protocol with AI Techniques

Traditionally, Safe distance vector routing is intended to be accomplished through the Secure Efficient Distance Vector protocol. Employs cryptographic techniques to prevent attacks, such as a sequence number attacks and routing loops. But artificial intelligence methods, especially deep learning, can be improve SEAD even more by making it more resilient and able to adapt to threats that get more complicated [26]. SEAD can be prepared with deep learning models like neural networks to examine routing patterns and find out any anomalies that could indicate malicious activity. A deep learning model for a instance may be used to examine and monitoring the network's routing behavior over time spotting anomalies that might mention potential avenues for routing attacks such as black hole or wormhole attacks.

5.4 *Addressing Security Issues Through Deep Learning*

Deep learning is a very powerful tool for dealing with various security concerns in VANETs, particularly through the development of advanced Intrusion Detection Systems. These Systems make use of deep learning's potential to handle and evaluate big volumes of data to find out and mitigate security threats in real-time.

5.4.1 Intrusion Detection Systems Using Deep Learning

Utilizing intrusion detection systems is a very crucial to maintaining an eye on and defending VANETs from malicious activity. Conventional intrusion detection systems have difficulty identifying fresh or developing dangers since they frequently depend on pre-established rules or signatures to identify recognized threats. However, deep learning-based intrusion detection systems make better use of both recognized and unidentified threats by learning from historical and real-time network data through the use of neural networks and other machine learning models. The convolutional neural networks and recurrent neural networks for example are deep learning models that are able to examine trends in network traffic and spotting anomalies that can point to intrusions like spoofing, denial-of-service attacks or unauthorized access attempts [27]. These models can be trained on labeled datasets of legitimate traffic and malicious to acquire unique properties.

6 Quality of Service

Is a critical aspect that ensures the network can support diverse application needs by managing traffic in a way that meets varying performance requirements?

6.1 Definition and Importance of QoS in VANETs

Quality of Service in VANETs is defined as the capability to deliver different service levels to network traffic based on the unique requirements of the applications, including bandwidth, latency, jitter and packet loss considerations. QoS is crucial for supporting a wide array of applications—ranging from safety–critical notifications to infotainment services—by making certain that sufficient network resources are allocated to communications that are urgent and high priority. Effective QoS management in VANETs improves communication reliability and performance, thereby supports the seamless functioning of every application that require stringent service levels [28].

6.2 Traditional QoS Protocols

For management of network traffic to ensure the performance of critical applications Quality of Service protocols are very important. Traditional QoS protocols focusing on only setting priorities certain types of traffic to meet particular performance criteria, such as latency, packet loss and jitter.

6.2.1 VEQMAC

The Vehicular Enhanced Quality of Service Medium Access Control (VEQMAC) protocol addresses QoS requirements by controlling media access in a manner consistent with the demands of different applications. VEQMAC prioritizes high-priority traffic, such as safety alerts, by dynamically adjusting transmission intervals and contention window length. In this type of approach ensures that critical communications gain priority over communications that are not as critical, thus meeting the QoS requirements for different types of vehicular traffic and maintaining optimal network efficiency [29].

6.2.2 QOLSR

Quality of Service Optimized Link State Routing (QOLSR) extends the OLSR protocol by using QoS measurements into its routing algorithm. By evaluating different parameters like delay, link reliability and bandwidth, QOLSR selects routes that fulfill the requirements for various apps' QoS. This protocol increase application routing efficiency with strict performance criteria, ensuring that critical traffic is given first priority and gets the as per required network resources [30].

6.3 Deep Learning-Based QoS Enhancements

Deep learning has introduced noticeable developments in a different domain, including network management and Quality of Service. Traditional QoS protocols often depends on static rules and predefined policies, which may not adapt well to dynamic network conditions. Deep learning-based QoS improves leverage machine learning models to optimize, predict and manage network resources more effectively.

6.3.1 Dynamic Resource Allocation Using AI

In VANETs Dynamic resource allocation can be considerably improved with AI methods such as deep learning. Models like reinforcement learning and neural networks predict traffic patterns and allotment of resources in real-time to maintain desired QoS levels. These types of different techniques adapt such type of parameters like transmission power and channel selection based on the real-time data, improving resource utilization and reducing latency. For instance, reinforcement learning can accept to different changes in traffic density and network topology, learning optimal strategies for resource allocation to ensure consistent QoS [31].

6.3.2 Predictive QoS Management with Neural Networks

Predictive QoS management forecasts network traffic using neural networks conditions and adjust QoS settings proactively. By analysis of historical data and current traffic data, these models predict future condition network states such as congestion or high demand periods. This foresight allows networks to proactively adjust

QoS regulations, such as reserving bandwidth or prioritizing traffic, thereby maintain optimal performance even under different conditions. Neural networks' predictive capabilities improve the ability of VANETs to provide required resources to critical applications during periods of high demand, guaranteeing dependable and effective service delivery [32].

7 Data Dissemination Protocols

In Vehicular Ad-hoc Networks which ensure reliable and timely communication among vehicles and infrastructure for efficiently data dissemination is important part. This section explores the different methods which are used in traditional techniques, data distribution and the role of deep learning in enhancing these protocols.

7.1 Effective Data Distribution in VANETs

In VANETs involves sharing knowledge between vehicles and infrastructure efficiently, considering the network's dynamic and movable nature is a data dissemination Effective data distribution ensures that critical information such as emergency alerts, traffic updates and road conditions are disseminated accurately and promptly to improve the overall traffic management and safety for persons [33].

7.2 Traditional Data Dissemination Techniques

Traditional data dissemination techniques have been crucial for sharing data among various types of networks, ensuring efficient and reliable data delivery.

7.2.1 Geocasting

Geocasting is a targeted data dissemination technique where the messages are send to vehicles within a particular geographical area in a network. In this type of method ensures that only intended vehicles receive the information, reducing unnecessary network congestion and optimizing resource utilization. Geocasting is very successful for applications requiring localized data dissemination, such as road hazard warnings and intersection collision avoidance in traffic [34].

7.2.2 Broadcasting and Data Aggregation

Broadcasting is nothing but sending messages to all vehicles within the communication area, ensuring widespread dissemination of information. Data aggregation techniques are often used to consolidate same messages before the broadcasting, conserving network resources and decreasing redundancy. Broadcasting is very important for urgent notifications and emergency alerts that required immediate attention from all vehicles in the vicinity [30].

7.3 Deep Learning for Data Dissemination

Deep learning has revolutionized various domains, including data dissemination and network management.

7.3.1 Enhancing Geocasting with Machine Learning

In the Machine learning techniques, including deep learning, increases geocasting by enhancing the delivery of the messages depends on real-time traffic and environmental data. Deep learning models examine past traffic trends and uses vehicle behavior to forecast the best regions for distribution dynamically. By accepting to changes in network conditions, these models increase the efficiency and accuracy of geocasting, making certain that vital information reaches the targeted recipients promptly.

7.3.2 Optimized Data Aggregation and Broadcasting Using AI

In the Artificial intelligence techniques optimize data aggregation and broadcasting by the clever management of message priority and transmission scheduling. AI algorithms analyze traffic dynamics and congestion in the large network to identify the most effective broadcasting techniques. Reinforcement learning for example can optimize broadcast scheduling based on feedback from network conditions, effectiveness of data dissemination and improving the reliability in VANETs [31].

8 Applications and Case Studies

This section explores practical applications case studies and approaches to incorporate deep learning into Vehicular Ad-hoc Networks, Highlighting different developments in traffic control, safety applications, best practices for implementation and autonomous vehicle coordination in VANETs.

8.1 Practical Applications of VANET Protocols Enhanced by Deep Learning

VANET protocols improved by deep learning play a pivotal role in improving a range of useful uses. These type of protocols leverage machine learning techniques to improve data dissemination, optimize network efficiency and ensure reliable communication in dynamic vehicular environments. Deep learning models analyze real-time traffic patterns, optimize resource allocation, predict road conditions and predict road conditions, thereby increasing the performance of applications such as collision avoidance, autonomous vehicle coordination and traffic management systems [33].

8.2 Case Studies of Successful Deep Learning Implementations in VANETs

8.2.1 Real-Time Traffic Management Systems

Deep learning has revolutionized which is real-time traffic management systems in VANETs by enabling adaptive decision-making and predictive analytics. For instance, deep neural networks analyze historical traffic data and current sensor inputs to optimize signal timings, forecast traffic flow and suggest route diversions dynamically. This type of proactive approach reduces congestion, reduces travel time and contributing to improved urban mobility and driver experience, enhances overall traffic efficiency [29].

The VANET-Cloud architecture, as depicted in Fig. 2, introduces a multi-layered approach to improve traffic management through improved utilization of road infrastructure and reduction of data anomalies. This architecture encompasses three primary layers: The Infrastructure as a Service Layer and VANET-Cloud Layer. The Traffic Data Collection Layer, Traffic Data Collection Layer gathers real-time traffic information using connected detectors and transmits it to the cloud. It consist of sensing traffic data and communication interfaces to optimize data accuracy and service performance, benefiting applications like route planning and congestion management. The Infrastructure as a Service Layer comprises physical servers for computing and storage, along with a virtualized sub-layer to handle varying demands and ensure consistent performance despite high congestion. The VANET-Cloud Layer highlights on real-time processing and prediction of traffic data, enhancing the performance of vehicular communication applications through dynamic optimization [35].

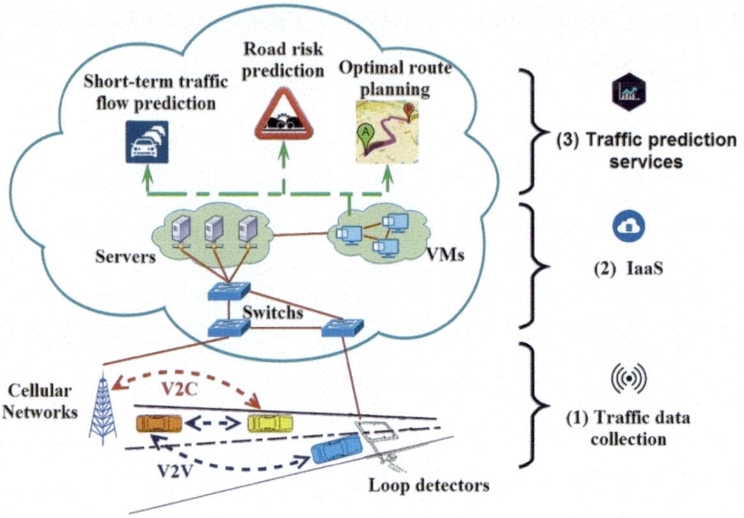

Fig. 2 VANET network with cloud computing [35]

8.2.2 Collision Avoidance and Safety Applications

In collision avoidance and safety applications, deep learning models analyze sensor data, vehicle dynamics and environmental factors to predict potential collisions and trigger timely alerts. These models increase the accuracy of collision prediction and enable vehicles to take preventive actions autonomously, such as adjusting speed. By leveraging deep learning, VANETs improve road safety by mitigating risks and assisting methods for cooperative collision avoidance among vehicles [30].

The research identifies two primary types of collisions that occur in VANETs.

Access Collisions which occur when multiple vehicles attempt to use the same time-slot for communication and merging Collisions which happen when vehicles come too close to each other while using the same time-slot, particularly in high-density traffic situations. A Fig. 3 shows the Access Collision. A frame in a TDMA-based VANET system is typically separated into a service channel and a control channel. Every vehicle produces and sends out a safety message containing its vehicle information. The control channel is used to periodically convey this message to other 1-hop vehicles. Emergency messages, audio, and video communications are among the non-secure messages that are sent over the service channel. The quality of service of the application that each vehicle's service channel provides is impacted by a resource collision in the control channel. More specifically, there are two types of collisions in a VANET protocol that uses TDMA. First, as seen in Figure, an access collision happens when several cars trying to obtain the same available time-slot simultaneously inside a two-hop communication range The simulation results in Fig. 4 demonstrating that the proposed algorithm significantly reduces collision rates compared to traditional methods. The performance improvements are particularly notable in scenarios with high traffic density and frequent changes in vehicle direction [36].

In this case study of Merging collision rate for a highway the proposed algorithm significantly reduces the merging collision rate by up to 55.1%, 40.8%, and 11.7% compared to VeMAC, CFR MAC and PTMAC, respectively shown in Fig. 4. The simulation results indicate that the vector-based prediction model performs better than slot partitioning-based algorithms. Despite both PTMAC and the proposed algorithm avoiding slot partitioning, the latter is more effective at preventing collisions between vehicles moving in different directions.

In Fig. 5 the proposed algorithm decreases the merging collision rate by up to 63.3, 46.9, and 25.2% compared to VeMAC, CFR MAC, and PTMAC, respectively. The reduction is more pronounced when the traffic light is red, as fewer vehicles are moving. However, with the light green, high mobility and various vehicle directions

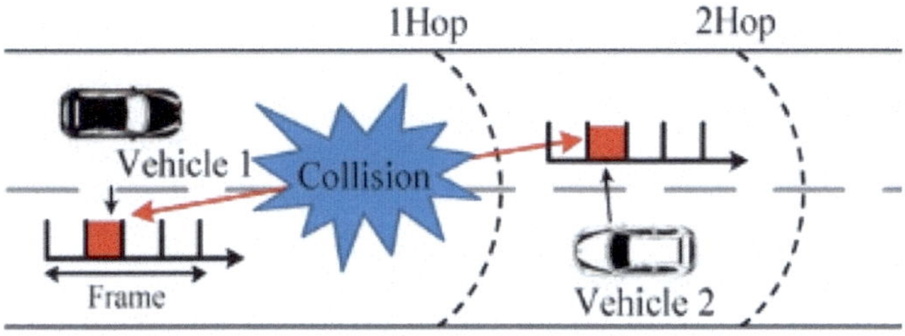

Fig. 3 Access collision [36]

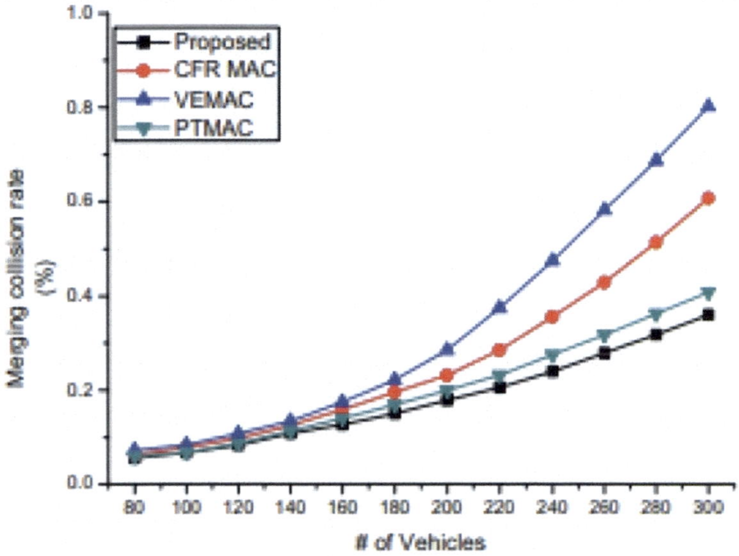

Fig. 4 Merging collision rate for a highway [36]

increase collision chances, where the proposed algorithm outperforms others by effectively predicting vehicle movements.

In Fig. 6 simulation result shows the proposed algorithm achieves a reduction in merging collision rates by up to 49, 25.8 and 14.9% compared to VeMAC, CFR MAC, and PTMAC, respectively. This performance improvement, though lower than on highways, is still notable and reflects the algorithm's ability to handle varying traffic densities and directions [36].

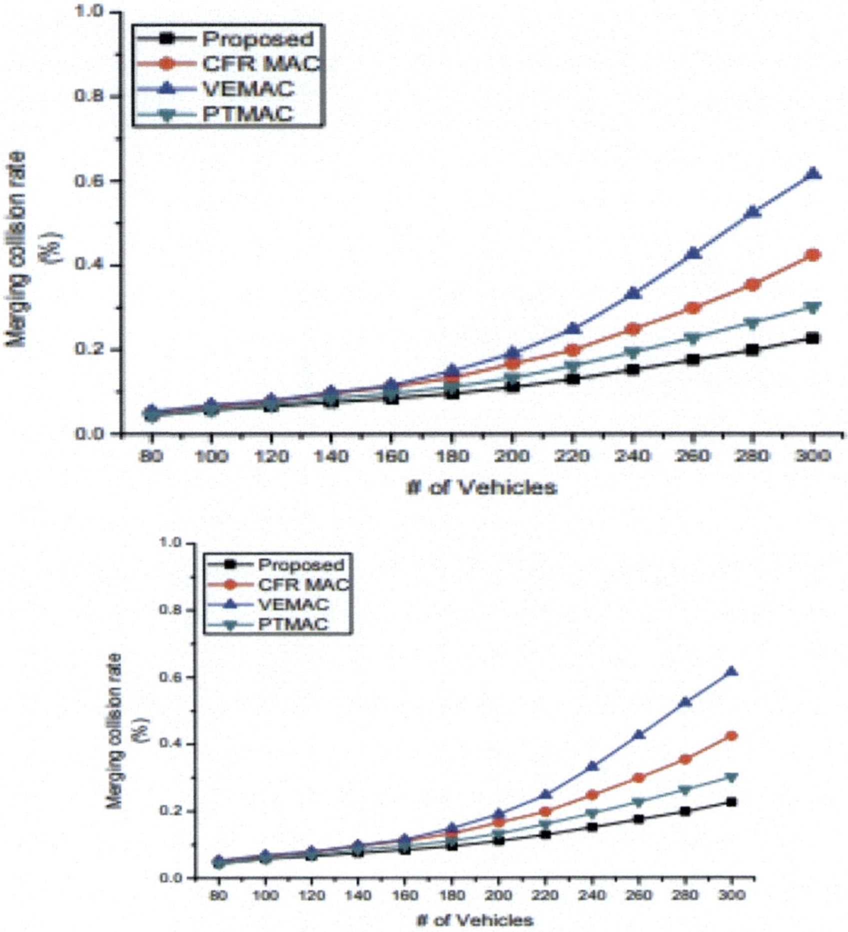

Fig. 5 Merging collision rate for an intersection with traffic lights [36]

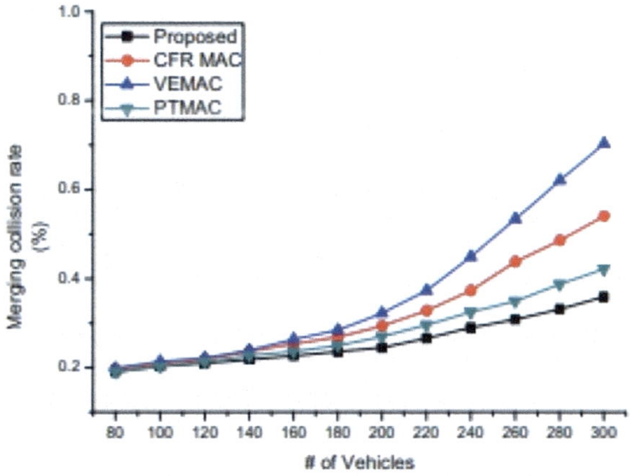

Fig. 6 Merging collision rate for a circuit intersection without traffic lights. [36]

8.2.3 Autonomous Vehicle Coordination and Communication

Deep learning facilitates efficient coordination and communication along autonomous vehicles in VANETs. Neural networks analyze communication signals and sensor data to predict the intentions of neighboring vehicles, coordinate maneuvers seamlessly and optimize route planning. This type of capability enables autonomous vehicles to navigate complex traffic scenarios safely, adapt to changes in road conditions collaboratively and maintain optimal spacing. Deep learning-driven communication protocols responsiveness and improves reliability, paving the way for widespread adoption of autonomous driving technologies [37].

9 Conclusion

The combination of deep learning into Vehicular Ad-hoc Networks significantly enhances various network protocols and applications. By advancing Medium Access Control (MAC) protocols, routing techniques, security measures, data dissemination, and Quality of Service (QoS) management, deep learning introduces predictive and adaptive methods that focuses the complex and dynamic nature of VANETs. The improvements in MAC protocols and collision forecasting improves channel access and reduce network congestion. Deep learning-driven routing and security protocols effectively manage dynamic topologies and mitigate potential threats, ensuring robust communication. Enhancement in QoS protocols and data dissemination methods optimize resource allocation and application performance. Practical implementations, such as in collision avoidance, traffic management and autonomous vehicle coordination, demonstrate how deep learning can revolutionize VANETs. Continued research in this area, standardization and collaboration are essential for realizing these advancements, addressing security and interoperability challenges. Overall, the deep learning offers promising opportunities to elevate the performance and reliability of

VANETs, paving the way for smarter and safer vehicular communication networks which is safer for human.

10 Motivation and Research Contribution

10.1 Motivation

- Addressing the limitations of traditional VANET protocols, including inefficient handling of network congestion, security vulnerabilities, and inadequate QoS management.
- Enhancing the adaptability and real-time performance of VANET systems through advanced techniques like deep learning.
- Improving data dissemination methods and QoS management to better support critical vehicular applications and autonomous driving technologies.

10.2 Research Contribution

- Introduction of deep learning-based enhancements to Medium Access Control (MAC) protocols, leading to improved channel access and reduced collision rates.
- Development of adaptive and predictive routing and security protocols that manage dynamic topologies and evolving threats more effectively.
- Optimization of resource allocation and data dissemination using deep learning to boost application performance and network efficiency.
- Demonstration of practical implementations in traffic management, collision avoidance, and autonomous vehicle coordination, showcasing the transformative impact of deep learning on VANETs.

Organization of the Paper

- Introduction: Overview of VANETs and the need for advanced techniques.
- Taxonomy of VANET Protocols and Enhancements: Classification of protocols and methods.
- Deep Learning in VANETs: Application of deep learning techniques to various VANET aspects.
- Case Studies and Practical Implementations: Examples of deep learning applications in real-world scenarios.
- Conclusion: Summary of findings, contributions, and future directions.

References

1. Zhou R (2022) VANET architecture analysis and protocols. Int J Comput Appl 184(13):44–54
2. Bisht A, Khaitan V (2023) Reliability analysis of 5G—VANET using cloud-fog-edge based architecture. RAIRO Oper Res 58:129–149
3. Yeferny T, Hamad S (2021) Vehicular ad-hoc networks: architecture, applications and challenges. arXiv abs/2101.04539
4. Zhou R (2022) VANET architecture analysis and protocols. Int J Comput Appl 13:44–54
5. Ahmed TH, Tiang JJ, Mahmud A, Gwo-Chin C, Do D (2023) Proposed CtCNet-HDRNN: a cornerstone in the integration of 5G mmWave and DSRC for high-speed vehicular networks. IEEE Access 11:126482–126506
6. Albattah W, Habib S, Alsharekh MF, Islam M, Albahli S, Dewi DA (2022) An overview of the current challenges, trends, and protocols in the field of vehicular communication. Electronics 11(21):3581
7. Badole MH, Thakare AD (2024) Cluster-based multicast optimized routing in VANETs using elite knowledge-based genetic algorithm. Knowl Based Syst 294:111773
8. Mahmood J, Duan Z, Yang Y, Wang Q, Nebhen J, Bhutta MNM (2021) Security in vehicular ad hoc networks: challenges and countermeasures. Secur Commun Netw 9997771.1:9997771.20
9. Lochert C, Scheuermann B, Wewetzer C, Luebke A, Mauve M (2008) Data aggregation and roadside unit placement for a vanet traffic information system. In: Proceedings of the Fifth ACM International Workshop on Vehicular Inter-NETworking, VANET'08, San Francisco, CA, USA, 15 September 2008; Association for Computing Machinery, New York, NY, USA
10. Karunathilake T, Förster A (2022) A survey on mobile road side units in VANETs. Vehicles 4(2):482–500
11. Pierce A (2011) Crash survival systems + car-to-car communication = a next gen crash avoidance system. Technol Today 2011:8–9
12. Dharsandiya AN, Patel RM (2016) A review on MAC protocols of vehicular ad hoc networks. In: 2016 International Conference on Wireless Communications, Signal Processing and Networking (WiSPNET), pp 1040–1045
13. Kenney JB (2011) Dedicated short-range communications (DSRC) standards in the United States. Proc IEEE 99:1162–1182
14. Huang Y, Chen Y (2020) Autonomous driving with deep learning: a survey of state-of-art technologies. ArXiv abs/2006.06091
15. Liu L, Xia W, Shen L (2009) An adaptive multi-channel MAC protocol with dynamic interval division in vehicular environment. In: 2009 First International Conference on Information Science and Engineering, pp 2534–2537
16. Zerrouki H, Moussaoui S, Boualouache A (2022) Prediction-based congestion avoidance MAC protocol for IoV (PCA-MAC). In: 2022 First International Conference on Computer Communications and Intelligent Systems (I3CIS), pp 123–127
17. Hartenstein H, Laberteaux LP (2008) A tutorial survey on vehicular ad hoc networks. IEEE Commun Mag 46:164–171
18. Rawat P, Singh KD, Chaouchi H, Bonnin JM (2014) Wireless sensor networks: a survey on recent developments and potential synergies. J Supercomput 68:1–48
19. Luong NC, Hoang DT, Gong S, Niyato D, Wang P, Liang YC, Kim DI (2018) Applications of deep reinforcement learning in communications and networking: a survey. IEEE Commun Surv Tutor 21:3133–3174
20. Amalia A, Pramitarini Y, Perdana RH, Shim K, An B (2023) A deep-learning-based secure routing protocol to avoid blackhole attacks in VANETs. Sensors (Basel, Switzerland) 23(19):8224
21. Jiang Y, Zhu J, Yang K (2023) Environment-aware adaptive reinforcement learning-based routing for vehicular ad hoc networks. Sensors (Basel, Switzerland) 24:40
22. Blum JJ, Eskandarian A, Hoffman LJ (2004) Challenges of intervehicle ad hoc networks. IEEE Trans Intell Transp Syst 5:347–351

23. Liu X, Fang Z, Shi L (2007) Securing vehicular ad hoc networks. In: 2007 2nd International Conference on Pervasive Computing and Applications, pp 424–429
24. Lu N, Cheng N, Zhang N, Shen X, Mark JW (2014) Connected vehicles: solutions and challenges. IEEE Internet Things J 1:289–299
25. Chen Y, Wu K, Zhang Q (2015) From QoS to QoE: a tutorial on video quality assessment. IEEE Commun Surv Tutor 17:1126–1165
26. Hu YC, Johnson DB, Perrig A (2002) SEAD: secure efficient distance vector routing for mobile wireless ad hoc networks. In: Proceedings Fourth IEEE Workshop on Mobile Computing Systems and Applications
27. Yin C, Zhu Y, Fei J, He X (2017) A deep learning approach for intrusion detection using recurrent neural networks. IEEE Access 5:21954–21961
28. Hafeez KA, Zhao L, Liao Z, Ma BN (2010) Performance analysis of broadcast messages in VANETs safety applications. In: 2010 IEEE Global Telecommunications Conference GLOBECOM 2010
29. Sherin C, Mary Anita EA (2018) A survey on MAC protocols for vehicular ad-hoc networks
30. Badis H, Al Agha K (2005) QOLSR, QoS routing for ad hoc wireless networks using OLSR. Eur Trans Telecommun 16:427–442
31. Cui Y, Shi H, Wang R, He P, Wu D, Huang X (2024) Multi-agent reinforcement learning for slicing resource allocation in vehicular networks. IEEE Trans Intell Transp Syst 25:2005–2016
32. Sciancalepore V, Costa-Perez X, Banchs A (2019) RL-NSB: reinforcement learning-based 5G network slice broker. IEEE/ACM Trans Netw 27:1543–1557
33. Hafeez KA, Zhao L, Liao Z, Ma BN (2010) Performance analysis of broadcast messages in VANETs safety applications. In: 2010 IEEE Global Telecommunications Conference GLOBECOM 2010, pp 1–5
34. Sherin C, Mary Anita EA (2018) A survey on MAC protocols for vehicular ad-hoc networks
35. Abdelatif S, Derdour M, Ghoualmi-Zine N, Marzak B (2020) VANET: a novel service for predicting and disseminating vehicle traffic information. Int J Commun Syst 33
36. Bang J-H, Lee J-R (2020) Collision avoidance method using vector-based mobility model in TDMA-based vehicular ad hoc networks. Appl Sci 10(12):4181
37. Kumar AS, Zhao L, Fernando X (2023) Task offloading and resource allocation in vehicular networks: a Lyapunov-based deep reinforcement learning approach. IEEE Trans Veh Technol 72:13360–13373

Integration of Deep Learning Algorithms into VANET

Prateek Pandey, Deepak Sharma, and Mehul Sweeti Agrawal

Abstract Vehicular ad hoc networks (VANETs) overhauled the way the modern transportation system works changed the way. This chapter is designed to target the key elements and several issues faced by the systems using VANETs. This chapter explores the key components and challenges of VANETs, which includes but not limited to security concerns, protocols for communication, use of deep learning and machine learning for analyzing traffic flow, etc. This chapter also addresses the deployment requirements of the VANET architectures with special attention being given to the energy efficiency, resource utilization and randomness of the VANET environment. Since vehicles swiftly change their locations, which results in interactions with different topologies and network dynamics. Since regular Machine Learning algorithms, which often are based on a pre-defined deterministic model doesn't adapt well with dynamic VANET environments, deep learning algorithms are used along with the reliable and secure mechanisms to protect the VANET system against malicious attacks. All these requirements demand a comprehensive framework customized specially for VANETs. This chapter proposes a novel framework for VANETs that addresses intricacies of the VANET environment while leveraging the potentials of communication technology and deep learning to address the underlying challenges.

Keywords Vehicular networks · Machine learning · Deep learning · Communication technology and protocols · Safety–critical · Security · Simulation

P. Pandey (✉) · M. S. Agrawal
Department of Computer Science Engineering, Jaypee University of Engineering and Technology, Guna, India
e-mail: pandeyprat@yahoo.com

M. S. Agrawal
e-mail: mehulsagrawal33@gmail.com

D. Sharma
Department of Electronics and Communication Engineering, Jaypee University of Engineering and Technology, Guna, India
e-mail: deepak2.sharma@juet.ac.in

J. Bhatia et al. (eds.), *Deep Learning Based Solutions for Vehicular Adhoc Networks*, Studies in Computational Intelligence 1207,
https://doi.org/10.1007/978-981-96-5190-0_3

1 Introduction

VANETs (see Fig. 1) are a peculiar case of Mobile Ad hoc Network MANETs where the nodes involved are vehicles that may establish communication with each other called V2V Communication and with the Road Side Unit (RSU) (V2I Communication). Here, the vehicles have a circuitry known as OBU which can communicate with RSU. The prime units and their components of VANET are on board unit (OBU) and road side unit (RSU) and communication technology as shown in Fig. 1. The On-Board Unit is used for data collection from different sensors and communication with other OBUs and RSU. Road side unit (RSU) provides an infrastructure which enables communication to the outside network. In order to communicate through wireless access in vehicular environment (WAVE) the technology used by the system is multi-hop wave communication between the vehicles. VANETs can majorly be used to provide safety and non-safety (relating to comfort) services to drivers.

The major characteristics that VANET must possess are spontaneous data exchange between vehicles, key component for any intelligent transportation system (ITS), to follow a predictable pattern i.e., along roads, providing very high mobility, even much faster than found in MANETs. The advantages that VANET may provide now-a days are co-operative driving, information sharing and media Streaming, value added services like Internet access and navigation, post-crash notification, lane change assistance, blind spot warning and many more. VANET uses both infrastructure-based and infrastructure-less wireless networks where, inter-Vehicular Communication (V2V) is an infrastructure-less network and Communication with RSU (V2I) is an infrastructure-based network. V2V Communication: It is fully natural ad hoc network and capable to perform itself without any other infrastructure

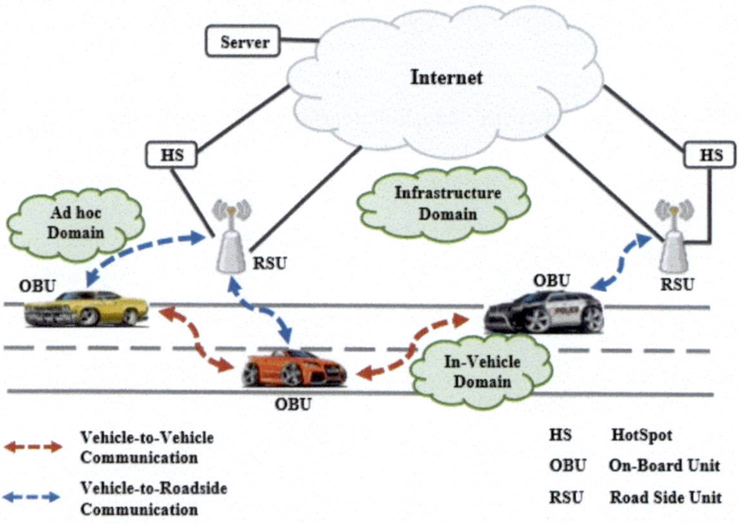

Fig. 1 VANET architecture [7]

support during its operation. In V2V communication, the vehicles communicate to each other. One vehicle can send and receive the messages from neighboring vehicles.

The motivation behind V2V communication is to keep drivers informed about the neighborhood vehicles. V2I and I2V imply the direction of information transfer from vehicle to infrastructure and infrastructure to vehicle, respectively. VANET-Hybrid Communication encompasses both: V2V and V2I. The hybrid communication is used to enhance the coverage area of roadside infrastructure. It uses multi-hop forwarding technique to send the information to the vehicles; those are out of coverage of roadside infrastructure. Similarly, the hybrid communication becomes more efficient for opportunistic forwarding of data for the roadside infrastructure with the assistance to the intermediate vehicles working as a message forwarding node [16].

The chapter is organized in 6 major segments. These segments are as follows: Sect. 2 presents a detailed review of the literature available on the VANET, contemporary architectures and its implementation details and deployment. Section 3 details the methodology for designing, developing and implementing components of VANET. Section 4 describes deep-learning algorithms implementation in context of VANET. Section 5 explains the integration details of the individual components, and the last Sect. 6 concludes the chapter.

1.1 Motivation

The sale figures of automobiles are all time high, which results in a greater number of automobiles on the roads that the road infrastructure can handle. On the positive side, there is a significant drop in the size and cost; and massive increase in the computation power of the onboard devices. Thus, a network of moving vehicles and roadside infrastructure, like traffic signals and toll plazas, can be created to inform drivers about their surroundings, potential collisions, and traffic jams in advance, so that a safe journey can be accomplished. VANET promises a real-time data-sharing solution among the participant vehicles in the network. The motivation behind writing this chapter is manifold, which includes, but is not limited to:

- Minimizing road accidents by generating collision avoidance alarms and contingency responses.
- Regulating traffic, thereby avoiding congestion by keeping drivers informed about existing or possible road jams.
- Proposing a fuel economy plan to drivers by suggesting routes that minimize fuel consumption.
- Investigating new avenues for VANET by exploring the possibilities of autonomous driving, smart cities, and drone-based delivery systems.

1.2 Contribution

This chapter proposes a novel framework for VANETs that addresses intricacies of the VANET environment while leveraging the potentials of communication technology and deep learning to address the underlying challenges.

2 Related Work

In this section, the VANET features and parameters (see Table 1), based on the related works on VANET [1–5, 8, 12–15, 18–20, 22, 25–28], are presented that offer a snapshot of the characteristics of a typical VANET system. The parameters are: cost, mobility, range, reliability, change of network topology, bandwidth, density of nodes, and volatile node patterns. Each parameter is an indicator of the VANET performance. For example, high mobility indicates that vehicles are constantly moving, impacting network connectivity; high cost suggests a significant investment in infrastructure and vehicle equipment; high mobility indicates that vehicles are constantly moving, impacting network connectivity; a range of 600 miles is substantial, implying good coverage but potential for handover challenges; high reliability is crucial for safety–critical applications in VANETs; frequent changes imply dynamic network conditions due to vehicle movement; 1000 Kbps suggests sufficient capacity for basic applications but might be limited for high-definition content; dense node density indicates a many number of vehicles in the network, potentially leading to congestion; and regular volatile nodes pattern suggests predictable vehicle movement patterns.

These parameters are the vital part of any complex VANET environment with its own set of challenges to develop superior system by improving such enlisted parameters. For example: the high cost might limit widespread adoption, especially in developing countries; these factors make network management and routing complex; congestion and interference can degrade network performance; and sufficient bandwidth is essential for various applications, but efficient utilization is crucial.

Table 1 VANET network parameters

Parameter	Performance metric
Cost	Expensive
Mobility	High
Range	Equal to 600 miles
Reliability	High
Change of network topology	Frequent
Bandwidth	Thousand Kbps
Density of nodes	Dense
Volatile nodes pattern	Regular

A number of challenges that the scientist's community still find difficult to address [16] are exploring ways to reduce the cost of VANET deployment without compromising performance; developing routing algorithms that can adapt to rapidly changing network topologies; implementing mechanisms to mitigate congestion in dense urban areas; efficiently utilizing available bandwidth for different types of traffic; and addressing security challenges in a highly dynamic environment with a large number of nodes.

Table 2 summarizes the contemporary related work in the area of VANET based algorithms to counter security threat by various researchers, the techniques adopted, the robustness offered by the proposed algorithms and their limitations. In the subsequent sections, a novel VANET framework is proposed, the underlying methodology is outlined, and the integration of the system is explained.

Table 2 Analysis of VANET algorithms on security threat

References	Techniques/methods	Robustness against attacks	Limitations
Singh et al. [33]	Symmetric cryptography, Hash function, digital signature, ID registration technique	DOS routing attack, Replay attack, Fake information attack	–
de Fuentes et al. [8]	Vehicular public key infrastructure, certificate validation	Eavesdropping effect, Identity revealing, Location mapping and tracking, DOS attack	Faces privacy issue due to RF fingerprinting
Tzeng et al. [11]	Initialization, anonymous identity generation, message verification using signing	Forgery attack	Useful to avoid forgery attack but not effective for other issues
Al Hasan et al. [2]	Public k key, ID based cryptography, Certificate revocation	For security and privacy	Enhance privacy and minimize overheads, fail to resolve attack
Pandey et al. [27]	Vehicular public key, ECC, Certificate authenticity, group signature	Fabrication attack, DOS attack, alteration and replay attack	Not useful in Eavesdropping
Raw et al. [30]	SEAD, ARAN, secure message transmission, non-disclosure method	Impersonation, replay attack, DOS routing attack, location tracking effect	Privacy and authentication requirement, Confidentiality missing

3 Proposed Framework and Methodology

The methodology defines a systematic technique for designing, developing, and implementing an OBU system in vehicles. It encompasses the entire process, from initial requirements analysis to deployment and maintenance, including software and hardware design, system integration, testing, and optimization. The OBU is a vital component for vehicle navigation, communication, and data management, integrating features like GNSS for positioning, a CAN bus for vehicle system communication, and V2X communication for real-time data transfer with external entities. The technique ensures that the OBU system is robust, reliable, and capable of enhancing vehicle safety and efficiency through advanced functionalities such as collision avoidance, remote diagnostics, and infotainment.

3.1 Working of On-Board Unit

An OBU is an embedded system-based hardware device that needs to be installed inside vehicles. The OBU consists of read–write memory used to store and retrieve information, and a processor for decision-making and computational capabilities, as shown in Fig. 2 for a four-wheeler system and Fig. 3 for a two-wheeler system. Additionally, it includes a crucial device called the GUI, which helps exchange information between drivers (part of the Application Unit (AU)). Apart from that, a GPS device is also installed to track vehicle positioning, along with a wireless interface that works as a radio antenna.

The radio communication sub-system is a critical component responsible for establishing wireless communication links between the vehicle and external entities, such as roadside infrastructure [19]. The complete block diagram of the communication that takes place among the OBU and the application ECU is depicted in

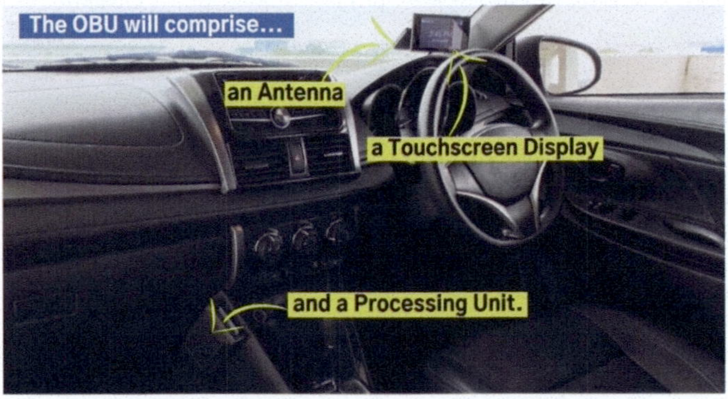

Fig. 2 Singapore based ERP 2.0 OBU for car [4]

Fig. 3 Singapore based ERP 2.0 OBU for a two-wheeler [35]

Fig. 4. The Global Navigation Satellite System (GNSS) antenna captures signals from GNSS satellites (such as GPS, GLONASS, Galileo, or BeiDou). It is generally placed on the roof, dashboard of the car, or integrated into other external parts of the vehicle to ensure a clear line of sight to the satellites. The GNSS receiver, on the other hand, processes these signals to determine the vehicle's precise location, speed, and time for various required applications. It is integrated within the vehicle's head unit, navigation system, or a dedicated telematic control unit [37].

The GNSS receiver is capable of performing various tasks, such as precision positioning, navigation, routing, and safety. Precision positioning includes location determination and real-time tracking. Location determination is achieved by

Fig. 4 Interaction of OBU with application ECU [9]

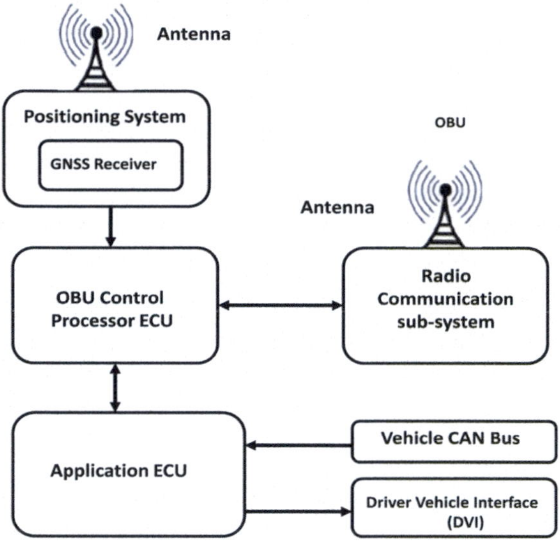

the GNSS receiver, which calculates the exact position of the vehicle by receiving signals from multiple GNSS satellites working together and independently [3] (such as GPS, GLONASS, Galileo, or BeiDou). Real-time tracking provides continuous monitoring of the vehicle's location, essential for navigation and other location-based applications.

Navigation and routing include turn-by-turn navigation, where the GNSS system integrates with the vehicle's navigation system to provide accurate directions. Route optimization helps optimize paths based on real-time traffic data received through V2X communication. In terms of safety, collision avoidance precisely detects vehicle positioning to help prevent collisions. This is primarily based on accurate distance localization and speed calculation algorithms. Geo-fencing aids applications by triggering alerts or actions when a vehicle enters or exits a predefined geographic area.

To perform these tasks, location-oriented services are activated to provide real-time data for various applications, such as finding nearby services, emergency assistance, and fleet management. Time synchronization is also essential, offering precise time coordination for the vehicle's systems, crucial for coordinated actions in V2X communication.

The Application ECU is a specialized control unit (see Fig. 4) within a vehicle responsible for running specific applications that enhance the vehicle's functionality and connectivity. It plays a key role in data processing. Sensor data integration allows it to aggregate and process data from various vehicle sensors (e.g., cameras, radar, LiDAR) to support applications. Decision-making helps analyze this data to make real-time decisions that improve vehicle safety and efficiency.

The Application ECU performs the following functions:

- Traffic Alerts: Receives real-time traffic updates from other vehicles and infrastructure, warn the driver about congestion and suggesting other possible routes.
- Collision Avoidance: Analyze data from vehicle sensors and V2V communication to identify potential mishaps, alerting the driver or automatically taking evasive action.
- Infotainment: Manages the vehicle's infotainment system, providing navigation assistance, music services, and other internet-oriented services, while ensuring these applications do not conflict with critical safety measures.
- Remote Diagnostics: Generate diagnostics alert by monitoring automobile health and contact service center for appointment.

The touch screen display application unit (AU) is not directly connected to the OBU processor but communicates through the Application ECU. The Application ECU accesses data from various sensors mounted on the vehicle using the CAN Bus. The Controller Area Network (CAN bus) is a message-based protocol that allows Electronic Control Units (ECUs) to communicate in modern automobiles based on a priority-driven mechanism [34].

The CAN bus visualization is shown in Fig. 5. Devices equipped with the CAN bus are called "nodes." These nodes consist of a main unit's CPU, a CAN controller, and a transceiver. The transceiver receives and modulates the signal levels of both

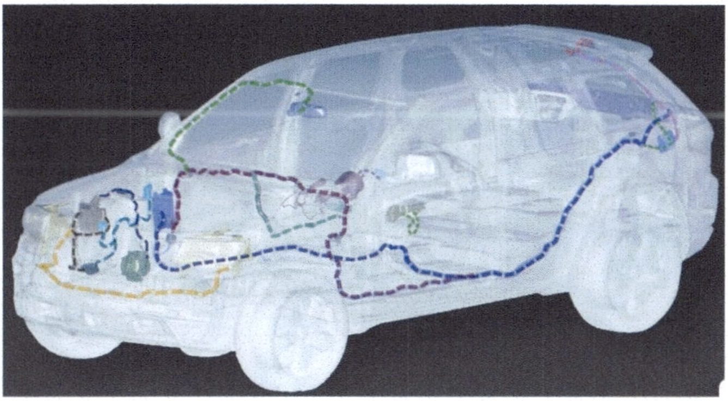

Fig. 5 Interaction of OBU with application ECU [6]

transmitted and received data for each node [12]. Every node is capable of transmitting and receiving data, but in a half-duplex manner, meaning data cannot be transmitted and received at the same time. Instead of transmitting data directly to each other, nodes send their data onto the network [18]. If data is available for a node and it is addressed for communication, the node processes the data accordingly.

The CAN protocol is activated and employs bitwise arbitration to resolve contention issues on the bus. All nodes are synchronized to sample data on the network simultaneously [5]. However, data is not transmitted with clock signal (timing) data, so CAN is not a truly synchronous bus, unlike EtherCAT. CAN sends all data in the form of frames, which are of four types: data, remote, error, and overload frames. Data frames transfer information to one or many receiver nodes as per requirements and instructions. Remote frames request data from other nodes. Error frames are used to report any detected errors, while overload frames report overload conditions caused by excessive data traffic.

Messages are transmitted in two lengths: either the standard length or the extended length [13]. The CAN protocol uses two data message formats: standard and extended CAN, as depicted in Fig. 6.

3.2 CAN Node Architecture

The CAN node architecture comprises a controller, transceiver, buses, and multiple nodes, as represented in Fig. 7. It collects and processes data from various sensors connected to the CAN node. In addition to this, it executes control algorithms for functions such as engine management, braking systems, and body controls [1].

The multiple functions performed by the CAN node architecture include data processing, application logic, memory management, real-time system control, diagnostics, and monitoring.

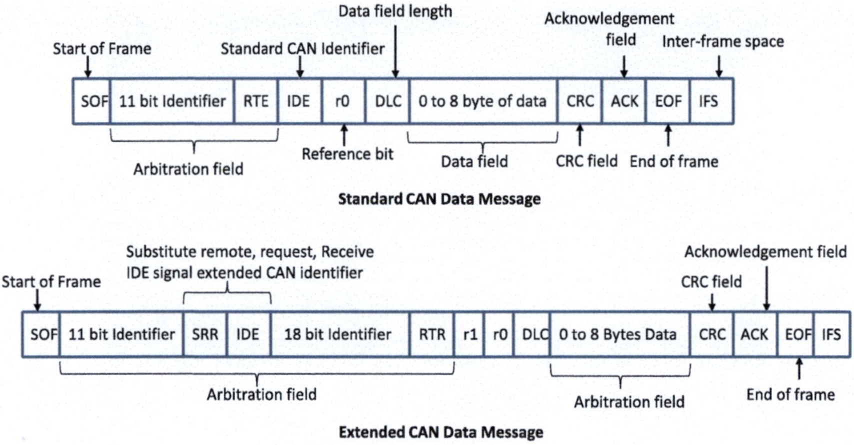

Fig. 6 CAN data message formats [29]

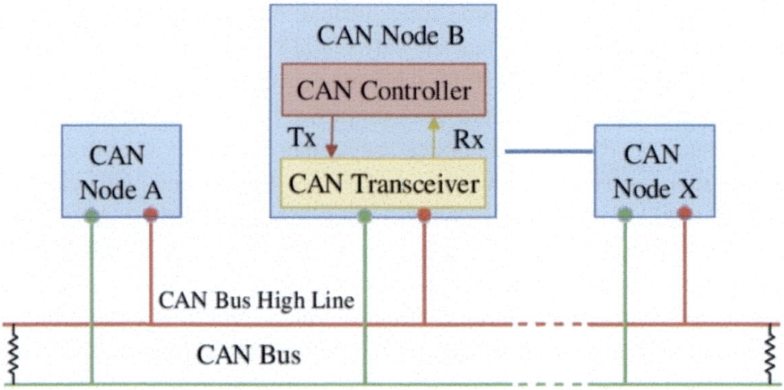

Fig. 7 CAN node architecture [20]

Data Processing: This function collects and processes data from distinct sensors connected to the CAN node and executes control algorithms for functions like engine management, braking systems, and body controls (e.g., lighting, climate control) [10].

Application Logic: This runs the main application code that dictates how the node should behave in response to data inputs. It also manages higher-level operations, such as adjusting the fuel injection in response to sensor data or activating ABS when wheel slippage is detected.

Memory Management: RAM is used for temporary data storage, including flash memory for the temporary storage of programs, and EEPROM for non-volatile data storage.

Real-Time Control: This activates to execute time-critical tasks, ensuring prompt responses to inputs and events. It utilizes interrupt handling to quickly respond to high-priority tasks, such as urgent sensor data or critical control commands [32].

Diagnostics and Monitoring: This function continuously monitors the status of the node's hardware and software. It applies diagnostic protocols (e.g., OBD-II) for vehicle health and performance monitoring. The CAN controller performs message framing, arbitration, error detection, and message filtering.

- Message Framing: Handles the framing of data into CAN messages, which include the identifier, control field, data field, CRC, and acknowledgment field. It also ensures the correct formatting and packaging of data for transmission on the CAN bus [36].
- Arbitration: Manages bus access to ensure that the highest priority message is transmitted first when multiple nodes attempt to send messages simultaneously. It uses a non-destructive bitwise arbitration process to resolve conflicts and determine which node gets bus access.
- Error Detection: Implements various error-checking mechanisms, such as CRC checks, to detect errors in transmitted messages. It also handles error signaling, error frames, and manages error counters to ensure robust communication.
- Message Filtering: Filters incoming messages based on their identifiers to determine if they are relevant to the node. It only passes relevant messages to the microcontroller, reducing the processing load.

The CAN transceiver comprises signal conversion, interfacing, bus driving, and fault tolerance mechanisms within itself. These functions are briefly described below [28]:

- Signal Conversion: Converts the digital data from the CAN controller (usually in the form of TTL logic levels) into differential signals suitable for transmission on the CAN bus [14].
- Physical Layer Interface: It facilitates the physical connection to the CAN bus, and ensure that the node can send and receive data over the network. It implements the electrical specifications of the CAN standard (e.g., ISO 11898) to facilitate reliable data transmission and reception [26].
- Bus Driving Unit: It drives the CAN bus lines, CAN_H and CAN_L, to represent the dominant (represented by 0) and recessive (represented by 1) states as required by the CAN protocol. It handles the differential signaling that enhances noise immunity and performs robust communication over long distances [21].
- Fault Tolerance: Includes mechanisms to detect and handle bus faults, such as short circuits or wiring issues. It may include features like automatic bus-off management to isolate faulty nodes from the network, preventing disruption [17].

Integration in a CAN Node: The microcontroller, CAN controller, and CAN transceiver work together. The microcontroller runs the main application, processes data from sensors by providing input signal data, and controls actuators that send and receive high-level commands and data to/from the CAN controller.

The CAN controller converts data from the microcontroller into CAN messages, handling message framing, arbitration, error checking, and filtering. It passes relevant messages to the microcontroller and prepares data for transmission received from the microcontroller [15]. The CAN transceiver converts the CAN messages from the CAN controller into differential signals for the CAN bus. It sends these signals onto the CAN bus and converts them back into digital data for the CAN controller [23]. It also ensures proper electrical characteristics and fault tolerance on the CAN bus.

4 Deep Learning Algorithms for VANET

In VANETs, deep learning algorithms are increasingly in demand for various purposes to enhance safety, efficiency, and intelligence within the transportation ecosystem. The important deep learning algorithms used in VANETs and their respective purposes are as follows:

1. Object Detection and Recognition: Object detection and recognition is efficiently achieved by using specialized machine learning technique called deep learning [23]. The objective is to detect and assign a class to the identified object, which may be vehicles, passerby, bicyclists, signs and other road blocks. This would help in improving awareness of the automobile driver to his or her surroundings.
2. Lane Detection and Semantic Segmentation: CNNs and semantic segmentation networks like U-Net, are used for lane detection and tasks of semantic segmentation. The objective is to accurately mark lanes, road edges, and regions of importance within the scene to assist in lane-keeping, and path planning.
3. Behavior Prediction: Long Short-Term Memory (LSTM) networks and Recurrent Neural Networks (RNNs) models are used to predict the future pathway of vehicles and the deterministic and/or random moving patterns of other drivers, including passerby and cyclists.
4. Cybersecurity and Intrusion Detection: Deep learning models are used again for intrusion detection purpose by identifying unusual patterns in the network. CNN and RNN may be deployed for the said purpose. The purpose here is to avoid incidents of spoofing and eavesdropping.

4.1 Strategies to Deploy Deep Learning Algorithms in VANET

The researcher fraternity have expressed a great deal of interest towards VANETs owing to its potential to impact the transportation industry in a significant way. At the same time, the deep learning algorithms may play a vital role to develop vehicle intelligently and enabling, and equipping them with all the qualities to be classified as an intelligent vehicle that are capable to take real-time decisions, enhancing safety, customize as per necessity, efficiency, and enhance jolly driver experience in the next-gen of automobile industry. This segment focuses on and suggests two key strategies for deploying deep learning algorithms on OBUs, the processing core of vehicles within a VANET.

Strategy-1: Deep learning algorithms in VANET are applied as software running on the OBUs processor. Steps to apply the algorithm in VANET are as follows:

I. Software Implementation: These algorithms are typically implemented by using software frameworks like PyTorch, TensorFlow, or Keras. These frameworks provide high-level APIs for defining, training, and deploying deep neural networks.
II. Embedded Processors: OBUs are equipped with embedded processors such as ARM Cortex-A series or specialized processors like NVIDIA Jetson modules. These processors are capable to provide sufficient computational power to run deep learning algorithms very efficiently [22].
III. Optimization: Deep learning models further need to be optimized for deployment on embedded processors. This optimization is required to ensure the performance of system under given resource constraints. This optimization techniques involves model quantization, pruning, and architecture design modifications to reduce computational complexity and memory footprint.
IV. Inference Engine: Once the deep learning model is trained and optimized, it is deployed on the OBU's processor as part of an inference engine but further it requires optimization and improvement for any new situations. The inference engine executes the model to perform inference tasks, such as object detection, semantic segmentation, or behavior prediction, using input data from vehicle sensors.
V. Real-Time Operation: The OBUs must operate in real-time to support time-sensitive applications like collision avoidance, lane detection, or pedestrian detection by using the deep learning algorithm [25].

Strategy-2: It is implemented by hardwiring algorithm in processor.

I. Define Requirements and Constraints

Identify Use Case: Clearly define the specific use case for the deep learning algorithm, such as object detection, lane detection, traffic prediction, or driver behavior analysis.

Hardware Constraints: Understand the computational, memory, and power constraints of the OBU hardware.

Performance Metrics: Establish performance metrics like latency, accuracy, throughput, and power consumption.

II. Select Suitable Hardware

OBU with Accelerators: Choose OBUs that come with hardware accelerators such as GPUs, TPUs, or dedicated AI chips (e.g., NVIDIA Jetson, Intel Movidius).

Embedded Systems: Ensure the selected hardware is rugged and suitable for automotive environments, capable of operating under varying temperature and vibration conditions [24].

III. Optimize Deep Learning Model

Model Selection: Select a proper deep learning model that balances accuracy and computational efficiency (e.g., MobileNet, YOLO, or EfficientNet for lightweight applications).

Quantization: It reduces model precision from floating-point to fixed-point (e.g., int8 quantization) but it is required to save on memory space and computation time.

Pruning: This is required to eliminate the redundant neurons and unnecessary layers in the model which help to reduce size of model and improve computation time and inference speed.

IV. Model Training and Testing:

Data Collection: Collection of data should be from diverse sources and standard representative for training and testing purpose for specific to the vehicular environment.

Training: Train the model using high-performance computing resources and big data set, such as cloud GPUs or specialized hardware.

Validation: Validate the developed model on real-world data to ensure it meets the desired performance for real time system.

V. Model Deployment

Conversion: For the deployment of OBU hardware convert the trained model to a format suitable e.g., Lite, ONNX, TensorFlow or TensorRT.

Integration: To integrate the model into the OBU's software stack may involve developing a lightweight inference engine tailored for the hardware system.

Optimization: The optimization is required for inference engine work with low latency and performs high in real-time environment under the constraints of the embedded system.

VI. Software and Firmware Development

Driver Development: Develop or adapt hardware drivers to interface with the deep learning model and accelerators.

Firmware Integration: Integrate the deep learning model and its dependencies into the OBU firmware, ensuring seamless interaction with other OBU functions.

VII. Testing and Validation

Simulation: Test the integrated system in a simulated environment to identify and address any issues without risking real-world failures.

Field Testing: Conduct extensive field testing under various conditions such as different weather, traffic situations to ensure robustness and reliability of system.

Performance Tuning: Continuously monitor and tune the system for optimal performance based on real-world feedback to up lift the technology over the period.

VIII. Deployment and Maintenance

Deployment: Deploy the OBUs with the integrated deep learning model across the suitable position in a vehicle.

Monitoring: Implement monitoring tools to track the performance and behavior of the deployed models in real-time system for avoiding any unfavorable conditions.

Updates: Develop a system for over-the-air (OTA) and take updates to continually improve the model and fix any issues if any occurs.

IX. Security and Privacy

Encryption: Implement data encryption for all communications involving the deep learning model to protect against eavesdropping and tampering.

Access Control: It ensures robust access control mechanisms that allow to prevent unauthorized access to the OBU and its deep learning capabilities.

Privacy Protection: Adhere and stick to privacy and compliance all regulations and follow best practices to ensure that sensitive data collected and processed by the model is adequately protected and no rules are violated.

X. Compliance and Standards:

Regulatory Compliance: It needs to ensure before the deployment of system it fulfills all relevant criterions of automotive and communication regulatory standards.

Interoperability: The solution is to be interoperable with other systems and infrastructure components should be match with the VANET ecosystem. During implementing deep learning algorithms on OBUs, vehicles should perform intelligent decision-making capability, able to tasks autonomously, leveraging data from onboard sensors and communicating with other vehicles and roadside infrastructure efficiently within the VANET system.

4.2 WAVE Architecture (Communication Technology)

Figure 8 represents the architecture of Wireless Access in Vehicular Environments (WAVE) framework, which is crafted to reinforce communication in VANET. The WAVE architecture works on IEEE 802.11p standard and developed according to the standards of IEEE 1609 family. It helps in step up the vehicle safety, traffic management and supports vehicle to infrastructure (V2I) and vehicle to vehicle (V2V) communication.

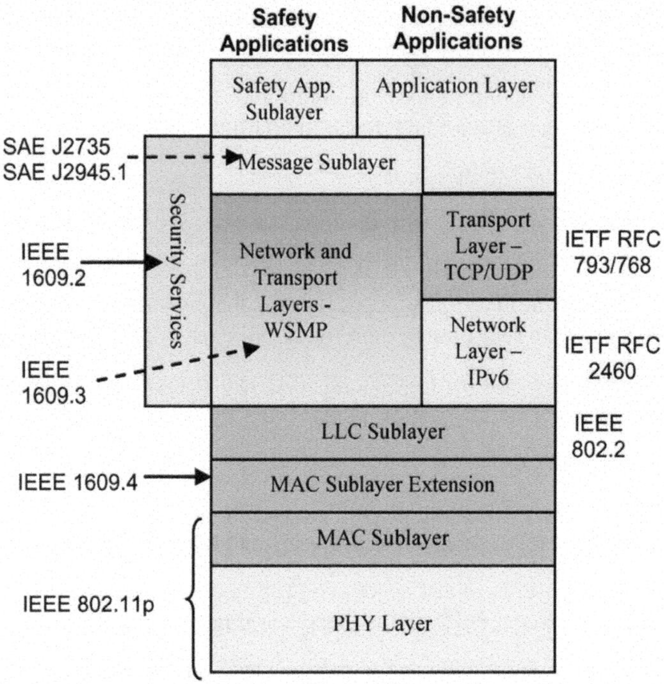

Fig. 8 WAVE architecture proposed by IEEE [3]

The WAVE Architecture (see Fig. 8) follows OSI Layered Architecture starting from Physical Layer to Application Layer. IEEE 1609.1 protocol defines access mechanism to the system resources that are a part of the OBU or RSU. It includes, processor, memory or UI. IEEE 1609.3: It's used for addressing and forwarding. It is an alternate protocol for Network Layer and Transport Layer. IEEE 1609.4: It is used for multi-channel operations. Its purpose is to provide simultaneous operational support for both safe and vulnerable applications.

The multi-channel environment provides us with a way to use safety and non-safety applications simultaneously. It also enhances the performance of the physical layer by supporting both applications at same time.

There are two types of channels—service and control, available to vehicle in a VANET architecture. Control channel paves the way for safety-related communications and service channel provides for either safe or vulnerable applications [21, 24].

IEEE 802.11p makes the physical layer of web architecture. Due to its cross-layered characteristics, it spreads through the physical and data link layer using OFDM scheme from IEE802.11a. The enhanced distributed channel access technology from IEEE 802.11e [31].

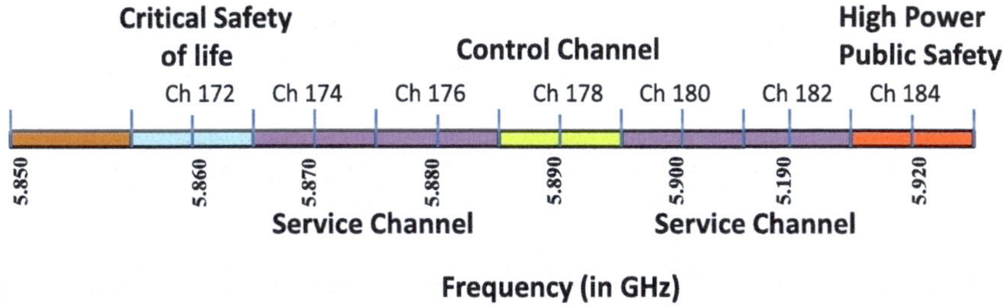

Fig. 9 Channel allocation [34]

IEEE 802.11p doesn't require the Basic Service Set (BSS) or central access point & this service would allow vehicles to exchange data at high speeds in a high-speed environment. Its beneficial as we have very small interconnect times between vehicles and between RSU and vehicles, so we don't want to spend too much time in authorization, authentication and permission from the base station, because then we miss the actual part which is data transfer [37].

SAE J2735 and SAE J2945.1 are at the top of WAVE Architecture and defines the semantics of the messages mostly related to safety like—roadside alerts signal status, emergency vehicle alert message etc. Multi-Channel Operation and working: It performs multi-channel operation for different range of applications. Its channel allocation is divided in three segments as shown in Fig. 9.

Figure 9 shows different features of channel allocation. The spectrum is made up of seven channels of 10 MHZ each. Control channel (labelled 178) announces WAVE service advertisement (WSA) and is used for safe applications. Channels from 172 to 182 are all reserved and are used for safety critical applications. Every channel is guarded by a 5 MHz guard band.

Service channels (174, 176, 180 and 182) are used for safe and vulnerable applications, both. Now, if a vehicle is interested in a particular service, it knows the channel number and its frequency, so it can tune its frequency to suit a particular channel number and can avail the service of that channel. Safety applications of VANET accessible through control channels such as in collision avoidance alerts the driver of any potential collision with the object ahead. Intersection collision warning: Provides warnings about potential collisions at intersections. Emergency electronic brake lights popup an alerts driver when a vehicle ahead is performing emergency braking, allowing them to react in time. Emergency vehicle notification generates alert for surrounding vehicles of indicating their presence to provide them way and enhancing reachability to needy place and people.

Curve Speed Warning is an important feature to advise the driver to reduce speed when approaching a curve at high speed. Road hazard notification notifies drivers of upcoming road hazards like debris, icy conditions, or roadwork. Intersection movement assists in gathering and provides information about the intersection status and recommends safe passage through the intersection.

Non-safety applications of VANET accessible through service channels are traffic management by using traffic signal controlling, dynamic routing guidance and infotainment services, electronic toll collection parking assistance, fleet management navigation etc. Dynamic route guidance helps us in finding the most efficient routes based on current traffic data, helping to reduce congestion. Infotainment services are provided through Internet access that allows Internet connectivity to passengers for browsing, streaming, and other online activities.

Multimedia streaming enables streaming of music, videos, and other multimedia content for passenger entertainment. Electronic toll collection allows automates toll payment processes, reducing delays at toll booths. Parking assistance helps us to identify parking availability and informs drivers of available parking spaces nearby. Automated payment facilitates automated payment for parking services as well. Fleet management gives real-time tracking and management of commercial vehicle fleets, improving logistics and operational efficiency.

Usage-based insurance is also an attractive feature that collects driving behavior data to offer insurance policies based on actual vehicle usage and driving patterns. Remote diagnostics and maintenance enable remote monitoring of vehicle health and predictive maintenance, alerting owners of potential issues before they become serious problems.

Nowadays, navigation based on augmented reality is being provided with vehicles where the map and other information gets displayed on the windshield, which improves the driver's awareness of the situation.

5 Integration of System

In this section, a brief about the working— communication, processing information and decision making, of the components of vehicular Ad-Hoc Network is given:

The On-Board Unit

The OBU orchestrates all communication that occurs within the vehicle and with external systems. Since the OBU is installed inside the vehicle, it must be sturdy enough to withstand the jerks and bumps experienced during vehicle movement.

Communication Antennas

Antennas are generally installed on the roof of vehicle to transmit and receive signals from the other vehicles and roadside installations using IEEE 802.11p and the 5.9 GHz band. This band is specially dedicated to vehicular communications.

Sensors and Data Gathering

A number of sensors are installed inside vehicles to gather data from other vehicles and roadside installations that are part of the larger network. LIDAR (for mapping distances), radar (for speed detection), ultrasonic sensors, and dashboard cameras (for visual information) are fitted in the vehicles and provide data to the OBU.

The Application Unit (AU)

The AU works in conjunction with the OBU. All decision-making occurs in the AU based on data received from sensors installed in vehicles, traffic lights, and other roadside installations. The AU is responsible for object detection, such as identifying other commuters, roadblocks, etc.

Deep Learning Algorithms

AU is the unit, which is responsible for running the deep learning algorithms. Deep learning models like convolutional neural network (CNN) performs image processing tasks such as detecting traffic signs, other vehicles and passerby. Recurrent neural network (RNN) or long short-term memory (LSTM) models may also be used by the AU for real-time and reliable decision making.

Embedding Algorithms

Powerful algorithms are embedded within the processor attached to the OBU. Since the size of the OBU is limited, the machine learning models are trained on huge data sets and further compressed and optimized to run smoothly on the OBU. Techniques of pruning and quantization are used to minimize the delay while relaxing the computational load of the models.

Safety and Control

The on-board unit continuously monitors incoming data from sensors and the AU to make real-time decisions. Algorithms detect potential collisions, prompting the OBU to signal the application of brakes or adjust the steering to prevent accidents. To achieve this, highly reliable algorithms and fast processors are essential, as even a small delay could be life-threatening.

Networking with Other Vehicles

The future of road transport lies in vehicles seamlessly communicating with each other, sharing information about roadblocks, potholes, crowds, or other points of interest as soon as one of them encounters it. This future envisions collective decision-making by vehicles operating within a network, working together in harmony. The integration of processing units, IoT devices, and advanced algorithms will enable efficient road transportation, saving both time and lives—this is the future that VANET promises.

6 Conclusion

As a concluding remark, it can be said that VANET provides a safe environment for transportation without negatively impacting efficiency and enriching user experience. The assembly of on-board units used in the vehicles and roadside units facilitate communications among vehicles and between a vehicle and infrastructure. Thus, on-board units are the core component of a VANET based node that collects information from sensors installed on the vehicles and roadside, and transmit the same to other participants in the network. Deep learning algorithms empower on-board units in decision making by identifying objects, lane detection, and prediction of behavior. WAVE architecture, meanwhile, provides a standard solution for communication by ensuring reliable exchange of data and supporting interoperability. VANET along with the ongoing advancement in the hardware and software technology coupled with the mighty deep learning models have the potential to provide the safest road transportation network, the humanity has witnessed to date.

As a scope for future work, the anomaly detection algorithms may be used in conjunction with VANET to identify unusual patterns in network traffic and thwart attempts at intrusion. Also, work on protocols to improve and define the interoperability standards in VANETs and connected devices is needed. Additionally, green algorithms that minimize power consumption while not compromising performance are also required to be developed. Work on user experience is also required to provide user-centric navigation, advanced infotainment systems, traffic alerts, etc.

References

1. Afzal Z, Kumar M (2020) Security of vehicular ad-hoc networks (VANET): a survey. J Phys Conf Ser 1427(1):012015. https://doi.org/10.1088/1742-6596/1427/1/012015
2. Al Hasan AS, Hossain MS, Atiquzzaman M (2024) Security threats in vehicular ad-hoc networks. In: Conferences on advances in Computing, Communication and Informatics, pp 22–24
3. Arena F, Pau G, Severino A (2020) A review on IEEE 802.11p for intelligent transportation systems. J Sensor Actuat Netw 9(2):22. https://doi.org/10.3390/jsan9020022
4. Change.org, https://www.change.org/t/erp-en-gb Last Accessed On: 30-July-2024
5. Cucor IB (2021) Outlines of vehicular ad-hoc networks. Transp Res Proc 55:1312–1319. https://doi.org/10.1016/j.trpro.2021.07.115
6. Dewsoft (2024) https://dewesoft.com/blog/what-is-can-bus. Accessed 30 Jul 2024
7. Dharmendra S, Pradhan S (2010) Data dissemination techniques in vehicular ad hoc network. Int J Comput Appl. https://doi.org/10.5120/1240-1725
8. de Fuentes JM, Gonzalez-Tablas AI, Ribagorda A (2010) Overview of security issues in vehicular ad-hoc networks. In: Handbook of research on mobility and computing, IGI Global
9. Feng Y, He D, Niu L, Yang M, Guan Y (2017) The overview of Chinese cooperative intelligent transportation system. In: Vehicular Communication Application Layer Specification and Data Exchange Standard 1
10. Ghori MR, Zamli KZ, Quosthoni N, Hisyam M, Montaser M (2018) Vehicular adhoc network (VANET): review. In: 2018 IEEE International Conference on Innovative Research and Development (ICIRD), pp 1–6. https://doi.org/10.1109/ICIRD.2018.8376311
11. Horang S, Tzeng S, Li T, Wang X, Huang P, Khan M (2015) Enhancing security and privacy for identity based batch verification schemes in VANET. IEEE Trans Veh Technol
12. Jantosova A, Dolnak I, Dado M (2019) An overview of vehicular ad hoc networks. In: 17th International Conference on Emerging ELearning Technologies and Applications (ICETA), pp 305–308. https://doi.org/10.1109/ICETA48886.2019.9040098

13. Karabulut MA, Shah AFMS, Ilhan H, Pathan A-SK, Atiquzzaman M (2023) Inspecting VANET with various critical aspects—a systematic review. Ad Hoc Netw 150:103281. https://doi.org/10.1016/j.adhoc.2023.103281
14. Kelotra A, Pandey P (2021) Adaptive optimization-based deep convolutional long short-term memory for bank NIFTY index prediction. Int J Wavelets Multiresolut Inf Process. https://doi.org/10.1142/S0219691320500873
15. Kelotra A, Pandey P (2020) Stock market prediction using optimized deep-ConvLSTM model. Big Data 8(1):5–24. https://doi.org/10.1089/big.2018.0143
16. Lim KG, Lee CH, Chin RKY, Beng Yeo K, Teo KTK (2017) SUMO enhancement for vehicular ad hoc network (VANET) simulation. In: 2017 IEEE 2nd International Conference on Automatic Control and Intelligent Systems (I2CACIS), pp 86–91. https://doi.org/10.1109/I2CACIS.2017.8239038
17. Litoriya R, Gulati A, Yadav M, Ghosh RS, Pandey P (2021) Social, ethical, and regulatory issues of fog computing in healthcare 4.0 applications. In: Fog Computing for Healthcare 4.0 Environments (pp. 593–609). Springer, Cham. https://doi.org/10.1007/978-3-030-46197-3_23
18. Malik S, Sahu PK (2019) A comparative study on routing protocols for VANETs. Heliyon 5(8):e02340. https://doi.org/10.1016/j.heliyon.2019.e02340
19. Martinez FJ, Toh CK, Cano J, Calafate CT, Manzoni P (2011) A survey and comparative study of simulators for vehicular ad hoc networks (VANETs). Wirel Commun Mob Comput 11(7):813–828. https://doi.org/10.1002/wcm.859
20. Mehedi SK, Anwar A, Rahman Z, Ahmed K (2021) Deep transfer learning based intrusion detection system for electric vehicular networks. Sensors 21:4736. https://doi.org/10.3390/s21144736
21. Pandey P, Litoriya R (2021) Software process selection system based on multicriteria decision making. J Softw Evol Process 33(2). https://doi.org/10.1002/smr.2305
22. Pandey P, Litoriya R (2020) Implementing healthcare services on a large scale: challenges and remedies based on blockchain technology. Health Policy Technol 9(1):69–78. https://doi.org/10.1016/j.hlpt.2020.01.004
23. Pandey P, Litoriya R (2020) Promoting trustless computation through blockchain technology. Natl Acad Sci Lett 44:225–231. https://doi.org/10.1007/s40009-020-00978-0
24. Pandey P, Litoriya R (2020) Securing E-health networks from counterfeit medicine penetration using blockchain. Wireless Pers Commun 117:7–25. https://doi.org/10.1007/s11277-020-07041-7
25. Pandey P, Litoriya R (2020) An IoT assisted system for generating emergency alerts using routine analysis. Wireless Pers Commun 112(1):607–630. https://doi.org/10.1007/s11277-020-07064-0
26. Pandey P, Litoriya R (2020) Ensuring elderly well-being during COVID-19 by using IoT. Disaster Med Public Health Prep 16(2):763–766. https://doi.org/10.1017/dmp.2020.390
27. Pandey M, Litoriya R, Pandey P (2019) Application of fuzzy DEMATEL approach in analyzing mobile app issues. Prog Comput Softw 45(5):268–287. https://doi.org/10.1134/S036176881950050
28. Pandey M, Litoriya R, Pandey P (2020) Applicability of machine learning methods on mobile app effort estimation: validation and performance evaluation. Int J Softw Eng Knowl Eng 30(1):23–41. https://doi.org/10.1142/S0218194020500023
29. Peserico G (2019) Università degli Studi di Padova Development of functional safety applications for Autec products. Study of protocols: CANopen, CANopen Safety, FSOE and ProfiSafe Candidato: Relatore. https://doi.org/10.13140/RG.2.2.31726.36161
30. Raw RS, Kumar M, Singh N (2013) Security challenges issues and their solutions for VANET. Int J Netw Secur Appl
31. Ruchita, Shankar R (2022) Electric vehicles charging infrastructure framework for smart transportation. Distrib Generat Altern Energy J. https://doi.org/10.13052/dgaej2156-3306.3815
32. Sharma U, Pandey P, Kumar S (2022) A transformer-based model for evaluation of information relevance in online social-media: a case study of COVID-19 media posts. N Gener Comput. https://doi.org/10.1007/s00354-021-00151-1

33. Singh A, Mishra R, Kumar R (2016) VANET security: issues, challenges and solution. In: International conference on Electrical Electronics and optimization techniques
34. Špitálová Z (2023) Vehicle-to-everything communication. Commun Sci Lett Univ Zilina 25(1):C24–C35. https://doi.org/10.26552/com.C.2023.017
35. StraitsTimes. https://www.straitstimes.com/singapore/transport/new-erp-system-to-start-2023-but-no-distance-based-charging-just-yet. Accessed 30 Jul 2024
36. Verma A, Nawaz S, Singh S, Pandey P (2021) Importance of 5G-enabled IoT for industrial automation. In: Blockchain for 5G-enabled IoT—the new wave for industrial automation. Springer, Cham
37. Zhang K, Olmsted A (2021) Examining autonomous vehicle operating systems vulnerabilities using a cyber-physical approach. In: 2021 IEEE International Intelligent Transportation Systems Conference (ITSC), pp 976–981. https://doi.org/10.1109/ITSC48978.2021.9564848

Impact of Deep Learning on VANET Security: An Analysis

Suja Cherukullapurath Mana, A. Clara Kanmani, and Arpita Chaudhuri

Abstract Vehicular Ad Hoc Networks (VANETs), which enhance traffic efficiency and road safety, are becoming a crucial component of intelligent transportation systems. However, the dynamic and decentralized nature of VANETs makes them susceptible to numerous security threats, such as Denial of Service (DoS), Sybil attacks, and data manipulation. Existing security solutions encounter challenges in extracting features and managing intricate network dynamics, mainly because they rely on conventional cryptography methods and machine learning models. Motivated by the need for more robust security protocols, this chapter investigates the application of deep learning (DL) methods for VANET security. This chapter presents a detailed analysis of deep learning framewch as DeepVCM and DeepADV for intrusion detection and anomaly detection, respectively, demonstrating their superiority over conventional methods. Additionally, few DL-based solutions for secure communication and trust management in VANETs, including a framework for Sybil attack detection and trust computation using deep neural networks are discussed in this chapter. The comparative study of mitigation strategies highlights the effectiveness of DL in automating feature extraction and improving detection accuracy. This chapter presents multiple methods of notable breakthroughs in VANET security by examining these DL procedures, ensuring a more reliable and safe vehicular communication environment.

1 Introduction

Worldwide Health Organisation (WHO) figures indicate that road traffic accidents cause between 1.15 million and around 50 million non-fatal injuries and deaths annually [1]. Governments have looked for feasible methods to lower the annual number of road accidents. Since autonomous driving systems in Intelligent Transport Systems (ITS) may reduce the impact of human elements on the entire traffic system, they become potential solutions for the advancement of the Internet of Things (IoT).

S. C. Mana (✉) · A. C. Kanmani · A. Chaudhuri
PES University, Electronic City Campus, Bengaluru, India
e-mail: cherukullapurathsuja@gmail.com

J. Bhatia et al. (eds.), *Deep Learning Based Solutions for Vehicular Adhoc Networks*,
Studies in Computational Intelligence 1207,
https://doi.org/10.1007/978-981-96-5190-0_4

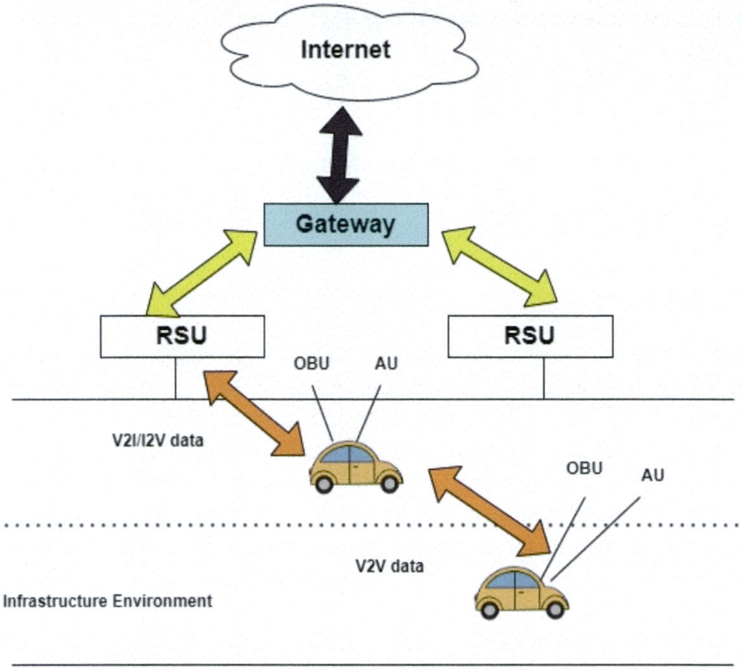

Fig. 1 Basic architecture of VANET

Furthermore, due to the close connection between autonomous driving and VANETs, VANET security has emerged as a crucial area of study in recent years [2]. Vehicle-to-vehicle (V2V) and vehicle-to-infrastructure (V2I) networks (Fig. 1) are commonly made up of wireless communication channels that allow information to be transmitted between vehicles and roadside devices. It facilitates the transmission of emergency notifications, improves traffic efficiency and road safety, and offers drivers practical services [3]. Because of the high degree of topological complexity, the high degree of vehicle mobility, and the diversity of communications, VANETs are susceptible to a wide range of cyberattacks, including wormhole, blackhole, DOS, and Sybil attacks [4–7]. Besides attacks, numerous aspects, including node collaboration, data security, user privacy, the quality of transferred information, authenticity, and so forth, influence the success of VANET applications linked to data dissemination.

Despite enormous research efforts, VANET research has not yet reached full commercialization, in part because of the issues that were previously described that still require resolution. Practical security and privacy are two of the most crucial needs that must be efficiently satisfied among the many obstacles in VANET. Authentication, privacy, confidentiality, trust and reputation, attack detection, misbehavior detection, and mitigation, and access control are just a few of the security features that have been thoroughly studied in VANET to date [6, 8–10]. Certain security issues are either directly or indirectly related to one another, even though they require separate procedures to counteract potential attacks in VANET. For example, there are relationships between authentication, reputation, trust, and misbehavior detection.

Numerous cryptographic approaches have been put forward in the past to address various security-related problems [11]. Vehicles in automotive networks can also be authenticated using certain popular traditional authentication methods, such as password protection, key-based authentication, and biometric security approaches. These methods, however, fail short of verifying whether the transferred value is authentic or fake. Furthermore, achieving high accuracy and using these techniques in low-powered vehicle security systems can be challenging. Machine learning (ML) has attracted a lot of attention recently as a means of addressing vehicle security and obtaining quicker and more precise attack predictions. In the realm of wireless networks nowadays, ML is one of the most promising technologies [12–15]. The literature uses a variety of machine-learning algorithms for various wireless applications [16]. In recent years, deep learning has gained attention in the solution of VANET security [7, 17–22] due to automatically identifying and extracting features from raw data without the requirement for complex feature engineering.

In the context of enhancing the security of VANET using deep learning-based approaches, this chapter summarizes the applicability of deep learning approaches in various aspects of VANET security. Moreover, the chapter highlights the importance of security in VANET, which is facing promising challenges. In addition, the comparative study mentioned in this chapter helps to get a quick idea regarding the differences between security attacks and the strategies for mitigating them.

The key contributions of this paper include:

- The paper investigates various deep learning methods to ensure secure vehicular communications in VANETs. Specifically, it explores how DL-based Intrusion Detection Systems (IDS), anomaly detection, secure communication protocols, and trust management can address cyber threats like DOS attacks, Sybil attacks, and data integrity issues.
- The paper introduces frameworks like DeepVCM and DeepADV that leverage deep learning for intrusion detection and anomaly detection. These frameworks are designed to efficiently detect network attacks and improve the accuracy and performance of vehicular network security.
- A comparative study is conducted on different types of security attacks and strategies, showing how DL-based models can outperform traditional methods in terms of accuracy and resource efficiency.

The rest of the chapter is organized as follows: Sect. 2 highlights the importance of security in VANETs. Application of deep learning in VANET security is discussed in Sect. 3. Finally, Sect. 4 concludes the chapter.

2 Importance of Ensuring Security in VANET

People drive their vehicles for extended periods of time these days. Autonomous vehicle networks or VANETs, are a subset of mobile ad hoc networks or MANETs. Smart cars are designed to make that time safer, more enjoyable, environmentally friendly, and rational congress in the US approved the creation of the Intelligent Vehicle High-way Systems (IVHS) program to improve road safety, lessen traffic jams, and clean up the air. In this sense, enhanced driving conditions and safety have been made possible by IVHS development. As the medium employed in VANET has drawbacks that can make the network vulnerable, there are many different kinds of security threats in VANET, including interference, jamming, and eavesdropping as media, which can make the network susceptible, has drawbacks. There are several VANET architectures. WAVE is among the most often utilized ones. The IEEE [3] suggested a system architecture called WAVE (wireless access in vehicle environment) (UZCATEGUI; SUCRE; ACOSTA-MARUM, 2009). WAVE adheres to the layered architecture of Open System Interconnection (OSI). It also includes two protocol stacks namely, the Wave short message protocol (WSMP) and Internet Protocol version six (IPv6). For Web-based applications, traditional IPv6 is used; for time-sensitive applications, the WSMP is suggested (IEEE Std 1609.3, 2010). Furthermore, a number of standard protocols for WAVE architecture have been specified by the IEEE 1609 working groups (Fig. 2).

From the Fig. 2 , it is inferred that the wave architecture is composed of many sublayers and various IEEE standards pertaining to it. The functions of each layer is discussed as follows:

- **Application Layer**: Supports different applications, including safety (e.g., impact evasion, traffic cautions), infotainment, and route administrations.
- **Transport Layer**: For non-security administrations, standard TCP/UDP protocols are utilized.
- **Network Layer**: Supports both WSMP and IPv6 for correspondence.
- **MAC Layer**: Essentially carries out the IEEE 802.11p Macintosh.
- **Physical l Layer**: Works in the 5.9 GHz band utilizing DSRC.

WAVE architecture is the foundation of VANET communication, empowering vehicles to securely and productively communicate with one another and framework. The requirement for security is one of the important topics to be discussed in the present scenario. Information is a vital asset for an organization. Computer security is the protection of information from any kind of harm, or theft and preventing unauthorized users from accessing resources. The basic security objectives for VANET are authentication, confidentiality, integrity, and availability as discussed in [23].

- **Authentication**: To avoid impersonation in vehicular communication, it is crucial to verify the identity of the message sender.
- **Confidentiality**: It is imperative that any organization's privacy be protected. Confidentiality should be ensured by using directional antennas and encrypted data.

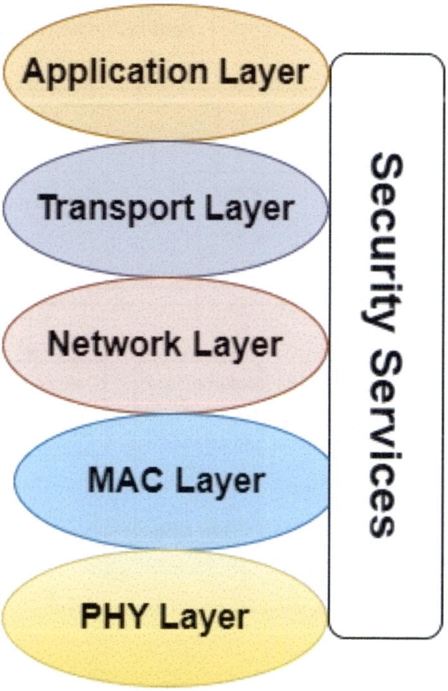

Fig. 2 WAVE architecture

- **Integrity**: To stop hackers from changing messages, the intactness of all messages needs to be safeguarded.
- **Non-repudiation**: By using it, we will be able to create the capacity to recognize the attack even after it has occurred. It keeps those who commit deception from denying their misconduct.
- **Availability**: The requirement that authorized users have access to network resources even while the communicating entities are being attacked via alternate means without impairing the network's functionality.

This section describes the security objectives and challenges in VANET, a list of cryptographic solutions proposed, and a comparative analysis of various solutions.

2.1 Security Objectives in VANET

There are a number of challenges when the proposed application for VANET is implemented. One of the foremost objectives is scalability in terms of creating a system to handle a larger number of nodes, especially handling a larger number of vehicles in a smaller area is a very common occurrence [24]. Hussein et al. [25] explored other issues such as fluctuating mobility, limitations on movement, frequent

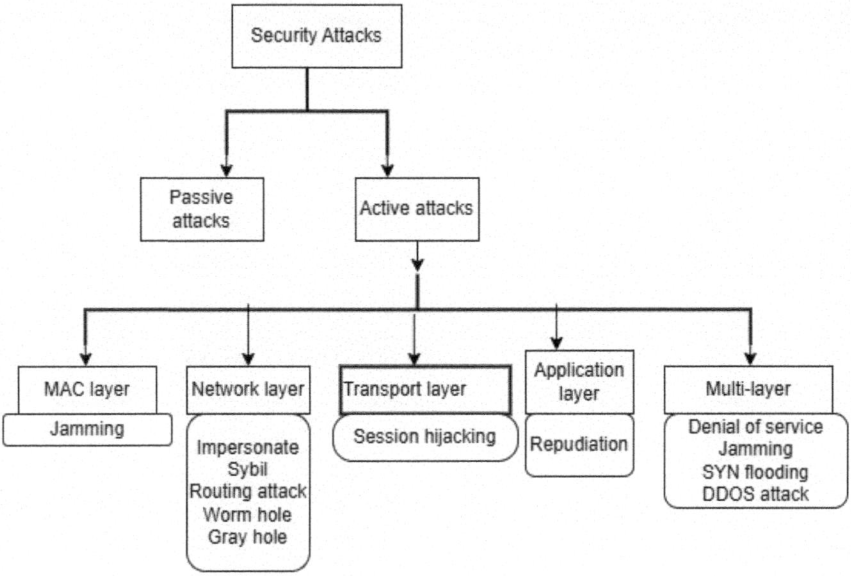

Fig. 3 Classification of attack based on layers

network fragmentation, heterogeneity, limitless power and computational resources, limited availability of spectrum, the accuracy of the information, low fault tolerance, and data privacy.

However, in the existing works [26, 27], the authors mentioned that every message in vehicular communication needs to be validated in order to guarantee its origin and manage the vehicles' authorization level. To accomplish this, the vehicles will assign a private key and certificate to each message. Once the recipient has received the message, they will check it for the key and certificate, and once this is done, they will verify the message.

There are some other attacks in VANET discussed in the paper [28], such as attacks on availability such as blackhole, sinkhole, broadcast tampering, DOS, and other attacks on authentication namely, message replay, Sybil, masquerading and location service spoofing. These attacks are also classified based on layers [29] and depicted in Fig. 3.

The above attacks shown in Fig. 3 are discussed briefly below.

- **Jamming**: A circumstance that an attacker intentionally sends a sequence of messages or signals to make vehicles incapable of using the wireless channel.
- **Impersonate**: An authenticity of the legitimate node is stolen by the attacker.
- **Sybil**: It is an attack on VANET in which multiple entities are stolen maliciously by the intruder and these identities are used to interrupt the functionality of VANET.
- **Routing attack**: An attempt to exploit the vulnerability in VANET architecture to split into multiple isolated components.

- **Wormhole**: It is a kind of DOS attack interrupting the routing of a network which affects multicasting and broadcasting routing.
- **Grey Hole**: This is a scenario where malicious nodes are constantly dropping random nodes which in turn decreases the network performance.
- **Session hijacking**: It is a scenario of taking control of a user's web session.
- **Repudiation**: This attack occurs when a malicious user manipulates the data on behalf of others and false data is saved in log files.
- **Denial of service**: In the context of VANETs, a denial-of-service (DoS) assault is any activity that impedes the availability or proper operation of a network [30].
- **SYN flooding**: It is one category of DOS attack where the server is unavailable to legitimate traffic.

2.2 *List of Cryptographic Solutions Proposed*

The fundamental structure of VANET, various security mechanisms, and recent work on security and privacy on VANET are discussed in the paper [31]. Kau et al. [32] presented a full outline of VANET for example, framework model, plan, conventions, and Challenging problems, after a review of numerous analyses on cutting-edge methods for managing security and protection in VANETs. Additionally, in order to offer safe communication, the security architecture needs to be upgraded with suitable verification mechanisms and authentication procedures. Systems that are now in place in VANET for detecting various types of attacks, including jamming, Sybil, black hole, and collision attacks, are discussed.

2.3 *Comparative Analysis of Various Solutions*

These attacks usually target the data that is transferred over the internet. For this reason, it's critical to be aware of potential threats so that you can respond appropriately. It is therefore important to be aware of possible attacks in order to react to them. A detailed view of various vulnerabilities in VANET is discussed in [33]. Security mechanisms for VANET vary from model to model. Most of the mechanism includes public key infrastructure, situation modeling-based mechanism, and ID-based cryptography [34]. Table 1 describes the mitigation strategies for various security issues and attacks on VANETs [31].

From the Table 1, it is observed that the use of strong cryptographic techniques, implementation of IDS, and end-to-end encryption are the widely used techniques for overcoming several attacks on VANET.

Al-Shareeda and Manickam [35] discussed VANET which is based on the VEINS framework. Filtering was done based on security considerations. The first category

Table 1 Mitigation strategies for security issue/attack on VANETs

Security issue/Attack on VANETs	Description	Mitigation strategies
Authentication and authorization	Ensures the identity of communicating entities and grants appropriate access rights	Use of strong cryptographic techniques for secure communication. Implementation of robust authentication protocols. Integration of access control mechanisms to restrict unauthorized access (RBAC, ABAC)
Sybil attack	Creation of multiple fake identities to deceive the network	Employ reputation-based systems to detect and isolate Sybil nodes. Collaborative approaches for detection involving neighboring vehicles. Use of secure positioning systems to mitigate identity spoofing
Denial of service attack	Intentional disruption of network services or communication	Implementation of intrusion detection systems (IDS) to identify and filter malicious traffic. Traffic prioritization for critical safety messages
Location spoofing	Manipulation of reported location information	Adoption of secure positioning systems. Time-synchronization mechanisms to detect replay attacks. Cooperative awareness for cross-verification of reported positions
Eavesdropping	Unauthorized interception of wireless communications	Implementation of end-to-end encryption for message confidentiality. Use of intrusion detection systems (IDS) to monitor for anomalous patterns. Continuous pseudonym changes to enhance privacy

consists of threats and vulnerabilities. Using the VEINS framework, the consequences of threats and vulnerabilities on VANETs are assessed, and several mitigation strategies are recommended. Articles about the application of blockchain, artificial intelligence, and Software-Defined Networking (SDN) techniques in VEINS-based VANET applications are included in the second category. The third category comprises the specifications required to satisfy the revocation, privacy, authentication, trust management, and dependability requirements of the VANET security-based VEINS architecture.The system concept, design, protocols, safety, and challenging concerns related to VANET security and privacy were all carefully studied in [36], An appropriate authentication system is required to address the security risks in the future.

3 Application of Deep Learning in VANET Security

High dependability and low latency are necessary for the operation of safety models based on VANETs. All trust-based security methods currently in use, however, have intricate iterations. These have the potential to increase network latency and reduce VANET performance. Apart from that, these conventional methods need a lot of training data and accurate category labeling by knowledgeable persons in order to handle a variety of difficulties in diverse and dynamic environments. As a result, machine learning and deep learning algorithms prove to be effective methods for identifying and analyzing security problems in VANETs. The model's capacity to manage the dynamism of vehicular networks determines how well ML/DL performs. Deep learning has been applied extensively in a number of domains recently to enhance detection performance. When compared to conventional machine learning techniques, deep learning can automatically identify and extract features from unprocessed data without the need for intricate feature engineering. This lowers the amount of resources required for testing and training while also increasing the accuracy of detection. Many VANET security features, including intrusion detection, anomaly detection, secure communication, and trust management, show great potential for deep learning. The use of deep learning algorithms in the four categories described above will be examined in the ensuing subsections.

3.1 Intrusion Detection Systems for VANET Using Deep Learning

VANETs are the underlying network for intelligent transport systems. VANETs are helpful in increasing security, minimizing road traffic congestion, and improving the safety of passengers and drivers. Still, due to the size and characteristics of VANET, it is prone to many types of attacks. An effective intrusion detection system is necessary to prevent many such attacks from happening. Attacks like DOS attacks, Sybil attacks, etc., can be prevented by using an efficient intrusion detection system. DL techniques can be used to implement intrusion detection systems(IDS) effectively. This section of the article studies some of the deep learning-based intrusion detection system implementations.

One of the DL-based intrusion detection systems is discussed by authors Zeng et al. [17]. Vehicles participating in a VANET consist of onboard units (OBU) and roadside units [17]. As discussed in the article [17], intrusion detection methods like misuse-based technologies and anomaly-based technologies have some shortcomings. Misuse-based technologies are only capable of detecting known attacks, and anomaly-based techniques using machine learning encounter difficulties in feature selection, data processing, etc. The proposed deep learning-based solution can address some of the above-discussed challenges [17]. Deep learning-based methods have efficient feature extraction capabilities and they are capable of attaining high performance with traffic data. It has low resource requirements also. The proposed DeepVCM analyzes the traffic going through VCM. The core components of this DL architecture are the convolutional neural networks (CNN) and Long Short-Term Memory (LSTM) unit.

The DeepVCM [17] proposed by Zeng et al. consists of mainly two functional modules namely the preprocessing procedure and the DeepVCM procedure. An overview of DeepVCM module [17] from the article is redrawn in Fig. 4. The first step is the package capture tool [37]. It will capture the flow of packets in both directions. Traffic cleaning is happening in step 2. In the traffic cleaning step inference data

Fig. 4 Overview of DeepVCM module

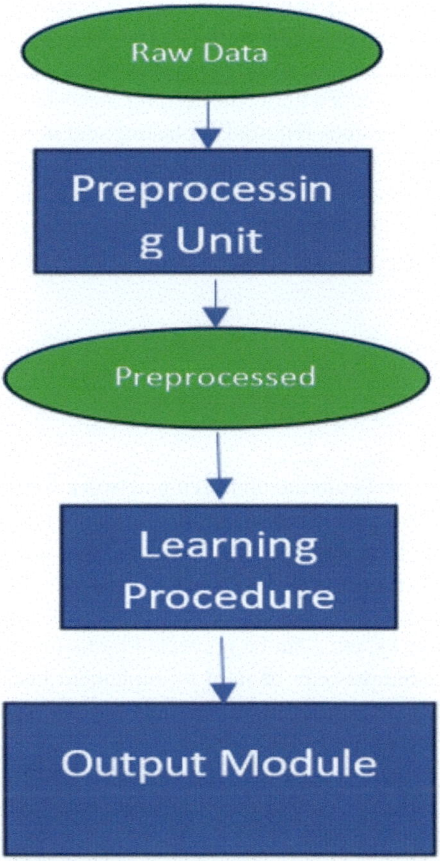

is removed. This data includes the header information like UDP, TCP header, and MAC addresses as it may interfere with the result produced. As discussed by the authors [17] the third step involves traffic purification which involves the removal of duplicate and empty files. The next step is a unification of length which will shorten files longer than 900 bytes. In the last step, it converts a uniform PCAP file into a two-dimensional IDX file. The steps in preprocessing are diagrammatically represented [17] in Fig. 5.

The DeepVCM procedure mainly consists of two components namely CNN [38] and LSTM [39]. As a first step, the information is extracted using a one-dimensional convolutional neural network. The reason behind selecting one-dimensional CNN over two-dimensional CNN is that most of the traffic data is a sequence organized by hierarchy. The second part of DeepVM is the LSTM. Time-dependent features are learned with the help of LSTM. Output from the first level can be given as input to the second level. DeepVM architecture is shown in Fig. 6 [40].

The initial part of the DeepVCM architecture contains two convolution layers, pooling layers and LRN layers [17]. ReLu activation function is used in the convolution layer. Convolution layers process data using 32 filters. The results of this convolution operation are given to the activation layer. ReLu activation function is used in this implementation. The next layer is the pooling layer. The Max pooling

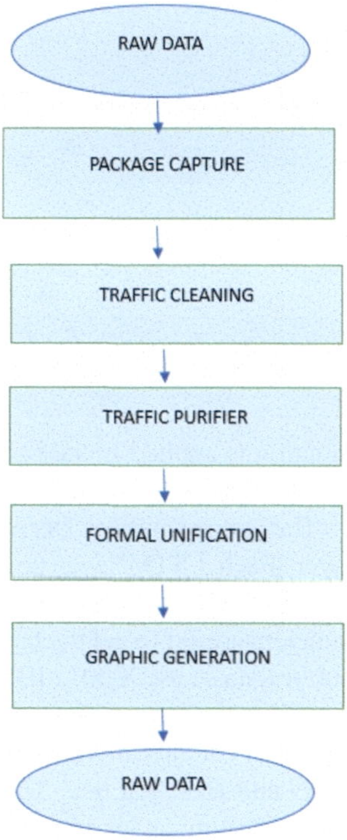

Fig. 5 Steps in preprocessing

Fig. 6 DeepPVCM architecture

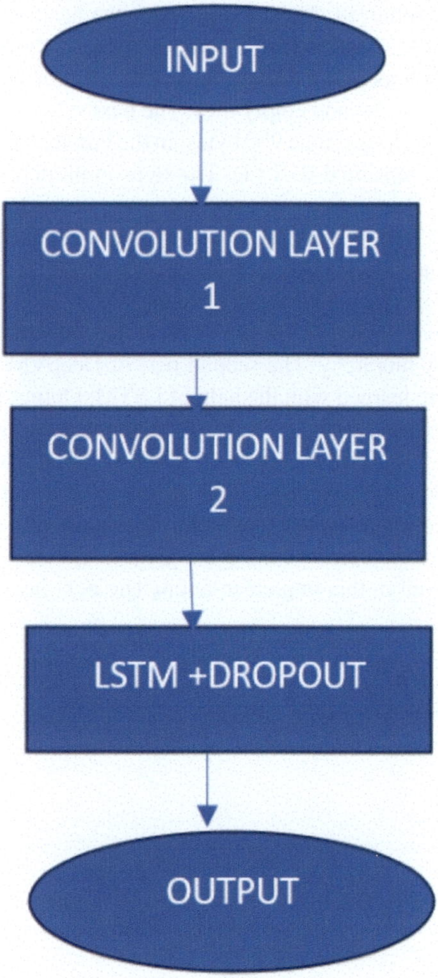

function is applied in the pooling layer with stride value 3. Then the input is given to the LRN layer.

The second part of DeepVCM architecture consists of LSTM and the dropout layer. Each LSTM layer consists of 256 LSTM cells. A densely connected layer follows the LSTM layer. At the end, the data will flow to a SoftMax layer. Hyperparameters are set to get the best results. The performance of DeepVCM is tested on a public dataset ISCX 2012 IDS dataset [40]. A wireless channel is selected for simulation. The performance is evaluated based on 3 measures namely precision, recall, and F1 score. Outputs are further compared with outputs of two other algorithms namely KNN and Decision tree. The tabular and graphical representation of the results of the Infiltrating attack is given [17] in Table 2. From the results obtained, it is clear

Table 2 Performance evaluation chart

Algorithm	Precision	Recall	F1 Score
KNN	89.21	87.23	0.882
Decision tree	86.52	87.83	0.871
DeepVCM	95.45	96.97	0.92

that the proposed DeepVM DL-based technique is highly successful in intrusion detection in VANETs.

3.2 Anomaly Detection in VANET Using Deep Learning

Deep learning can be effectively utilized for road anomaly detection. Deep learning algorithms like residual convolutional neural networks, VGG 16 [41] can be effectively utilized for anomaly detection. As the number of vehicles connected in a VANET is huge, the possibilities of anomalies also increase exponentially. In the work by authors Alladi et al. proposed an anomaly detection framework using deep neural networks. The deep neural network architecture is deployed on roadside units (RSU). This RSU receives the broadcast vehicular data and it runs anomaly detection tasks. The main component of this DNN-based anomaly detection framework is an anomaly detection framework capable of identifying anomalous data. Mainly three types of anomalies namely fault only, attack only, and a mix of fault and attack are considered in this. The dataset Kamel et al. [42, 43] is used for experimentation. Five steps involved in the proposed DeepADV [44] architecture are listed below.

1. In the first step, messages are extracted from the data exchanged in the network.
2. In the second step, data is cleaned and converted into sequences.
3. These sequences will be then passed into a sequence reconstruction unit in the third step.
4. In the next step, the error between the original sequence and the reconstructed sequence is calculated.
5. Finally, use the potential threshold value to classify sequences as anomalous or genuine.

The architecture diagram of the proposed DeepADV is given in Fig. 7. The performance comparison of the proposed model with some other existing models for all anomaly scenarios is given in the given Table 3 [44]. Here M1 is our proposed model. M2 is the CNN-LSTM bidirectional model, M3 is LSTM attention and M4 is the autoencoder model. The table shows that the proposed model outperforms in all aspects.

In order to discover network anomalies, Dhadhania et al. [18] integrate Graph Neural Network (GNN) with Software-defined networking (SDN), focusing on a variety

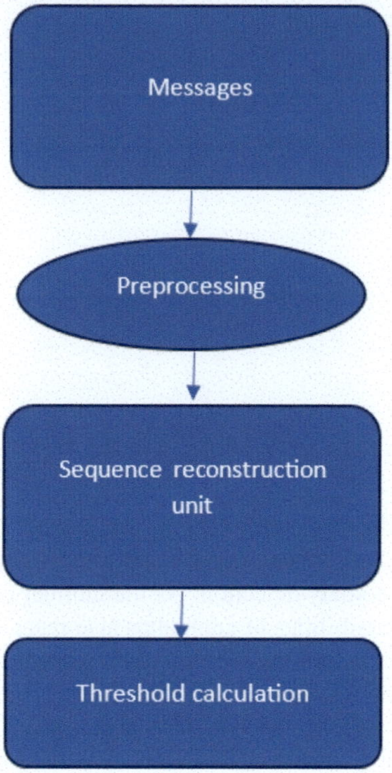

Fig. 7 DeepADV architecture

Table 3 Performance comparison chart

Model	Class	Precision	Recall	F1	Accuracy
M1	Genuine	0.969	0.997	0.983	0.980
	Anomaly	0.996	0.956	0.976	
M2	Genuine	0.959	0.996	0.977	0.973
	Anomaly	0.995	0.941	0.967	
M3	Genuine	0.938	0.998	0.967	0.960
	Anomaly	0.997	0.908	0.950	
M4	Genuine	0.938	0.995	0.965	0.958
	Anomaly	0.992	0.908	0.948	

of trends, concerns, and obstacles. They emphasized about the benefits of adopting SDN to provide network monitoring with programmable and fine-grained management. Simultaneously, GNN is capable of learning representations from graph-structured input and modeling network behavior as a graph. The limits of rule-based approaches and other conventional anomaly detection techniques in SDN are also covered by the authors, along with how GNN may be able to get around these drawbacks by utilizing its capacity to identify dynamic and non-linear patterns in network traffic. Additionally, a case study of DoS attack detection with SDN is presented in this research.

3.3 Secure Communication in VANET

Through Vehicle-to-Vehicle (V2V) and Vehicle-to-Infrastructure (V2I) communication, VANET enables vehicles to exchange messages for various applications including life safety, road efficiency, and infotainment. If a vehicle receives inaccurate information, it can have unfavorable effects, increasing the likelihood of accidents and traffic jams.

A useful framework for VANET security and Sybil attack detection was presented by Nova et al. [45]. There are five stages in the framework. Using the kernel K-harmonic means (KKHM) clustering technique, the VANET region was initially separated into several clusters in order to achieve a faster delivery ratio and a shorter propagation latency. In this case, kernel functions are included in the cost function of the k-means problem by kernelization in the feature space. It is thought that cluster centers are located in the feature space, a higher-level Hilbert space. In the subsequent, cluster heads are selected via the Floyd-Warshall algorithm (FWA). Using FWA, the shortest paths between every pair of vertices are quickly determined. A deep learning technique called The GBEHO-DNN is utilized to identify the malicious nodes. A method for optimizing Deep Neural Network parameters is called GBEHO. Finally, the data in the Cluster Head (CH) is securely sent to the cloud using the Modified Advanced Encryption Standard (MAES). The proposed architecture is shown in the following Fig. 8.

Shen et al. [46] developed a straightforward group key generation and verification procedure in addition to a lightweight authentication protocol to avoid intricate processing. In order to lower the cost of communication and the frequency of the group key updating phase, the authors have created a dynamic batch-based group key updating approach. In order to keep the process quick and safe, the authors modified a hash function and utilized it in place of a straightforward ID-based approach for authentication.

The next phase is proposing a dynamic time slot modification approach to monitor the vehicle making the request and the vehicle's departure. In order to manage dynamic time slots, an LSTM model is used. In the first step, the historical dataset, which includes distance, angle, and trend in its vicinity, is gathered for every RSU based on recent traffic flow. The dataset can be stored in the cloud within the allotted time, taking into account the Road Side Unit (RSU)'s storage capacity. Following the historical dataset's data normalization, RSU must further gather a sample of real-time queries. After that, they attempted to compute several similarity metrics and extract the modeling trajectories for every cloud technique using the real-time sample and dataset.

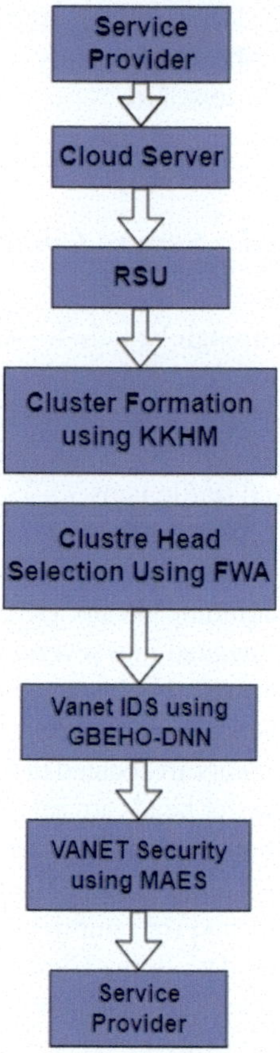

Fig. 8 GBEHO architecture

3.4 VANET with Trust Management

In VANET, modeling trustworthy peers in a road journey presents significant obstacles. There are two conditions that can be used to describe the significant obstacles that VANET faces. The network's nodes are mostly extremely dynamic and move within it constantly. A node traveling on the road usually travels at a speed of 100 to 120 km/h. Peers must be able to verify incoming information because responding to an impending event in real time at this high speed on the highway is crucial. Moreover, it would be expected that the VANET's node count could increase at any time. For instance, in urban locations, the number of nodes in the network is constantly ten thousand, and it will increase significantly during rush hours. This causes network congestion, which presents a number of problems. Additionally, because the VANET is a shared channel network, peers experience information overload during rush hour as a result of receiving a lot of information from other peers.

Thus, the possibility exists for intelligent vehicle communication systems to react in potentially hazardous circumstances. Since the network is an open, decentralized system, modeling confidence in it presents a third significant problem. Furthermore, there is no guarantee that the vehicle will interact with the same node in the future as it may join or exit the VANET at any time. Therefore, depending on mechanisms based on centralized systems-such as those utilizing Trusted Third Parties (TTP) and Central Certificate Authorities (CCA)-or relying on social networks to establish long-term relationships are not helpful.

A deep learning framework for trust computation and driver classification was presented by Tangade et al. [47]. They noted in this work that the cryptography-based security protocol for VANET is unable to stop fake messages from authorized senders from being broadcast. In the end, they developed a trust value by offering reward points that were determined through the use of a deep learning methodology (Fig. 9).

Taha et al. [48] proposed a strategy that combines a clustering algorithm and a Multi-Objective Task Distribution algorithm based on Differential Evolution (MOT-DDE) in order to guarantee stability and reliability in the VANET. The primary area of interest for the writers was distributing the task for the VANET, which allows host vehicles to offload subtasks to specific cars. Then, the Deep Learning model was used to estimate the cost and duration of the different subtasks(s). The DL model breaks the primary, complex task down into smaller, more manageable tasks according to variables including the amount of processing required, car numbers, task complexity

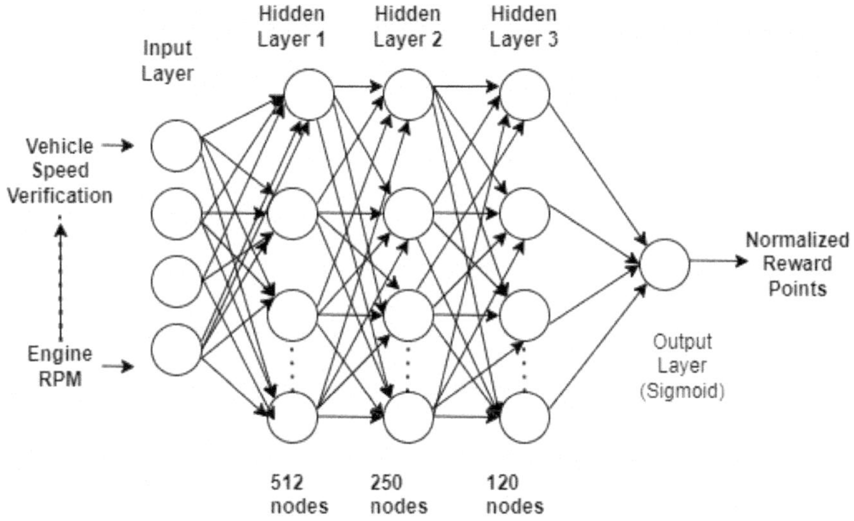

Fig. 9 GBEHO architecture

(i.e., size, policy length), and the state of the resources such as idle, moderate, or difficult. The system is able to distribute and assign the best tasks for the vehicles' resources in this way, ensuring that each application is completed on time and using the fewest resources possible for the vehicles involved in the process.

4 Conclusion

This chapter analyzes the impact of deep learning-based solutions in ensuring the security of VANETs. Deep learning based solutions have significant impact on improving the VANET security. As vehicle connectivity increases, the risk of security threats rises.The deep learning based solutions discussed in this chapter are capable of avoiding these threats effectively. Deep learning based frameworks like DeepVCM are effective in detecting intrusions happening to the network. Easy feature extraction properties of deep learning make it an effective solution in VANETs to ensure security. Frameworks like DeepADV are very effective in anomaly detection in VANET. DeepADV works based on the concept of calculation of reconstruction errors.It has proven to be highly effective in detecting anomalies in VANET. Deep learning techniques like GBEHO-DNN are highly effective in identifying the malicious nodes. Trust management in VANET is another area where deep learning architectures can contribute to increase efficiency. In the future, there is immense potential for the integration of deep learning with emerging technologies like blockchain, quantum cryptography, and edge computing to create even more secure VANET architectures. Research can focus on improving the scalability of DL models to handle larger networks with reduced latency. Additionally, hybrid approaches combining machine learning with reinforcement learning or federated learning could offer improved adaptability in dynamic vehicular environments.

References

1. World Health Organisation. https://www.who.int/news-room/fact-sheets/detail/road-traffic-injuries
2. Maalej Y, Sorour S, Abdel-Rahim A, Guizani M (2018) Vanets meet autonomous vehicles: multimodal surrounding recognition using manifold alignment. IEEE Access 21(6):29026–40
3. Benamar M, Benamar N, Singh KD, El Ouadghiri D (2013) Recent study of routing protocols in VANET: survey and taxonomy. In: Proceedings on 1st International Workshop Vehicles Network Telematics, pp 1–5
4. Sakiz F, Sen S (2017) A survey of attacks and detection mechanisms on intelligent transportation systems: VANETs and IoV. Ad Hoc Netw 61:33–50
5. Xiao L, Wan X, Dai C, Du X, Chen X, Guizani M (2018) Security in mobile edge caching with reinforcement learning. IEEE Wirel Commun 25(3):116–122
6. Lu Z, Qu G, Liu Z (2019) A survey on recent advances in vehicular network security, trust, and privacy. IEEE Trans Intell Transp Syst 20(2):760–776
7. Qi J. A prototype hybrid approach for defending Sybil attack in VANET. Doctoral dissertation, University of Auckland

8. Qu JF, Wu Z, Wang F, Cho W (2015) A security and privacy review of VANETs. IEEE Trans Intell Transp Syst 16(6):2985–2996
9. Isaac JT, Zeadally S, Camara JS (2010) Security attacks and solutions for vehicular ad hoc networks. IET Commun 4(7):894–903
10. He D, Zeadally S, Xu B, Huang X (2015) An efficient identity-based conditional privacy-preserving authentication scheme for vehicular ad hoc networks. IEEE Trans Inf Forensics Security 10(12):2681–2691
11. Sheikh MS, Liang J, Wang W (2019) A survey of security services, attacks, and applications for vehicular ad hoc networks (vanets). Sensors 19(16):3589
12. Jiang C, Zhang H, Ren Y, Han Z, Chen K, Hanzo L (2017) Machine learning paradigms for next-generation wireless networks. IEEE Wirel Commun 24(2):98–105
13. Kibria MG, Nguyen K, Villardi GP, Zhao O, Ishizu K, Kojima F (2018) Big data analytics, machine learning, and artificial intelligence in next-generation wireless networks. IEEE Access 6:328–338
14. Chen M, Challita U, Saad W, Yin C, Debbah M (2019) Artificial neural networks-based machine learning for wireless networks: a tutorial. IEEE Commun Surv Tutor 21(4):3039–3071
15. Nama M, Nath A, Bechra N, Bhatia J, Tanwar S, Chaturvedi M, Sadoun B (2021) Machine learning-based traffic scheduling techniques for intelligent transportation system: opportunities and challenges. Int J Commun Syst 34(9):e4814
16. Sun Y, Peng M, Zhou Y, Huang Y, Mao S (2019) Application of machine learning in wireless networks: key techniques and open issues. IEEE Commun Surv Tutor 21(4):3072–3108
17. Zeng Y, Qiu M, Zhu D, Xue Z, Xiong J, Liu M (2019) DeepVCM: a deep learning based intrusion detection method in VANET. In: 2019 IEEE 5th Intl Conference on Big Data Security on Cloud (BigDataSecurity), IEEE International Conference on High Performance and Smart Computing, (HPSC) and IEEE International Conference on Intelligent Data and Security (IDS), Washington, DC, USA, pp 288–293
18. Dhadhania A, Bhatia J, Mehta R, Tanwar S, Sharma R, Verma A (2024) Unleashing the power of SDN and GNN for network anomaly detection: state-of-the-art, challenges, and future directions. Secur Privacy 7(1):e337
19. Thorat SS, Rojatkar DV, Deshmukh PR (2024) A deep learning approach for sustainable ad hoc vehicular network. In: International Conference on Smart Computing and Communication 2024. Springer Nature Singapore, Singapore, pp 429–443
20. Ahsan SI, Legg P, Alam SM (2024) Privacy-preserving intrusion detection in software-defined VANET using federated learning with BERT. arXiv preprint arXiv:2401.07343
21. Bahramnejad S, Movahhedinia N, Naseri A (2024) An LSTM-based method for automatic reliability prediction of cognitive radio vehicular ad hoc networks. SN Comput Sci 5(3):291
22. Nandy T, Noor RM, Kolandaisamy R, Idris MY, Bhattacharyya S (2024) A review of security attacks and intrusion detection in the vehicular networks. J King Saud Univ Comput Inf Sci 1:101945
23. Raya M, Hubaux J-P (2007) Securing vehicular ad hoc networks. J Comput Secur 15(1):39–68
24. Lee M, Atkison T (2020) VANET applications: past, present, and future. Veh Commun 1:100310. https://doi.org/10.1016/j.vehcom.2020.100310
25. Hussein NH, Yaw CT, Koh SP, Tiong SK, Chong KH (2022) A comprehensive survey on vehicular networking: communications, applications, challenges, and upcoming research directions. IEEE Access 10:86127–86180. https://doi.org/10.1109/ACCESS.2022.319865
26. Raya M, Papadimitratos P, Hubaux JP (2006) Securing vehicular communications. IEEE Wirel Commun 13:8–15
27. Parno B, Perrig A (2005) Challenges in securing vehicular networks. In: Proceedings of HotNets-IV
28. Mejri MN, Hamdi M (2015) Recent advances in cryptographic solutions for vehicular networks. In: 2015 international symposium on networks, computers and communications (ISNCC). IEEE
29. Nagpal S, Aggarwal A, Gaba S (2022) Privacy and security issues in vehicular ad hoc networks with preventive mechanisms. In: Agarwal B, Rahman A, Patnaik S, Poonia RC (eds)

Proceedings of International Conference on Intelligent Cyber-Physical Systems. Algorithms for Intelligent Systems. Springer, Singapore. https://doi.org/10.1007/978-981-16-7136-4_24
30. Kim Y, Kim I, Shim C (2014) A taxonomy for DoS attacks in VANET. In: International Symposium on Communications and Information Technologies (ISCIT)
31. Asituha E (2023) Privacy and security issues surrounding vehicular adhoc network 24. World J Adv Res Rev 20(03):1449–1479
32. Kaur R, Singh TP, Khajuria V (2018) Security issues in vehicular ad-hoc network (VANET). In: 2018 2nd International Conference on Trends in Electronics and Informatics (ICOEI), Tirunelveli, India, 2018, pp 884–889. https://doi.org/10.1109/ICOEI.2018.8553852
33. Nema M, Stalin S, Lokhande V (2014) Analysis of attacks and challenges in VANET. Int J Emerg Technol Adv Eng 4:831–835
34. Bariah L, Shehada D, Salahat E, Yeun CY (2015) Recent advances in VANET security: a survey. In: IEEE 82nd Vehicular Technology Conference
35. Al-Shareeda MA, Manickam S (2023) A systematic literature review on security of vehicular ad-hoc network (VANET) based on VEINS framework. IEEE Access 11:46218–46228. https://doi.org/10.1109/ACCESS.2023.3274774
36. Sapala T, Penumallu RCSR, Kiran RS, Rajesh MV, Devi BSK (2022) A survey on VANET attacks and its security mechanisms. In: 2022 Seventh International Conference on Parallel, Distributed and Grid Computing (PDGC), Solan, Himachal Pradesh, India, pp 435–440. https://doi.org/10.1109/PDGC56933.2022.10053295
37. Orebaugh A, Ramirez G, Beale J (2006) Wireshark and ethereal network protocol analyzer toolkit
38. Krizhevsky A, Sutskever I, Hinton GE (2012) Imagenet classification with deep convolutional neural networks. Adv Neural Inf Process Syst 1097–1105
39. Hochreiter S, Schmidhuber J (1997) Long short-term memory. Neural Comput 9(8):1735–1780
40. Shiravi A, Shiravi H, Tavallaee M, Ghorbani AA (2012) Toward developing a systematic approach to generate benchmark datasets for intrusion detection. Comput Secur 31(3):357–374
41. Bondre SV, Bondre VD, Yadav U, Thakare B, Nanwani J (2023) Automated road anomaly detector in VANET by using deep learning. In: 2023 OITS International Conference on Information Technology (OCIT), Raipur, India, pp 963–968. https://doi.org/10.1109/OCIT59427.2023.10431273
42. Kamel, J, Wolf M, van der Hei RW, Kaiser A, Urien P, Kargl F (20220) VeReMi extension: A dataset for comparable evaluation of misbehavior detection in VANETs. In: Proceedings on IEEE International Conference Communication, pp 1–6
43. VeReMi Extension (2020) https://github.com/josephkamel/VeReMi-Dataset. Accessed Sep 2020
44. Alladi T, Gera B, Agrawal A, Chamola V, Yu FR (2021) DeepADV: a deep neural network framework for anomaly detection in VANETs. IEEE Trans Veh Technol 70(11):12013–12023. https://doi.org/10.1109/TVT.2021.3113807
45. Nova K, Umaamaheshvari A, Jacob SS, Banu G, Balaji MSP, Srithar S (2023) Floyd-Warshalls algorithm and modified advanced encryption standard for secured communication in VANET. Measurem Sensors 27:100796
46. Shen X, Huang C, Pu W, Wang D (2022) A lightweight authentication with dynamic batch-based group key management using LSTM in VANET. Secur Commun Netw
47. Tangade S, Manvi SS, Hassan S (2019) A deep learning based driver classification and trust computation in VANETs. In: 2019 IEEE 90th Vehicular Technology Conference (VTC2019-Fall). IEEE, pp 1–6
48. Taha MB, Talhi C, Ould-Slimane H, Alrabaee S, Choo KKR (2022) A multi-objective approach based on differential evolution and deep learning algorithms for VANETs. IEEE Trans Veh Technol 72(3):3035–3050

An Intrusion System for Vehicular Ad Hoc Networks Using Deep Learning

Mohit Kadwal, Deepali Sharma, Nishu Singh, Lalita Pawar, and Sangeeta Jain

Abstract The increasing use of the Internet in vehicles has made travel more convenient. However, there are several technological weaknesses that intelligent vehicle attackers can exploit, creating several security issues. These security concerns have drawn attention to the safety protection technologies of the in-vehicle system. We investigated the application of deep learning-based recurrent neural networks (RNNs) for enhanced auto encoder networks and intrusion detection in automobiles. To effectively comprehend the vehicle's boundary behaviour and identify invasive behaviour, we merged two algorithms. The correctness and efficacy of the proposed model were verified through an analysis of actual vehicle data. The experiment's results show that integrating the two technologies can successfully and dependably identify aberrant border behaviour. The model's parameters are updated iteratively using the time-based backpropagation technique. We verified that the suggested model in the study may attain a detection accuracy of more than 96%.

Keywords Time-based back propagation algorithm · Powerful auto-encoder · The web of vehicles · Detection of intrusion system · Recurrent neural network · Safety protection technology

---

M. Kadwal (✉)
Assistant Professor AI and Data Science, Prestige Institute of Engineering Management and Research, Indore, India
e-mail: mohitkadwal.mk@gmail.com

D. Sharma
Assistant Professor, Shri Vaishnav Institute of Management and Science, Indore, India

N. Singh
Department of Computer Application, Medicaps University, Indore, India

L. Pawar
Department of Computer Science and Engineering, Medicaps University, Indore, India

S. Jain
Govindram Seksaria Institute of Management and Research, Indore, India

J. Bhatia et al. (eds.), *Deep Learning Based Solutions for Vehicular Adhoc Networks*, Studies in Computational Intelligence 1207,
https://doi.org/10.1007/978-981-96-5190-0_5

1 Introduction

The advent of smart vehicles in recent years, which fuse Internet technology with mechanical manufacturing, has made it easier to build important data services to facilitate travel and daily commutes [1]. The development of these intelligent network vehicles must ensure optimal security and ongoing information access. Vehicle computer systems used to function in isolation from outside networks, which meant that their safety was neglected. But as recent events have shown, hackers can infiltrate vehicles that make use of networks of computers [2–4]. The initial attacks targeting vehicles via the On-Board Diagnostics-II (OBD-II) port required direct connection to a laptop, limiting the feasibility of sustained attacks across different vehicle types [5]. However, recent advancements in computer technology have led to the availability of affordable encryption tools for OBD-II connections, applicable to various vehicles irrespective of their make, model, or production year [6]. These tools, accessible to malicious hackers, pose a serious risk to drivers', passengers', and pedestrians' safety. As intelligent connected vehicles continue to evolve, the escalating concerns surrounding automobile safety underscore the critical importance of robust automotive security measures. To guarantee vehicle safety, several artificial intelligence-based network security systems have been put into place [7]. As the main line of defence for security on the Internet, firewalls are placed among the inside and outside networks and are intended to restrict packet flow according to pre-established criteria. While encryption techniques, Public Key Infrastructure (PKI) structures, and identity authentication increase security in certain network environments, they become less useful in complex communication scenarios like vehicles, where data frames have special properties [8]. Conventional approaches and detection of intrusion technology, a popular approach to network security, face comparable difficulties. To address these complications, researchers have lately proposed a multitude of real-time intrusion detection technologies that are based on statistical frameworks derived from user behaviour or behaviour norms.

But putting intrusion detection models into vehicles is difficult for several reasons:

1. Restricted computation and Storing Competencies: Due to the restricted computation and storing competencies of vehicles, it is difficult to monitor and analyse actions in detail. As a result, effectively allocating the resources required for vehicle intrusion detection systems is a major difficulty.
2. Complex Communications Conditions: The interaction environment in vehicles is influenced by a wide range of subjective and objective factors, including the weather. Since these considerations were often ignored in the initial designs of intrusion detection systems (IDS), many of the existing models are not suitable for direct installation in vehicles.
3. Large Erroneous Alert Rate: A large erroneous alert rate is a prevalent problem with intrusion detection systems used today. The complex communication topology of vehicles exacerbates this issue in non-standard environments. For instance, Control Area Network (CAN) bus lost packets rate may spike sharply during events like heavy rain or high traffic, leading to false alarms.

Currently, anomaly-based intrusion detection systems wrongly label certain situations as aberrant behaviour, which leads to a higher percentage of false alerts. Consequently, when designing an IDS concept for a vehicle, the following factors must be considered:

1. The IDS model should be designed to minimise communication load and storage space consumption when deployed on platforms with limited computing and storage capabilities. It must to possess sufficient adaptability to conform to the dynamic topology and rigorous real-time processing demands of automotive networks.
2. The intricate communication architecture of vehicles needs to be considered. For instance, the IDS should employ learning techniques to identify and correctly categorise aberrant activity that is not hazardous in and of itself.
3. Through independent learning processes, it is possible to achieve a lower rate of false alarms for benign anomalies and a greater alarm rate for recognised forms of attacks by differentiating between non-malicious anomalous behaviour and attacks.

1.1 Motivation

Our goal was to improve the detection efficacy within vehicular system and overcome the lapses in current Intrusion Detection Systems (IDS) that are limited in their ability to identify certain kind of attacks [9, 10]. This article presents a novel IDS model that uses deep learning's sophisticated automatic encoder and RNN approaches for in-vehicle systems. Higher-dimensional data utilisation in automotive information systems is not viable due to the limited storage and processing capabilities of vehicle systems and the need to reduce operational costs. We decreased the CAN bus's data dimension by using a sophisticated automatic encoder network with matching sparse term constraints to lessen this. The sophisticated auto encoder skill fully learns alternate representations of data at higher dimensions through supervised learning. Following post-dimension reduction with the sophisticated automatic encoder system, the data maintain all relevant information, like data that has been noise-filtered. Activation functions and matrix operations were used to recover the data. The sophisticated auto encoder efficiently reduces invalid noise and preserves relevant information in the processing of data, allowing transmission in lower dimensions for learning to the classifier. By reducing the expenses associated with training models, reasoning, and storage, this method improves the accuracy of models [11]. We optimised the hardware resources of the vehicle terminal, considering the constraints of embedded technology in vehicle systems. Model classification capabilities were maximised without unnecessary overhead by employing a RNN in conjunction utilising the SoftMax classifier to aid feature data categorization. Processing was made quick as a result. Sequence processing of data is a strong suit for the RNN because of its equal-time sequencing of the data on the CAN bus. The RNN also benefits from a simple Global Shared Parameter (GSP) technique [12]. Furthermore, it requires less data context backdrop during training than some RNN variants and traditional Convolutional Neural Networks (CNNs), i.e., Long Short-Term Memory (LSTM) networks; this saves training time [13]. The IDS model, which picks up on the characteristics of reservations on its own, accurately discerns the atmosphere of driving data properties. To ensure effective training and address any issues with gradient disappearance during training—a common occurrence in deep models such as RNNs—we employed at times-based Backpropagation Through Time (BPTT) technique. Compared to conventional intrusion detection methods, our suggested methodology achieves greater effectiveness and detection precision by taking timing and vehicle mechanics into account.

Organization of the Chapter

In the subsequent second segment, we review relevant literature on Internet of Vehicles and investigate state-of-the-art studies on RNN and CNN in intrusion detection systems (IDS). The final section describes the methods used in this research and describes how the automotive system's IDS makes use of the automatic encoder and an RNN. The fourth component makes use of actual testing to compare our suggested models' performance with that of conventional models employing vehicles. Finally, we provide a summary of the issues brought up and talk about potential directions for this research.

2 Related Work

A review of Deep Learning and Internet of Vehicles (IOV) will be provided in this section.

2.1 The (IOV) Internet of Vehicles' Architecture

To establish a complete web connection inside the vehicle as well as between people and vehicles, roads and vehicles, and vehicles and services, an entirely novel type of wireless communication technology is being used, which is known as the "Internet of Vehicles. Figure 1 shows the interconnected diagram IOV.

An intelligent connected vehicle connects its nodes for the electronic control unit (ECU) by forming a bus network within the vehicle via the bus communication protocol. A condensed representation of the schematic diagram for intelligent network communication is presented in Fig. 2. The electronic control unit (ECU) serves as both the central component of the vehicle's overall communication system and the primary communication unit within the vehicle. To vehicle out the designated command action, the node gets distinct communication data from the bus [14]. The bus protocol must be implemented in the vehicle if various ECU nodes need to

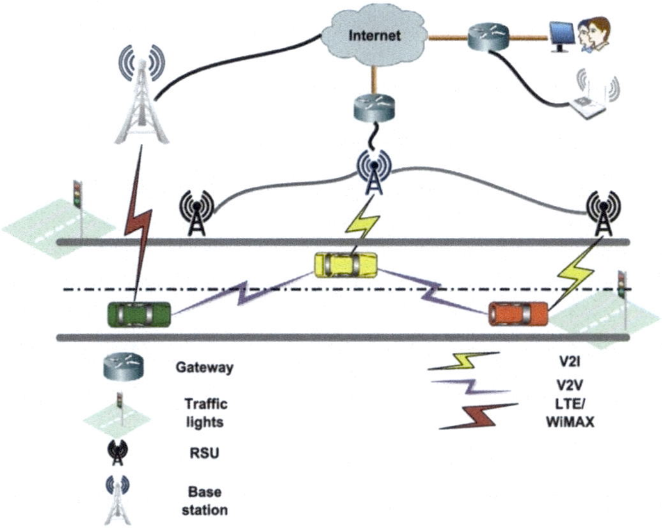

Fig. 1 Topology of a network of vehicles communicating

interact with one another. The CAN protocol is the most well-known bus protocol utilised in automobiles.

The CAN protocol is the most well-known bus protocol utilised in automobiles. The internal bus system of the vehicle, which can supply the ECU with sufficient communication data, uses the CAN standard. According to Seo et al. [5], the CAN bus is an affordable and dependable sequential bus of the vehicle network. Every ECU node communicates by sending data in a competitive manner to the bus. At this point, the order of priority region in the data frame is used to determine the bus control priority. If an ECU node has a small number in the arbitration field, prior to the other

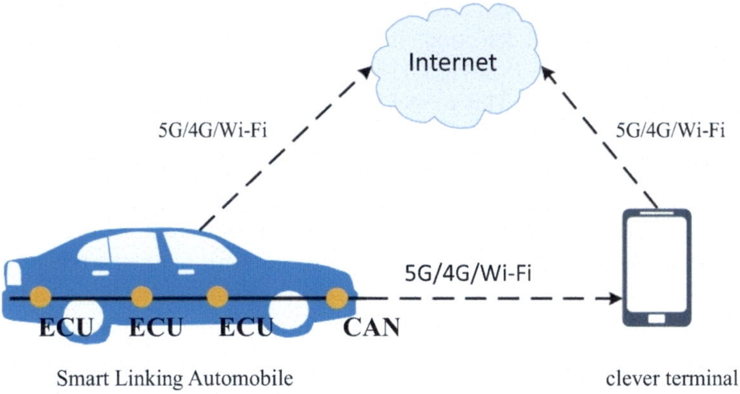

Fig. 2 An intelligent connected vehicle's simplified communication diagram

nodes being able to compete again, they will transmit the data first and wait for the network to idle. CAN was developed by Robert Bosch GmbH6 in the 1980s because this broadcasting method enhances the real-time operation of data connections. More than 50 ECUs are found in modern vehicles, allowing for communication speeds of up to 1 Mega bit/sec between them [15]. By providing data in a competitive manner to the bus, each ECU node can connect with the others. The bus control priority is now determined by sorting the priority regions in the information frame. When the bus is idle, the remaining nodes will be forced to wait to compete once more, and the ECU node with the smaller arbitration variable value will have priority in delivering the data. Robert Bosch GmbH6 invented the protocol known as CAN in 1980 because it enhances the data's real-time performance transmission through broadcasting.

2.2 Deep Learning in Intrusion Detection

In the current environments, vehicle has an IDS is often installed as software and hardware for vehicular. By performing the required analytical process and capturing information acquired from every ECU node or bus, it could guarantee the safety of travellers as well as drivers by quickly informing an emergency response assistance mechanism in the event of any aberrant activity [16]. However, due to the limiting technologies of a single-chip microcomputer, the ECU placed in vehicles during the early stages of the creation of the Intelligence Connected Vehicle (ICV) only had a little storage capacity and poor computational capability.

Given that the vehicle's system cannot directly use the current IDS, numerous academics have suggested an IDS depending on the in-vehicle system. In 2015, Liu and Li [17] introduced an intrusion detection methodology based on anomalies. In this model, data mining is used to detect the outlines, and the networking layer and MAC addresses of the layer are recognised to analyse the typical behaviour patterns of the mobile nodes. Used NS2 to validate this model to confirm their findings. The suggested technique has a high efficiency level for anomaly identification. Nevertheless, the model's overall detection efficiency declines as the number of detecting nodes rises. Based on AWISSENET [18] created a Distributed Vehicle Intrusion Detection system (DVIDS). This model looked through relevant paths and reliable services. A heterogeneous test platform was used to evaluate the proposed IDS model, and the results demonstrated the model's adaptability to various wireless networks. A vehicle-ad hoc network (VANET)-based anomalous intrusion detection model was introduced by [19]. In this approach, the network application layer that represents the contextual environment of the interacting nodes is recognised to facilitate the learning of the attributes of the attack behaviours. Additionally, the model makes use of the global behaviour maximum function and the density function to detect anomalous behaviour.

However, the model continues to have a significant false alert rate when exposed to certain atypical traffic circumstances. In 2010, Banković et al. [20] suggested a DVIDS for wireless networks of sensors that relies on reputation detection. In this approach, the trained model assigns a reputation to every node within the networks to the identification of malicious nodes. According to the accompanying trials, the suggested intrusion detection system prevented harmful activity from spreading by separating the target nodes from the network. The model can still identify rogue nodes by updating their reputation, nevertheless, with a high alert rate. For wireless sensor networks, Cho et al. [21] created a local IDS model. To lower the model's storage and computation needs, Howard Bloom's Bloom filter is introduced, which reduces the number of dimensions of the data coding. According to experimental results, by simulating the relevant wireless communication environment, the suggested method can identify possible denial-of-service attacks. Numerous researchers have discovered that feature science and technology and feature selection are frequently prioritised when utilising protocols pertaining to the network layer or cryptography to use IDS for malicious behaviour detection. However, these approaches are unable to adequately address the issue of enormous intrusion classification of data in real-world network environments [22]. Achieving the independent acquisition of attack attributes to identify similar attacks is difficult when the attack approach is changing [23, 24]. As a result of the widespread success that machine learning and deep learning have had, researchers worldwide are increasingly fusing these methods with intrusion detection systems to effectively detect both internal and external harmful activity.

A wireless sensor network-based intelligent intrusion detection model was presented by Medhat et al. [25]. The IDS model is trained using a combination of supervised and unsupervised learning techniques. Sensor nodes are trained using supervised learning, while base station and convergence nodes are trained using unsupervised learning. During this time, a binary tree that classifies normal and abnormal behaviour will be formed by a set of rules that have been learned. The outcomes of the experiment demonstrate that the model can achieve a reduced time complexity and a more precise attack data detection. A distributed intrusion detection system based on the Naive Bayes was presented by [26] for wireless networks. The suggested model is highly robust and capable of detecting numerous threats on its own. An artificial neural network was used by [27] to construct an IDS. The required understanding of traffic indicators is achieved by this model through supervised learning, allowing for the identification of abnormal traffic. But our model's accuracy was only 82% because the UDP distributed denial of service parameters and regular traffic values are comparable. Anzer et al. [28] published an artificial neural network-based abnormal intrusion detection algorithm. The model can use a neural network with full connectivity for modelling the network traffic data. Empirical studies show that the model developed in this study can acquire data with features more efficiently than certain traditional machine learning methods, such as SVM, ad boosting, and random forest. It also has a lower chance of alarms that are false and more accurate detection when compared to specialised anomaly detection techniques based on traditional deep learning [29]. A multi-layer perceptron-based neural network was proposed by Pavani et al. [30] as a means of detecting malicious behaviour in ad hoc mobile networks (MANET). Their trial results showed that the proposed model can distinguish between attacks from black holes and grey holes [31] to developed an intrusion detection system based on the Responsive Proportional Overlapping

Score (POS). By using the information gathered via the VANETs' trace file, their IDS model eliminates unnecessary features to enhance the autonomous vehicle's performance and security. Using the features extracted from the tracking files, it can distinguish between regular and abnormal cars. In addition, the model learns from audit data and uses a neural network that is artificial to detect attack behaviour. Tests demonstrated that their suggested model, which learns the data properties of the trace files, can identify black hole assaults efficiently. Unfortunately, these technologies have a lot of drawbacks. For example, to satisfy accurate recognition needs, these technologies are interacted with human experts at a highest level and require significant amount of domain expert knowledge when process data [32]. Deep learning can examine network data more thoroughly and swiftly spot any anomalous data as well as associated trends. Additionally, artificial intelligence boasts a strong black box component. As a result, attackers find it challenging to alter the detection system's internal structure. The IOT, particularly the IOV, has made use of deep learning for intrusion detection. The fundamental idea is to create a comparable classifier based on the boundary features identified by using deep learning to realise the learning of vehicle boundaries behaviour features. The effective categorization and identification of anomalies of an entity's behaviour data can be accomplished with this classifier. Many sophisticated and effective deep learning algorithms cannot be directly implemented into automobiles due to the restricted storage and processing capacity of conventional automotive ECUs [33]. Nonetheless, the advancement of intelligent vehicle information systems has increased the on-board ECU's processing speed for real-time operations as well as its calculation and storing capacity. A deep development of deep machine learning is provided by the auto encoder based on scant item restrictions and the RNN. A multi-layered was developed with learning capabilities to anticipate and recognise unknown behaviour by acquiring knowledge of the boundary features. The model's parameters were optimised using the appropriate optimisation procedure. We discovered via trials that the deep learning-designed model is more robust than the standard IDS based on signatures or statistics [34].

3 Deep Learning-Based Research on Intelligent Vehicle Intrusion Detection Systems

3.1 The Intrusion Detection Algorithm's Structural Design

Figure 3 displays the detection flow chart for the improved automatic encoder and RNN-based in-vehicle intrusion detection model. Three steps make up the suggested model: (1) Pre-processing of the data; (2) extraction of correlation characteristics between the data by the advanced auto encoder; and (3) classification of the corresponding data using a combination of RNN and Soft Max classifier. Given that the pertinent experimental data gathered from the vehicles consisted of both continuous and discrete data types, we had to apply certain pre-processing techniques to

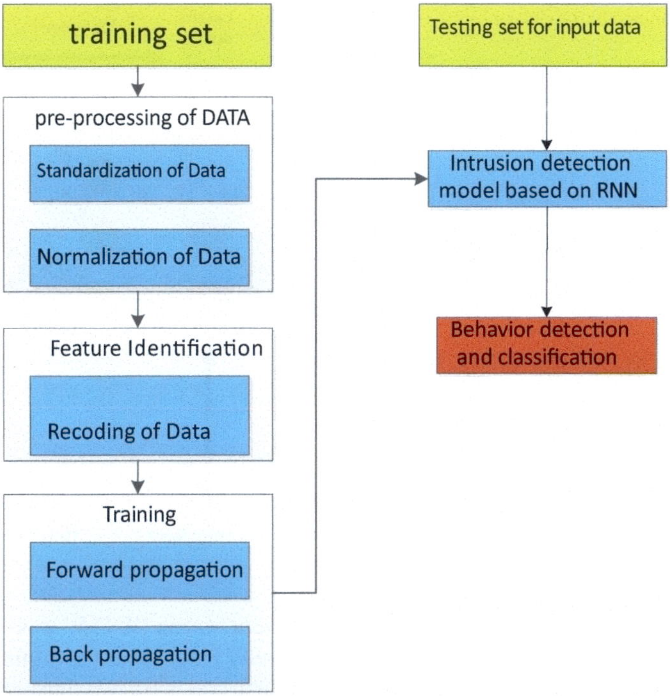

Fig. 3 Deep learning-based intrusion detection system flow chart

normalise the data. The fourth section provides a description of the specific data introduction. To finish the data extraction of features procedure, the conventional format data was fed into the sparse auto-encoder following the data pre-processing. Data with significant sparse characteristics were obtained. These highly sparsely feature data were fed into a RNN classifier. The relevant feature data was then learned and classified using Soft Max and RNN. Ultimately, we determined the pertinent CAN bus vehicle data was anomalous using the sorted findings as the output.

3.2 Methodology

A type of neural network structure called an auto encoder is used to Fig. out how to encode a piece of data. To accomplish decrease in dimension processing and associated learning of features of high-dimensional data, its main goal is to acquire knowledge about highest-dimension complicated information which is extract from an appropriate coding mode [35]. An Auto-Encoder's diagram shown in Fig. 4. In which the network's structure consists of two components: The equation $h = f(Wx + b)$ represents the data encoder, and the function $x = g(WTh + b)$ represents the

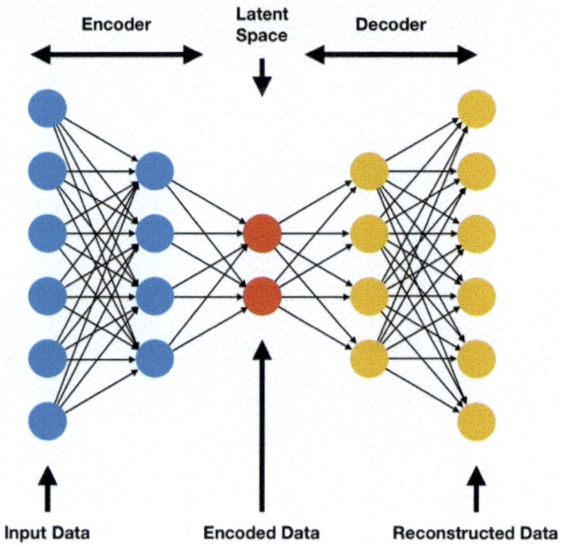

Fig. 4 Auto-encoder model

Auto-Decoder for production. The constraint Weight Matrix (WT) W was optimised using the unsupervised learning algorithm, and the WT was rebuilt to minimise error between the input and output models, resulting in $x(i) = x'(i)$. We finished the data dimension reduction process by limiting the number of features recovered from the data using an Auto-Encoder with restrictions on regular sparse terms. The successful extraction of feature was accomplished through the addition of regular L1 phrase constraints to the basic loss of function, which resulted in the sparse auto encoder. process 1 displays the precise steps of the process.

Algorithm 1 Note that the variables *ti*, *Ti*, and *sl* denote the training iterations, total training iterations, and layer l node count, respectively.

1: Coding:
$h = \mathrm{W}x + b$

2: While $t < T$:

3: For each $i < n$:

4: Calculate $\Delta \boldsymbol{W} l = \Delta \boldsymbol{W} l + \nabla Wl \cdot [J\ (W, b) + \beta \cdot \sum sl\text{^}2\ KL(\boldsymbol{\rho} \| \boldsymbol{\rho}')]$, where
$J\ (W, b) = 1 \cdot \sum n\ (1 \cdot \|h(xi) - yi\|\text{^}2) + \boldsymbol{\lambda} \cdot \sum sl - 1 \sum sl\ Wl\text{^}2$
and $KL\ (\boldsymbol{\rho} \,\|\, \boldsymbol{\rho}') = \boldsymbol{\rho} \log{(\boldsymbol{\rho}/\boldsymbol{\rho}')} + (1 - \boldsymbol{\rho}) \log{((1 - \boldsymbol{\rho}) / (1 - \boldsymbol{\rho}'))}$,
$\boldsymbol{\rho}' = 1 \cdot \sum n\ [a\text{^}2 \cdot x(i)] / (1 - \boldsymbol{\rho})$.

5: Update Δbl similarly.

6: Update Wl using: $Wl = Wl - \alpha\ [(1 \cdot \Delta Wl) + \boldsymbol{\lambda} Wl]$.

7: Update bl using: $bl = bl - \alpha\ [(1 \cdot \Delta bl)]$.

8: Increment i and t.

9: Decoding:
$x' = l(W^\wedge Tx + b)$

Following several iterative trainings of the auto encoder using Algorithm 1, the reconstruction matrix and optimal restriction weighting matrix were produced. The difference in error between the original data and the data acquired after dimension reduction is minimised when these two matrices are used. After obtaining a low-dimensional depiction of the high-dimensional characteristics of the data, the resulting data are fed into a series of RNN classifiers.

Three components typically make up an RNN: an input layer, a hidden layer, and an output layer. The main features of this method are the Uni-Direction flow of data, that takes significant time series data from the input layer, via the Internal layer that is hidden, and finally to the output layer. As soon as the present data reaches the appropriate hidden layer, it can determine the potential behaviour state for the next state by combining calculations from both the present data flow and the previously preserved digital data stream. As a result, we typically think of the network's hidden layer unit as the structure's storage. It is employed to compute the state data for the following behaviour and to save the state data from the preceding portion of the behaviour. Figure 5 displays the equivalent structure of networks diagram of the active network.

It is evident from Fig. 5 that adding a ring structure aids in "remembering" and applying the pertinent prior data to the present output computation. An RNN has a different structure than a conventional CNN. The output result of a hidden layer is belonging to sequence results which is computed by the matching hidden layer unit in

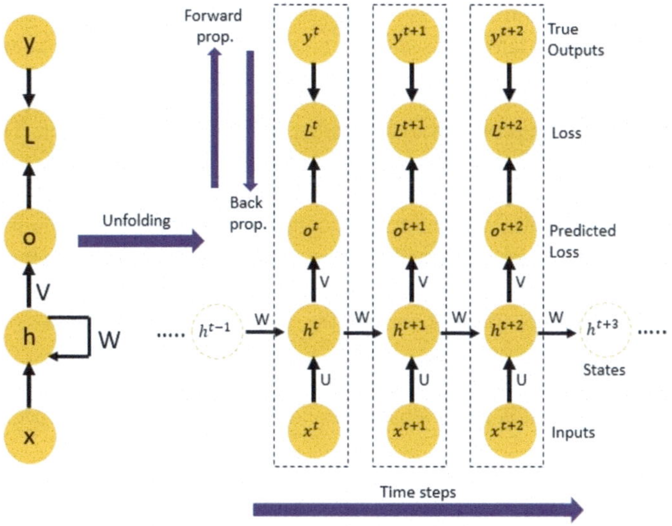

Fig. 5 The structure of a foldable RNN

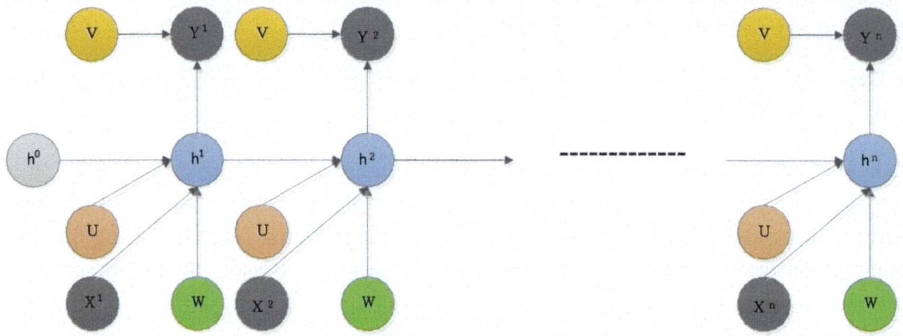

Fig. 6 The construction of an unfurled RNN

the currently working unit, and there is a particular information exchange procedure among each hidden layer unit's neurons. To finish learning the necessary features, we reduced the data dimensions using the sophisticated auto encoder created in the previous section.

The division of our RNN training procedure in-to two comparable stages—forward-propagation training and back-propagation—is also evident from Fig. 3. To finish this process in our study, we employed the BPTT algorithm. The anticipated value for each specimen under the appropriate weight matrix for a particular sample is determined by forward propagation. On the other hand, back-propagation computes the accumulated residual by updating the pertinent weight matrix utilising the differential.

The RNN's fully unfolded structure is shown in Fig. 6. We dissected the basic RNN structure into subsequent 3 components:

(1) Samples set for training $x_{i,}$ i is belongs to 1 to Infinity.
(2) Corresponding hidden layer's state units' sequence h_i i is belongs to 1 to Infinity.
(3) Set of anticipated outcomes y_i i is belongs to 1 to Infinity.

The following are the pertinent structure that are used for computation. The hidden layer unit's connection weight between its prior and current states is represented by U. V represents the link weight connecting the associated layer's hidden layer's unit to the resultant layer unit. W is the weight connection in-between the hidden layers and the corresponding input layers. The appropriate training procedure for the RNN depicted in BPTT algorithm. The procedure is displayed in Algorithm 2 and 3. For every input RNN model training sample (x_i, y_i), we calculated the loss using the following objective function:

$f(\theta) = L\ (y_i, y\square_1)$, wherein L is able to assess the real departure separation value between the label y_i that is really observed and the label $y\square_1$ that is expected.

Algorithm 2 Algorithm for Forward Propagation

Input: x, W (for $i = 1, 2, \ldots, n$)

Output: yi

1. For i from 1 to n, do:
   1. $si = Uhi - 1 + wxi + b$
   2. $hi = \sigma(si)$ (where $\sigma\sigma$ is the hyperbolic tangent activation function)
   3. $zi = Vhi + c$
   4. $yi = \text{SoftMax}(zi)$

End

Algorithm 3 Time-Based Backpropagation

Input: $yi,y\hat{}i$ (for $i = 1, 2,\ldots,n$)

Output: $\theta = \{W',U',V',b',c'\}$

1. For i from 1 to n, do:
   1. $\partial L\partial V = \partial Lt\partial V = \partial Lt\partial zi = (y\hat{}i - yi) \cdot(hi)T$
   2. $\partial L\partial c = \partial Lt\partial c = \partial Li\partial zi = y\hat{}i - yi$
2. Update V and c:
   1. $V' = V - \eta\sum i = 1n(y\hat{}i - yi) \cdot(hi)T$
   2. $c' = c - \eta\sum i = 1n(y\hat{}i - yi)$

Note: Parameters W, U and b are shared across time steps and affect both the output at time t and the input $st + 1$ for hidden layers on a particular time $t + 1$. That's why, their gradients must be computed in a step-by-step manner from the last time step backward.

3. For i from 1 to n, do:
   1. $\partial L\partial W = \partial L\partial hi\cdot\partial hi\partial si = (\partial L\partial zi + \partial L\partial hi + 1\cdot\partial hi + 1\partial si + 1) \cdot\partial hi\partial si$
      - $= (VT\cdot(y\hat{}i - yi) + UT\delta i + 1\odot\sigma'(zi + 1)) \cdot\partial hi\partial si$
      - $= (VT\cdot(y\hat{}i - yi) + UT\text{dig}(\delta i + 1) \cdot\sigma'(zi + 1)) \cdot\partial hi\partial si$
      - $= (VT\cdot(y\hat{}i - yi) + UT\text{dig}(\sigma'(zi + 1)) \cdot\delta i + 1) \cdot\partial hi\partial si$
      - $= (VT\cdot(y\hat{}i - yi) + UT\text{dig}(1 - hi + 12) \cdot\delta i + 1) \cdot\partial hi\partial si$
      - $= \text{dig}(1 - hi2) \cdot(VT\cdot(y\hat{}i - yi) + UT\text{dig}(1 - hi + 12) \cdot\delta i + 1) \cdot(xi)T$
   2. $\partial L\partial b = \partial L\partial hi\cdot\partial hi\partial si = \text{dig}(1 - hi2) \cdot(VT\cdot(y\hat{}i - yi) + UT\text{dig}(1 - hi + 12) \cdot\delta i + 1)$
   3. $\partial L\partial U = \partial L\partial hi\cdot\partial hi\partial si = \text{dig}(1 - hi2) \cdot(VT\cdot(y\hat{}i - yi) + UT\text{dig}(1 - hi + 12) \cdot\delta i + 1) \cdot(hi - 1) T$
4. Update W, U and b:
   1. $W' = W - \eta\sum i = 1n\text{dig}(1 - hi2) \cdot(VT\cdot(y\hat{}i - yi) + UT\text{dig}(1 - hi + 12) \cdot\delta i + 1) \cdot(xi)T$
   2. $U' = U - \eta\sum i = 1n\text{dig}(1 - hi2) \cdot(VT\cdot(y\hat{}i - yi) + UT\text{dig}(1 - hi + 12) \cdot\delta i + 1) \cdot(hi - 1) T$
   3. $b' = b - \eta\sum i = 1n\text{dig}(1 - hi2) \cdot(VT\cdot(y\hat{}i - yi) + UT\text{dig}(1 - hi + 12) \cdot\delta i + 1)$

   End

4 Discussion and Simulation Findings

4.1 Data Set Description

We collaborated with a domestic vehicle cyber security lab to conduct simulation experiments on actual intelligent vehicles to confirm that the model can effectively detect vehicle behaviour. Typically, there are two methods available for gathering the vehicle's internal parameters:

1. Use the OBD—diagnosis interface to quietly watch information stored in the Electronic Control Suspension network (ECS), or use the normal communication protocol to retrieve information from the vehicle via the OBD.
2. By accessing the vehicle CAN bus through the CAN converter to finish the CAN bus monitoring.

We opted for the second method, which involves simply attaching the CAN and USB converter to the OBD port to get data.

There are two causes:

1. After immediately accessing the CAN bus using the USB-to-CAN converter, we were able to get the necessary data with ease through passive monitoring.
2. It is frequently required to transmit request parameters to acquire the correct data when collecting data via the OBD port.

This is a complex procedure. Using statistical analysis, we discovered that Table 1: On every bus, certain important parameters broadcast. When the vehicle is operating regularly, these settings are sent without the need for a request process. Weighing first, we decided to collect information on (ECM) Engine Management Module bus since the sender bus has certain crucial information. For instance, the ECU calculates the engine speed directly and shares data via other available modules.

The data collection configuration is depicted in Fig. 7, which is the result of attaching the data collection of the vehicular that houses the ECU (Fig. 8). The Kvasses CAN Leaf LightV2 was the CAN acquisition device that was used, and Vehicle Spy, an open-source programme, was used to collect data on the computer. We gathered around 300,000 pieces of vehicle-related data. A portion of information that we gathered is displayed in Fig. 9. We shaded certain areas of the graph because the arbitration domain and a few other characteristics require the experimental centre and the automaker to exchange private information.

Table 1 On every bus, certain important parameters broadcast

	Parameter	Velocity	Per minute revolutions	Position of accelerator pedal	Pressure Intake	Status of brake pedal	Angle of steering	Position of gear
Bus category	Meters for motor	O	O	X	X	O	O	O
	System of convenience	O	O	X	X	O	X	X
	Power system	O	O	O	O	O	X	X
	Control engine module	O	O	O	O	O	X	O
	Stability of electronic controller	O	O	O	X	O	O	O

Fig. 7 Experiment's vehicular for data

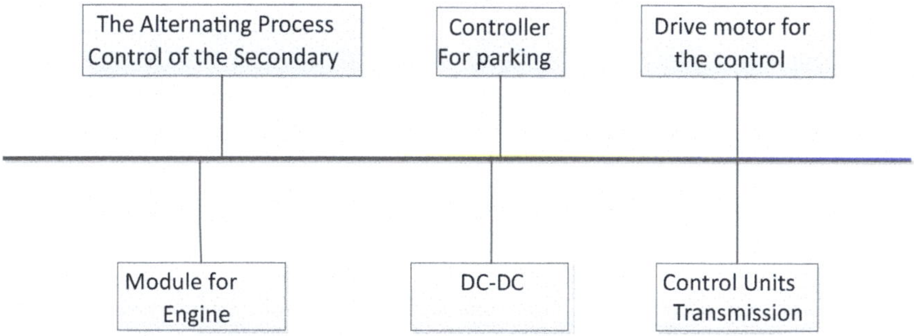

Fig. 8 Engine's CAN bus architecture

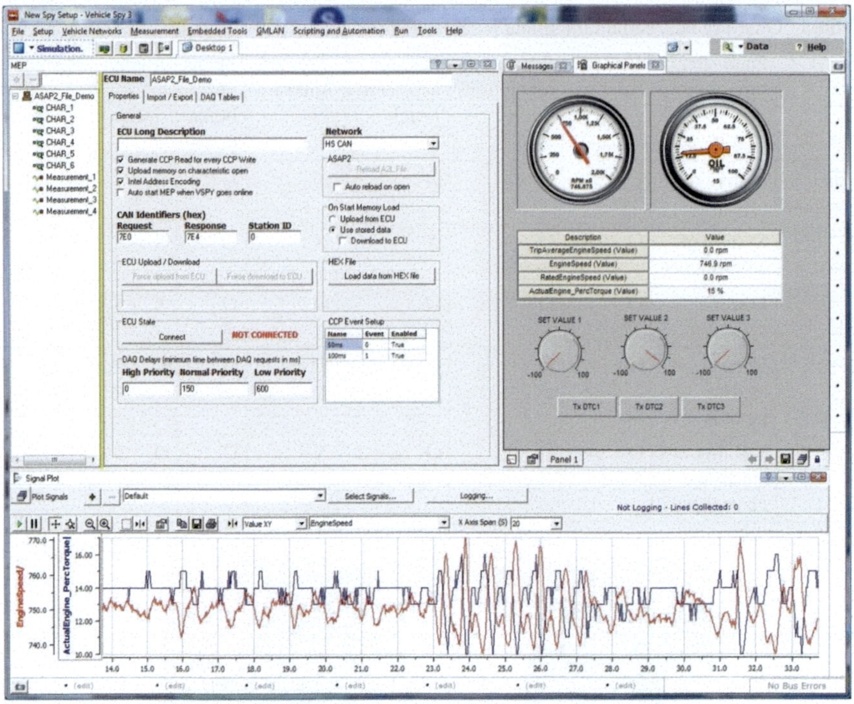

Fig. 9 Sample information gathered by a vehicle spy

4.2 *Preparing the Data*

4.2.1 Extracting and Calibrating the Data

We initialised the data in Python using the sklearn library. Figure 10 shows that the data sent across the CAN buses were sent in the most initial format. Multiple IDSs in a message can communicate one or more parameters. To illustrate, consider 8-byte information of CAN. The high or low 8 bits of information about the data are represented by bytes 1 and 2, respectively, while both the low and high 8 bits of speed are represented by bytes 3 and 4. Furthermore, the possible range for these data is 0 ×000–0×FFFF. Therefore, we had to convert them, usually using Eq. (1), to obtain actual speed.

$$v = ax + b$$

It was not possible to send the parameters that Vehicle Spy obtained straight to the computer network for training. We also needed to do the related data analysis procedure to obtain the variables that we required. As illustrated in Fig. 11, we

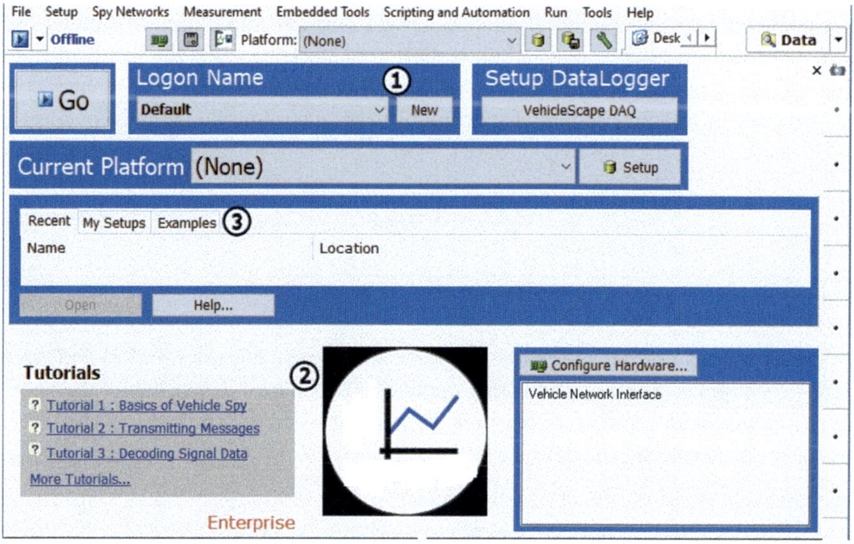

Fig. 10 The vehicle spy log file format

380504.9	577	0.021255	0.02149	0.023041	0.169643	0.000419287	0.096
380504.9	577	0.021255	0.02149	0.023041	0.169643	0.000419287	0.096
380508.9	527	0.021255	0.02149	0.023041	0.169643	0.000419287	0.096
380509.9	269	0.021255	0.02149	0.023041	0.169643	0.000419287	0.096
380518.9	527	0.021255	0.02149	0.023041	0.169643	0.000419287	0.096
380519.9	269	0.021255	0.02149	0.023041	0.169643	0.000419287	0.096
380524.9	577	0.021255	0.02149	0.023041	0.169643	0.000419287	0.096
380529.9	527	0.021255	0.02149	0.023041	0.169643	0.000419287	0.096
380529.9	269	0.021255	0.02149	0.023041	0.169643	0.000419287	0.096
380538.9	577	0.021255	0.02149	0.023041	0.169643	0.000419287	0.096
380538.9	527	0.021255	0.02149	0.023041	0.169643	0.000419287	0.096
380539.9	269	0.021255	0.02149	0.023041	0.169643	0.000419287	0.096
380545.9	577	0.021255	0.02149	0.023041	0.169643	0.000419287	0.096
380548.9	527	0.021255	0.02149	0.023041	0.169643	0.000419287	0.096
380549.9	269	0.021773	0.02149	0.023041	0.169643	0.000419287	0.096
time_in_ms	Msg_ID	AccPedal	RPM	Speed	Throttle	IntakePressure	AirFlowRate

Fig. 11 Pre-processed data

extracted the parameters from the Vehicle Spy CAN log using a script created in the Python sklearn package and translated them into CSV files.

4.2.2 Normalisation of Data

These values vary in different ranges, which explains in Fig. 11 (subsequent col 3) range from 0 to 1. The data normalisation approach is the standard. To address the discrepancies between the data indicators, data normalisation aims to standardise the data of various dimensions and units. Different data sets are at the identical level after normalisation, which is useful for thorough comparison analysis.

There are numerous methods for normalisation, including the Min–Max (Eq. 3) and Z-score (Eq. 2):

$$XX* = x - \mu/\sigma\sigma \tag{1}$$

$$XX* = x - \text{min} / \text{max} - \text{min} \tag{2}$$

The Min–Max normalisation method was chosen since it is the simplest and there were no unique requirements for this application scenario. Remarkably, the same set of maximum and lowest values must be utilised when we applied the Min–Max approach for normalise the data.

4.2.3 Interpolating Data

Figure 11 illustrates, these parameters were sent repeatedly via the CAN datagram. Numerous interpolation techniques exist. As seen in Fig. 12, we decided to use the forward interpolation method. In figure, the parameter is vertical axis and time is on horizontal axis. The values in the figure that are underlined and italicised represent the real received parameter value; the remaining values were inserted using the forward interpolating.

Air Velocity	Thermal Resistance
0	373.0
200	156.1
400	113.6
600	93.1
800	81.5
1000	73.7
1200	67.2
1400	62.4
1600	58.3
1800	55.0

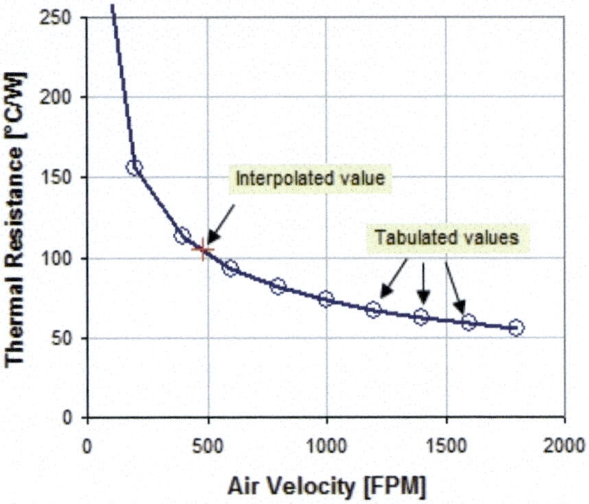

Fig. 12 Interpolation data sample

4.2.4 Data Collection

We analyzed the data and discovered that there was a significant amount of redundancy. The data interpolation was cause of redundant data. Because of the inherent unpredictability in the CAN channel arbitration system, data received were not dispersed equally across time. These factors made sampling the data necessary. Here, the sampling strategy used was to sample every row of data in milliseconds, rounding to the closest 10 ms. The training data set might be greatly trimmed by selecting the samples that were nearest to the total of 10.

4.3 Features of the Data and Correlation Study

The following two methods are typically used to finish the Co-Relation analysis and properties of information in ICV.

(1) Examine data flow produced by the vehicle while it is being driven normally. Analyzing data packet transfer rate, shift rate, and bus utilization CAN rate under typical driving circumstances, for instance, completes the analysis process.
(2) The vehicle's operating principles are used to analyses of vehicle in various states as well as the Co-Relations between them (e.g., the relationships between the vehicle's air intake and normal driving conditions and RPM (revolutions per minute) and speed). In the experiment, we analyzed the data gathered and identified the set of data with the strongest link using the Pearson correlation coefficient. The general pattern of the information is significant, as Fig. 13 illustrates.

By means of analysis, the data transmitted over a main bus may typically be categorised into two groups based on obvious distinctive relationships: the data pertaining to mechanically obvious laws, like the vehicle's pedal position and its speed and acceleration. Alternatively, data that does not seem to follow any pattern, like engine air intake or brake pedal condition. The fact that the driver must constantly modify the predetermined settings corresponding to these data to account for changing road circumstances is the reason why these data lack transparent characteristic relationships.

An equivalent normal threshold range will also be presented for various vehicle parameters with matching relationships under specific normal driving conditions. Examples of these parameters include the vehicle's driving speed, automatic gear, and engine speed under specific driving conditions. The average threshold range's upper bound is their changing rate. Under typical driving conditions, there is a restricted shift rate between the vehicle's speed, the engine's speed, and the gear speed, as seen in Fig. 14. We may readily observe a black curve as the related anomalous data (Fig. 15) an increased frequency, as illustrated in Fig. 16. There is a breakpoint in the associated vehicle speed when the combustion speed is repeated at a greater rate, and this has a specific effect on the vehicle's speed.

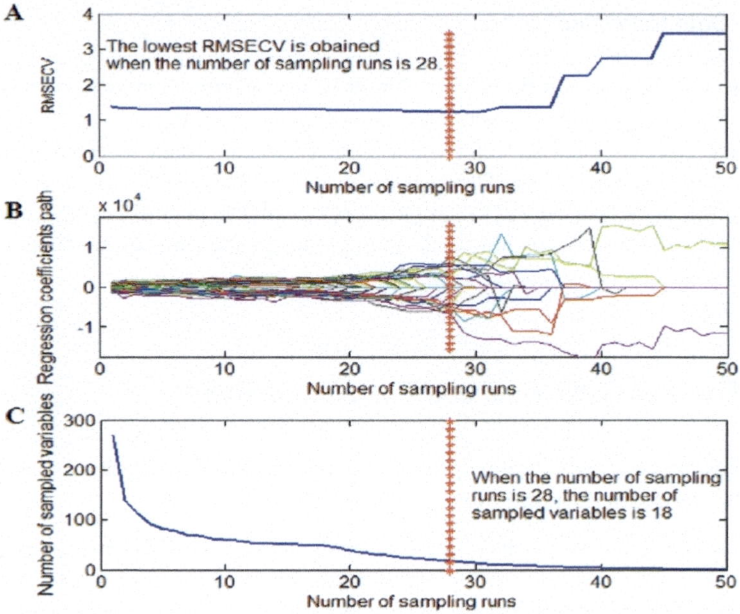

Fig. 13 The automobile parameters' changing trend

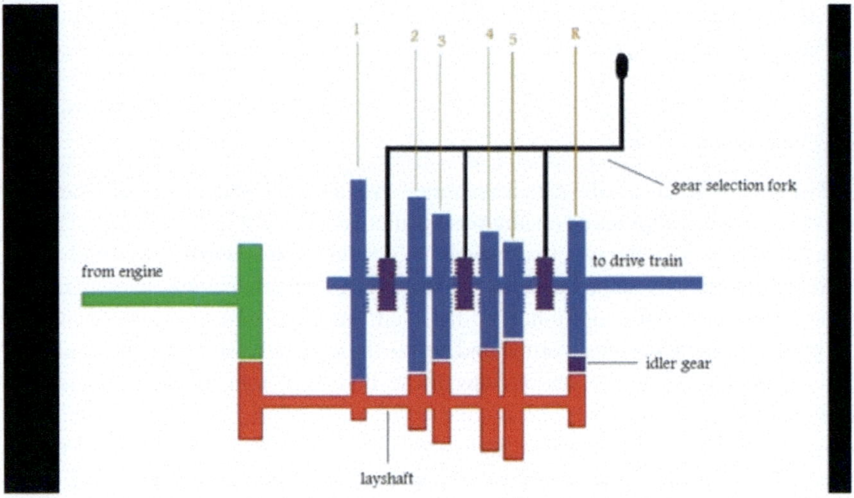

Fig. 14 The link between the vehicle's RPM, gear position, and driving speed under typical driving circumstances

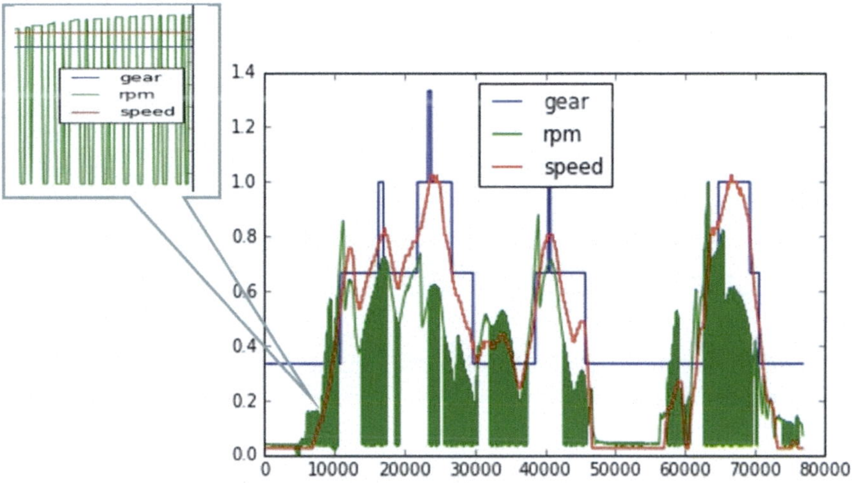

Fig. 15 The link between the vehicle's RPM, gear position, and driving speed throughout the replay attack

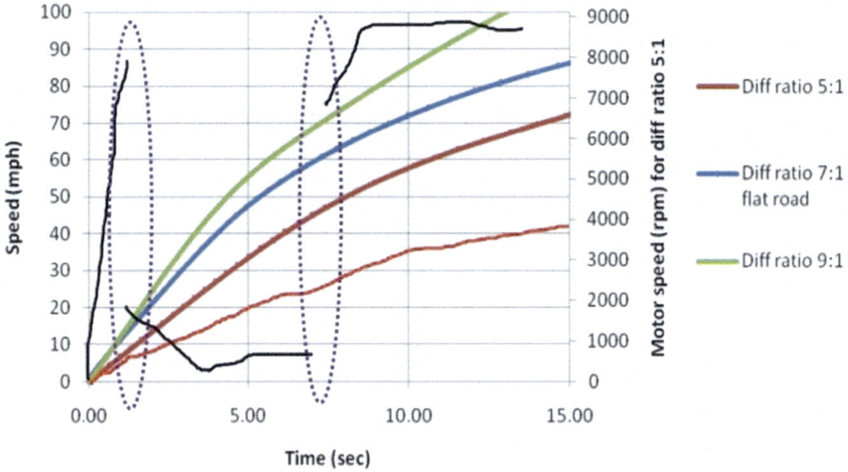

Fig. 16 After transmitting anomalous data at several frequencies, the vehicle's speed changes

4.4 Selecting Features

The car's speed, engine speed, engine intake pressure, accelerator pedal position, and automatic transmission gear are just a few of the auto parameters that we chose as the data vectors for the entire model after examining the strong correlations of the data in Section C. There are distinct features to the time correlation of these factors and how they relate to one another.

Table 2 Selected parameters' co-variance matrix

	RPM	Speed	MAP	MAF	AccPedal	Throttle
RPM	1	0.793538	0.525027	0.313592	0.438199	0.565193
Speed	0.793538	1	0.495519	0.365531	0.215287	0.494654
MAP	0.525027	0.495519	1	0.832216	0.654406	0.484693
MAF	0.313592	0.365531	0.832216	1	0.496407	0.553881
AccPedal	0.438199	0.215287	0.654406	0.496407	1	0.38837
Throttle	0.565193	0.494654	0.484693	0.553881	0.38837	1

We used Eq. (4) to determine the covariance coefficient of the chosen parameters, as indicated in Table 2, to confirm that there is a substantial correlation between the parameters we chose. These metrics exhibit a positive correlation, as can be seen.

$$\sum n(Xi - X)(Fi - F)\mathrm{cov}(XX, Y)/n - 1 \text{ where n} = \text{i} - 1$$

4.5 Analysis of the Experiment

This study aim is to increase the resolution speed of the complete network using BPTT and the detection efficiency utilising the RNN method and sparse feature-based auto encoder, respectively. To assess the overall effectiveness of the suggested methodology, we looked at the alarming rate and the percentage of false alarms. Equations (5) and (6) were utilised to compute the alert frequency and false alarm rate, respectively. The number of records classified as abnormal is called a (TP) True Positive, the number of records classified as normal is called (FP) False Positive, the number of records classified as normal is called a (TN) True Negative, and the number of records classified as abnormal is called (FN) False Negative:

$$TPR = (TP/(TP + FN)) * 100\%$$
$$FPR = (FP/(FP + TN)) * 100\%$$

To finish the model's training, we employed the widely known deep learning framework Kera's. We used a laptop to finish the matching experiment. The experimental system included an ASUS F1 8000 u, an Intel i7-8550 u the central processing unit (Sixteen gigabytes of RAM), and a graphics vehicle that was not used for acceleration. To compare the schemes, we also created the matching comparative experiments. This was vehicle slice out to contrast the machine learning approach with the methodology suggested in this study [25].

In the experiment, the auto encoders (one-hot encoding coding technology) were utilised to convert the 16-dimension data features into 48-dimension data features.

Consequently, the RNN classifier comprises 2 for output and 48 nodes for input. To achieve an optimal procedure, we confit a neural network to have 40, 60, 80, and 100 hidden layer nodes. Additionally, we set the learning rate to 0.05, 0.1, 0.01, during the phase of training. Table 3 displays the RPM's anticipated outcomes according to the proposed model.

Table 3 The rate of precision and convergence time of the model for various rates of learning and nodes in the hidden layer

	Training sets (%)	Test sets (%)	Time of convergence (s)	Detection time (ms)
Nodes that are hidden = 40 Rate of learning = 0.01	95.20	87.24	5340	6.47
Nodes that are hidden = 40 Rate of learning = 0.05	95.47	89.74	5041	6.12
Nodes that are hidden = 40 Rate of learning = 0.1	96.95	89.92	4885	5.94
Nodes that are hidden = 60 Rate of learning = 0.01	96.32	83.25	4991	6.03
Nodes that are hidden = 60 Rate of learning = 0.05	97.64	85.96	4764	5.76
Nodes that are hidden = 60 Rate of learning = 0.1	98.17	86.43	4683	5.21
Nodes that are hidden = 80 Rate of learning = 0.01	98.52	89.94	4954	5.84
Nodes that are hidden = 80 Rate of learning = 0.05	98.94	92.37	4892	5.32
Nodes that are hidden = 80 Rate of learning = 0.1	99.36	95.84	5103	4.97
Nodes that are hidden = 100 Rate of learning = 0.01	97.25	87.67	5701	6.15
Nodes that are hidden = 100 Rate of learning = 0.05	97.86	89.54	5394	5.87
Nodes that are hidden = 100 Rate of learning = 0.1	98.58	91.21	5248	5.63

Whether the set of training or set of testing is utilised, the experimental findings in Table 3 demonstrate that the model achieves a high accuracy whenever the learning rate has been customised to 0.1 and the layers hidden node to 80 nodes. Simultaneously, Table 3 displays the sparse auto encoder's independent feature learning capability as well as the RNN consideration of data temporal features. At the millisecond level, real-time aberrant data detection is achievable. Figure 17 displays the comparison findings utilising the suggested model between the actual engine speed values and the anticipated engine speed value. The data is accurately predicted when there is a tiny error between these values. Figure 18 presents the sequence of variance between the actual and anticipated values for engine speed, making the erroneous results in Fig. 17 simpler to see. Consequently, model classification along with data collection feature extraction can benefit from deep learning, which can help the model achieve excellent precision with a low false alarm rate. This would greatly increase the model's detection effectiveness.

The above graph produced whenever the vehicle information passes by using low-pass against the vehicle while it is operating normally is displayed in Fig. 19. We can observe that the computed results indicated the RPM forged data will affect the other vehicle characteristics and cause an anomalous behaviour on the other parameters' prediction after being injected into the CAN bus. We can now compute the judgement value index, which is the difference between the actual and anticipated values. We can identify the associated aberrant behaviour to estimated variance which is greater than given critical to the value for their money, including the unusual point in Fig. 19. In this way, our model can use the mechanical differences among the vehicle data to effectively detect the attack, regardless of what component of the vehicle is falsified or targeted by the attacker. These test findings allow us to take immediate action,

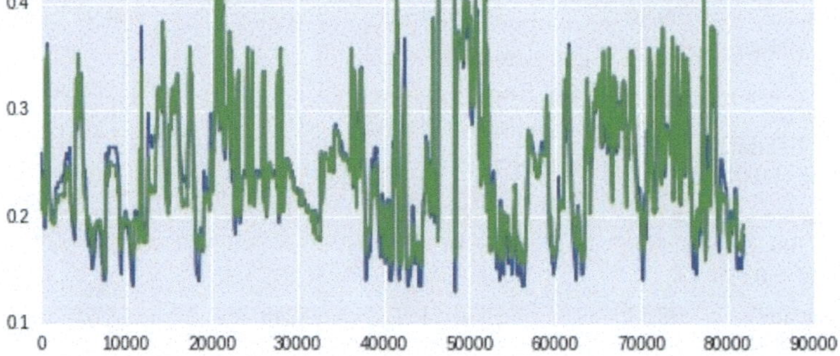

Fig. 17 The RPM's anticipated outcomes according to the proposed model

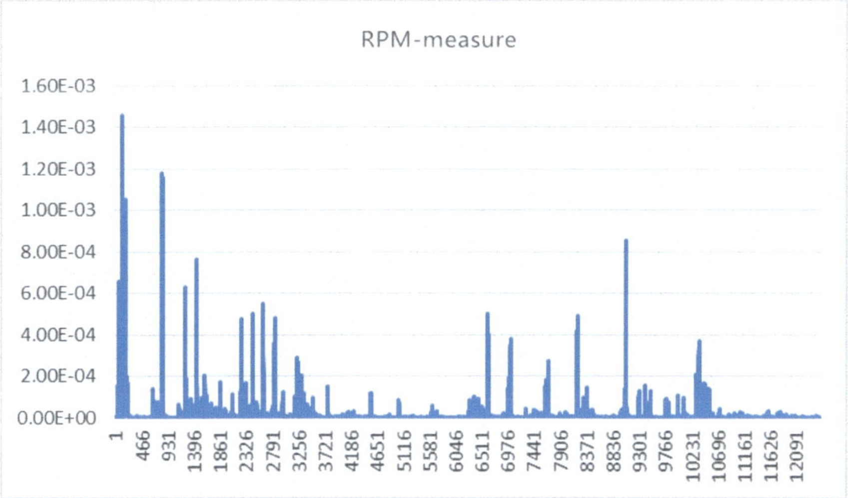

Fig. 18 Displays the sequence of RPM variance among anticipated and real values

such alerting the administrator to oversee the vehicles outside network safety and communication link management.

The matrix of confusion of the finding results for the simulated data set within the ideal parameter settings is displayed in Fig. 20 in terms of the suggested model. The experimental findings indicate that the model's accuracy is roughly 96%. Figure 21 that shows the experiment outcomes. As seen in image, the suggested technique achieves a lower false alarm rate, as it may attain about 96% in TPR index but only 2–s3% in the FPF. As seen in the picture, the model suggested in this work

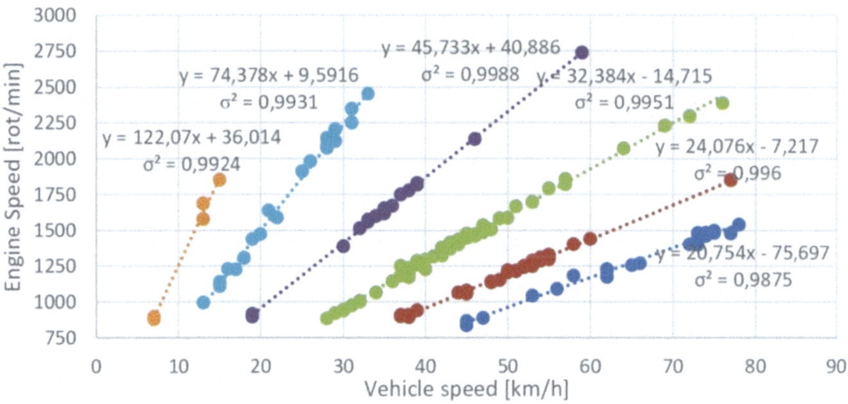

Fig. 19 Vehicle speed is affected by fictitious RPM data

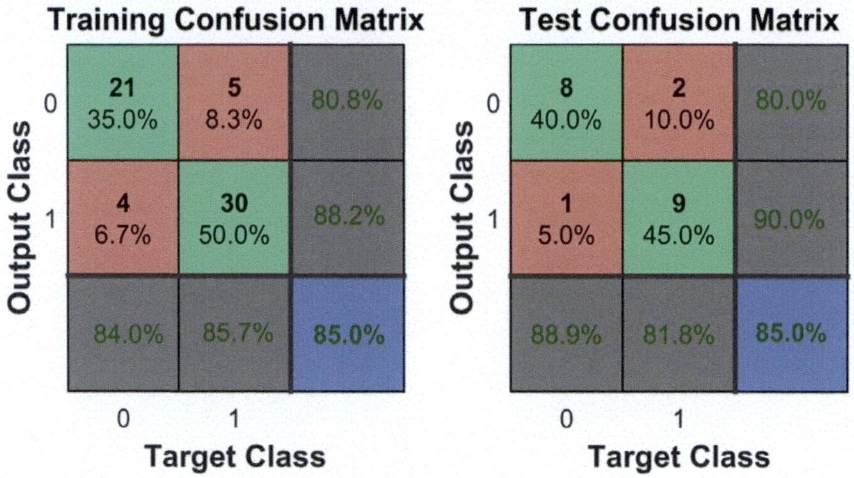

Fig. 20 Confusion matrix related with test findings

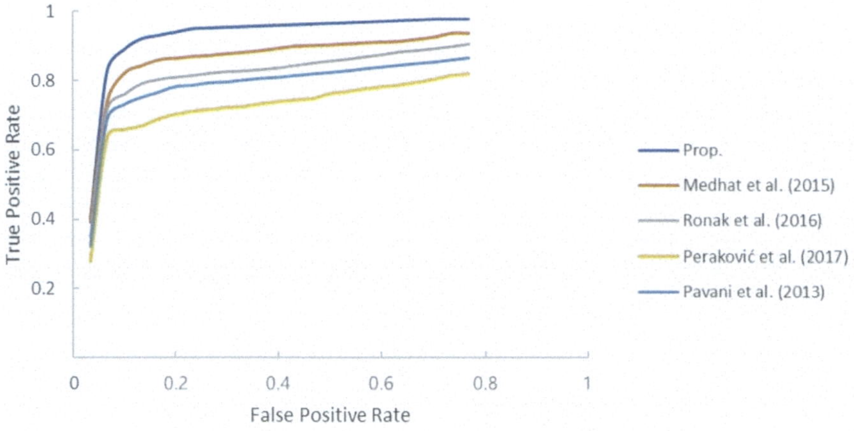

Fig. 21 ROC curves for the performance of vehicle intrusions detection

outperforms the deep learning-based IDS suggested by its predecessors in terms of detection efficiency and false alarm rate (Fig. 22).

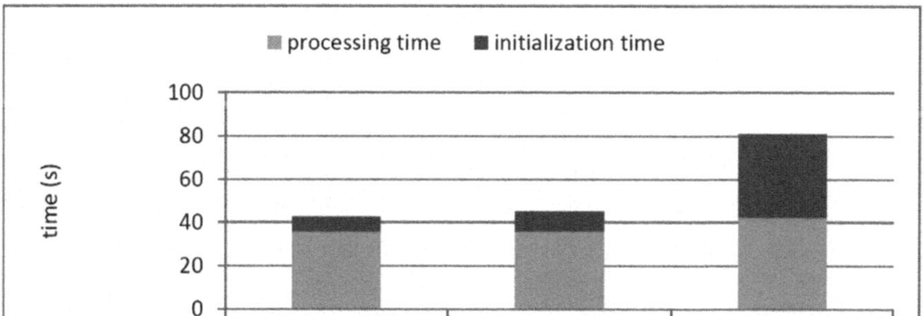

	Suricata CPU	Suricata GPU	Snort
initialization time	7,234	9,452	38,959
processing time	35,509	35,795	42,114

Fig. 22 A comparison of several IDS models' average processing times

5 Conclusion

Although our strategy has yielded great outcomes, we acknowledge that it is not perfect and believe there is still potential for development. The growth and convergence of information technology and automotive mechanical technology will make vehicles on the Internet (WWW) increasingly ubiquitous in people's lives. However, because of the special limitations of this kind of technology and the gaps in its existing security, government agencies and people everywhere should focus more on the security challenges surrounding the Internet of Vehicles (IOV). Considering the current security vulnerabilities with the Vehicle Network, we therefore recommend utilising RNN and sophisticated auto encoders to raise the rate of aberrant behaviours in the Vehicle System. Through studies that evaluated it using real ICV data, we found that our proposed model may be able to effectively categorise vehicle activity and considerably boost the security of the vehicle system. We plan to investigate the use of other deep learning technologies in the future to find a better solution and ensure the security of the in-car system. Additionally, we want to support the fusion of these advancements with AI methods for network security. We believe that our work may also contribute to the efficiency of network security challenges.

References

1. Contreras-Castillo J, Zeadally S, Guerrero-Ibañez JA (2017) Internet of vehicles: architecture, protocols, and security. IEEE Internet Things J 5(5):3701–3709
2. Sedjelmaci H, Senouci SM, Abu-Rgheff MA (2014) An efficient and lightweight intrusion detection mechanism for service-oriented vehicular networks. IEEE Internet Things J 1(6):570–577

3. Li X, Yang B (2015) Analysis of safety protection in vehicle networking. Mobile Commun 39(11):32–35
4. Tariq U (2019) Intrusion detection and anticipation system (IDAS) for IEEE 802.15.4 devices. Intell Autom Soft Comput 25(2):231–242
5. Seo E, Song HM, Kim HK (2018) GIDS: GAN based intrusion detection system for in-vehicle network. In: 16th Annual Conference on Privacy, Security and Trust, pp 1–6
6. Bernardini C, Asghar MR, Crispo B (2017) Security and privacy in vehicular communications: challenges and opportunities. Vehicul Commun 10:13–28
7. Samad A, Alam S, Mohammed S, Bhukhari MU (2018) Internet of vehicles (IoV) requirements, attacks and countermeasures. In: 5th International Conference on "Computing for Sustainable Global Development", pp 4037–4040
8. Woo S, Jo HJ, Lee DH (2014) A practical wireless attack on the connected vehicle and security protocol for in-vehicle CAN. IEEE Trans Intell Transp Syst 16(2):1–14
9. Sun X, Yan B, Zhang X, Rong C (2015) An integrated intrusion detection model of cluster-based wireless sensor network. PLoS ONE 10(10):1–16
10. Tyagi P, Dembla D (2014) Investigating the security threats in vehicular ad hoc networks (VANETs): towards security engineering for safer on-road transportation. In: International Conference on Advances in Computing, Communications, and Informatics, pp 2084–2090
11. Li L (2019) Research and implementation of intrusion detection technology for internet of vehicle based on machine learning (M.S. thesis). University of Electronic Science and Technology of China, China
12. Yang L, Wu Y, Wang J, Liu Y (2018) Research on recurrent neural network. J Comput Appl 38(s2):6–11, 31
13. Yuan F, Zhang L, Shi J, Xia X, Li G (2019) Theories and applications of auto-encoder neural networks: a literature survey. Chin J Comput 42(1):203–230
14. Zhao R, Yan R, Chen Z, Mao K, Wang P et al (2019) Deep learning and its applications to machine health monitoring. Mech Syst Signal Process 115:213–237
15. Martinelli F, Mercaldo F, Orlando A, Nardone V, Santone A et al (2018) Human behavior characterization for driving style recognition in vehicle system. Comput Electr Eng 4(31):1–14
16. Gao L, Li F, Xu X, Liu Y (2019) Intrusion detection system using SOEKS and deep learning for in-vehicle security. Clust Comput 22(6):14721–14729
17. Liu Y, Li Y, Man H (2015) Short paper: a distributed cross-layer intrusion detection system for ad hoc networks. In: First International Conference on Security and Privacy for Emerging Areas in Communications Networks, pp 418–420
18. Besson L, Leleu P (2016) A distributed intrusion detection system for ad-hoc wireless sensor networks: the AWISSENET distributed intrusion detection system. In: 16th International Conference on Systems, Signals and Image Processing, pp 1–3
19. Lauf AP, Peters RA, Robinson WH (2010) A distributed intrusion detection system for resource-constrained devices in ad-hoc networks. Ad Hoc Netw 8(3):253–266
20. Banković Z, Moya JM, Araujo Á, Fraga D, Vallejo JC et al (2010) Distributed intrusion detection system for wireless sensor networks based on a reputation system coupled with kernel self-organizing maps. Integr Comput Aided Eng 17(2):87–102
21. Cho EJ, Hong CS, Lee S, Jeon S (2013) A partially distributed intrusion detection system for wireless sensor networks. Sensors 13(12):15863–15879
22. Yin C, Zhu Y, Fei J, He X (2017) A deep learning approach for intrusion detection using recurrent neural networks. IEEE Access 5:21954–21961
23. Dong B, Wang X (2016) Comparison deep learning method to traditional methods using for network intrusion detection. In: 8th IEEE International Conference on Communication Software and Networks, pp 581–585
24. Mershad K, Artail H (2012) A framework for secure and efficient data acquisition in vehicular ad hoc networks. IEEE Trans Veh Technol 62(2):536–551
25. Medhat K, Ramadan RA, Talkhan I (2015) Distributed intrusion detection system for wireless sensor networks. In: 9th International Conference on Next Generation Mobile Applications, Services and Technologies, pp 234–239

26. Ronak A, Ganesh T, Akshay C, Shriganesh M, Deeplakshmi Z (2016) Distributed intrusion detection system using bayesian learning and apache mahout. J Adv Res Comput Commun Eng 5(1):361–364
27. Peraković D, Periša M, Cvitić I, Husnjak S (2017) Model for detection and classification of DDoS traffic based on artificial neural network. Telfor J 9(1):26
28. Anzer A, Elhadef M (2018) Deep learning-based intrusion detection systems for intelligent vehicular ad hoc networks. Adv Multimed Ubiquit Eng 518:109–116
29. Kwon D, Kim H, Kim J, Suh SC, Kim I et al (2017) A survey of deep learning-based network anomaly detection. Clust Comput 22:1–13
30. Pavani K, Damodaram A (2013) Intrusion detection using MLP for MANETs. In: Third International Conference on Computational Intelligence and Information Technology, pp 440–444
31. Leinmüller T, Held A, Schäfer G, Wolisz A (2014) Intrusion detection in VANETs. In: 12th IEEE International Conference on Network Protocols Student Poster Session, pp 482–486
32. Tan L, Li C, Xia J, Cao J (2019) Application of self-organizing feature map neural network based on k-means clustering in network intrusion detection. Comput Mater Continua 61(1):275–288
33. Kang MJ, Kang JW (2016) Intrusion detection system using deep neural network for in-vehicle network security. PLoS ONE 11(6):1–17
34. Qiao H, Li G, Chen Y (2018) Research on key technologies of vehicle networking system architecture. Electron Prod 352(11)
35. Li X, Zhong C, Chen Y, Zhang H, Weng J (2019) Survey of internet of vehicles security. J Cyber Secur 4(3):17–33

DLAR: Deep Learning-Based Adaptive Routing Strategies for Vehicular Ad-Hoc Networks (VANETs)

Harsh Pratap Singh, Hemlata Patel, Nagendra Singh, Pinky Rane, and Shubha Soni

Abstract The mobile ad hoc network, which allows cars to wirelessly connect to mobile devices as needed, served as a major source of inspiration for the development of the vehicle ad hoc network (VANET). The changing network topology and the automotive ad hoc network's constant mobility pose several challenges. Reliable communication between cars and infrastructure depends on efficient routing systems. Numerous intelligent solution alternatives for enhancing traffic flow, safety, and convenience are now feasible due to recent advancements in computing technologies, intelligent transportation systems, and communications. Many industries are increasingly utilizing artificial intelligence (AI) due to its significant enhancement of traditional data-driven methods. This chapter presents an intelligent deep route learning technique for estimating highway traffic density. Using network simulator 2.3, we simulate the proposed work and do a comparison study using different performance measurement factors, such as coverage ratio, packet delivery ratio (PDR), and control overhead. We compared the proposed methods with existing methods and found that our suggested route increases the aggregate packet delivery ratio (PDR) by

H. P. Singh (✉)
Department of Computer Science and Engineering, Shri Vaishnav Vidyapeeth Vishwavidyalaya, Indore, Madhya Pradesh, India
e-mail: drharshprataps@gmail.com

H. Patel
Department of Computer Science and Engineering, PIET, Parul University, Vadodara, Gujarat, India

N. Singh
Department of Electrical Engineering, Trinity College of Engineering and Technology, Karimnagar, Telangana, India

P. Rane
Department of Computer Science and Engineering, Medi-Caps University, Indore, Madhya Pradesh, India

S. Soni
Department of Computer Science and Engineering, Institute of Engineering and Technology, Sagar, Madhya Pradesh, India

J. Bhatia et al. (eds.), *Deep Learning Based Solutions for Vehicular Adhoc Networks*, Studies in Computational Intelligence 1207,
https://doi.org/10.1007/978-981-96-5190-0_6

23% over the current method. This is because it improves system overhead management and reduces latency, making data transfer more efficient. Overall, the simulation results demonstrate that IDRL routing is more complete than I-AREOR, AREOR, and ARIOR when it comes to routing over VANETs.

Keywords Mobility · Packet delivery ratio · Vehicular ad hoc network · Artificial intelligence · Adaptive · Reliable communication

1 Introduction

We tested the proposed adaptive routing algorithms in the real world and in simulations to see how well they work at improving communication reliability and lowering end-to-end latency in VANETs. The conclusions of our study show that deep learning approaches have the potential to overcome some of the unique challenges associated with routing in VANETs, paving the way for the development of more sophisticated and flexible vehicle communication systems. Major technology firms are primarily researching driverless technology to address the issue of autonomous vehicle control, but this does not imply that it is only applicable to intelligent vehicle control applications [1]. Sensors and communication devices can also integrate self-driving cars to follow predefined paths in the military theater. The utilization of driverless technology in the formation of a vehicle ad hoc network could potentially deter the enemy from causing fatalities during an attack on these vehicular systems. In harsh environments, such as deserts and disaster areas, autonomous technology can also establish a self-organizing network of vehicles for tasks such as sustainability data collection, data collection on plants and wildlife, rescue operations, disaster relief information interaction, and more. The vehicle network will send pertinent data to the terminal for decision-making and processing. This will reduce the need for human labor or protect against workplace mishaps [2].

Because car nodes move so quickly, vehicle ad hoc networks undergo substantial topological changes during network implementation. Furthermore, the network experiences fluctuations in the connection states of some vehicle nodes, disrupts the communication link, and struggles to establish a reliable communication path. As a result, routing technology is one of the primary technological problems with vehicle ad hoc networks. Furthermore, vehicles using fixed fuel in harsh environments or theaters may run out of fuel while completing missions and may not be capable of continuously providing electricity for on-board communication nodes. When there is not enough energy, node mortality will cause changes to the topology, disruptions to communication links, and interruptions to data transmission. Therefore, it is crucial to consider the energy concerns of the transmitting node during data transmission procedures. Depending on the node's transmission rate, an efficient routing node can sustain its energy for a while before successfully transmitting the data [3].

1.1 Motivation

The traditional routing protocols in VANETs are less effective to handle in an extremely dynamic environment because they are static or less adaptive in nature. The adaptive routing system has the ability to adjust its arrangement in real time to accommodate unstable network conditions, which is crucial for ensuring overall VANET performance and reliability. These approaches must endure unanticipated shifts in network topological structure, inconsistent traffic arrangements, and a diversity of environmental circumstances. We used deep learning, a type of algorithm capable of identifying intricate patterns and advancing inferences from data, in this explicit case. It has shown implausible prospects across multiple fields, including autonomous driving, natural language processing, and image recognition.

1.2 Contribution

In this chapter, the contribution is that we propose an intelligent deep route learning technique that helps in efficiently detecting traffic density on highways. We simulate this technique in the network simulator NS-2.3 and perform a comparative analysis using various measuring parameters such as PDR, network routing load, overhead, and coverage ratio. Upon analysis, we discovered that our suggested methods outperform the current ones.

1.3 Outline of Chapter

We have organized the chapters in this way: In Sect. 2, we briefly describe the vehicular ad hoc network, its characteristics, challenges, and issues. Section 3 discusses deep learning and its various techniques, which aid in the effective use of roadside units in vehicular ad hoc networks. In Sect. 4, we designate our proposed methodology in detail. Section 5 offers the result analysis, and the last section presents the overall conclusion of the chapter.

2 Vehicular Ad Hoc Network

Strategically placing vehicle units (VUs) adjacent to road segments forms vehicular-ad-hoc networks (VANETs), enabling effective announcement. Roadside units (RSUs) and adjacent vehicles can establish wireless communications with one another. VU users have the ability to access essential services via internet connectivity. The high mobility of VUs frequently leads to unstable linkages, causing dynamic changes in the network's design. According to the reference, both the transmission range and vehicle interference present limitations. Studies suggest that routing delays lead to intensification in the number of network hops that packets need to pass through during forwarding. Routing techniques effectively minimize delays by choosing route segments with a significant number of vehicles.

2.1 Design of a Vehicular Ad Hoc Network (VANET)

Vehicular Ad-hoc Networks (VANETs) are expanding rapidly in size due to their ad hoc network architecture [4]. The implementation of wireless communication technology between vehicles facilitates this. Vehicular ad hoc networks (VANETs) enhance traffic flow, alleviate congestion, and tackle navigation issues while also promoting safety [5]. As demonstrated in source [6], side of the road units (RSUs) as well as locally available units (OBUs) empower proficient declaration among vehicles in the two arrangements. The WAVE correspondence framework is basic for traveler security since it gives continuous updates on clog and vehicle data [7].

Furthermore, it improves vehicle and human mobility, resulting in improved traffic circulation and effectiveness, as well as increased protection for drivers and pedestrians [8]. The vehicular ad-hoc networks comprise RSUs, TAs, and OBUs. RSUs have the capability to run apps that facilitate smooth announcements with other devices and assist OBUs in connecting with Vehicular Units (VUs) to gather important data like location, speed, and fuel levels. Vehicles in the area receive this data through the wireless connection [9]. A wired network interconnects all RSUs and communicates with the Traffic Authority (TA) [10], responsible for ensuring authentication in VANETs.

- RSU: Equipped with considerable storage and powerful communication capabilities, the roadside unit is a static device. This provides localized connections for vehicles traveling along specific route segments. RSUs can interface with a variety of network devices across infrastructure networks using either wired or cable technology.
- On-board Unit (OBU): Introduced in the auto, an OBU occasionally associates remotely to Side of the road Units (RSUs) and different vehicles to protect security. It likewise utilizations input power, and prepares separately vehicle unit (VU) with a GPS and switched input sensor.
- Trusted Authorities (TAs) govern the maintenance of the VANET system, which involves the authorization of OBUs, RSUs, and car buyers.

To ensure VANET security administration, trustworthy authorities (TAs) supervise the identification of user IDs and the validation of OBU IDs for vehicles. Additionally, to detect dangerous communications emanating from vehicles, identify attackers, and provide full certification of whole vehicles, TAs requires tremendous computing power as well as substantial memory resources.

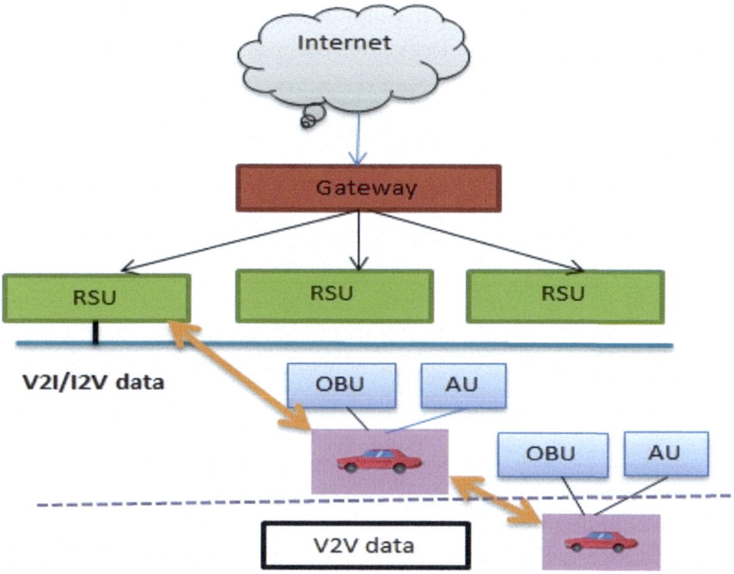

Fig. 1 Design of vehicular ad hoc network (VANET) [11]

The intelligent mobility system places a high focus on effective communication for road safety, including the promotion of secure movement (ITS). In tackling traffic congestion, Intelligent Transportation Systems (ITS) deliver dependable insights concerning traffic congestion and various other transportation-related challenges. Figure 1 represents the infrastructure of VANET.

2.2 VANET Characteristics

A roadside unit (RSU) or moving vehicle can be considered the node of a virtual area network (VANET), which involves physical infrastructure. It provides an integration of infrastructure-dependent modes as well as wireless medium approaches with the attributes of an ad hoc network, which employs a different topology for communication. One approach to summarizing the distinctive properties of VANET, a type of MANET implementation, is as follows [12].

High Mobility: The nodes in a VANET move swiftly, resulting in the network mesh narrowing inside the network. It becomes difficult to determine the vehicle's orientation and ensure node privacy.

Rapidly fluctuating network layout: As VANET nodes are extremely mobile and inconsistent in their vehicle velocity, their placements will fluctuate often. The topology is irregular and dynamic. It simplifies the process of initiating network attacks while making it harder to discover instances of wrongdoing.

Transmission medium accessibility: One can establish a VANET for a single town, a collection of cities, or an entire country. Therefore, location does not restrict the VANET's dimension.

Regular information transfer: VANETs tend to be ad hoc in nature. It encourages the nodes to collect data from nearby vehicles and roadside objects. The nodes need to regularly exchange information.

The concept of anonymity in transferring information through a wireless medium may be opaque. A transmitter, operating in the same frequency spectrum as broadcasting, allows anyone to escape the network's control and restrictions.

Restricted bandwidth: The VANET's traditional DSRC band, which had a width of 27 MHz, should be considered limited. There was a theoretical throughput of 27 Mbps.

Attenuations: According to digital transmission, the Doppler Effect, diffraction, dispersion, reflection, and other fading modes all cause losses in the DSRC band. Multipath reflections can be blamed for the propagation delays.

2.3 Major Concerns in VANET

Vehicular Ad-hoc Networks (VANETs) have special requirements and features, which pose a number of issues [13]. The following are some major VANET challenges:

- Network management: As a result of high mobility, channel conditions and network topology fluctuate quickly. We cannot deploy and maintain tree structures as quickly as changes in topology.
- The infinite network size also presents the challenge of congestion and collision control. Traffic is light at night in rural areas, even in metropolitan regions. As a result, network partitions frequently occur during rush hours, when heavy traffic overloads the network and makes it prone to collisions.
- MAC Design: Since VANETs often interact over shared media, MAC design is the main problem. There have been numerous proposals for strategies such as TDMA, SDMA, and CSMA. IEEE 802.11 accepted the CSMA-based Mac for VANET.
- Security: VANET offers life-critical traffic safety applications. Thus, it is necessary to ensure the security of these messages.

2.4 Critical Challenges in Routing

Each vehicle is appointed intelligent location by the different steering technique of the portable impromptu organizations. In any case, existing directing techniques don't make preparations for the peculiarity of numerous location tasks in auto frameworks. The VANET faces a few difficulties, including lacking street aspects for transmission inclusion, demography, the quantity of cars, versatility game plans, and the unpredictable campaign of vehicles entering and takeoff the organization [14].

Finding and keeping up with the best way for sending information parcels by means of transitional hubs is the primary objective of the directing calculation. Since the portable hubs are dynamic, VANETs' working is intricate and requires the utilization of explicit steering conventions from MANETs. For steering frameworks in view of geography, each taking part hub must have a remarkable location. In this manner, a method for relegating exceptional addresses is essential; on the other hand, these conventions don't ensure that address occupations stay extraordinary inside the organization. Therefore, we can reason that MANET techniques are not suitable for VANETs. Furthermore, certain VANET troubles, for example, versatility models, traffic volumes, quick changes, street widths, and organization geographies, are past the abilities of ordinary steering conventions. Table-driven directing, additionally ordinarily alluded to as proactive steering conventions, keeps up with steering data. This makes it simpler for each virtual client (VU) on the organization to interface with each other VU. The recently expressed conventions empower each hub to disseminate refreshes consistently while keeping a reasonable and reliable perspective on the organization structure. One advantage of proactive directing frameworks is that they can stay away from low idleness continuously conditions by proceeding to keep up with the objective way behind the scenes. Subsequently, this maintains a strategic distance from the development of underutilized information pathways, which might diminish the transmission capacity that is accessible.

3 Overview of Deep Learning

The phrase "deep learning" refers to artificial intelligence (AI) and machine learning that mimic how people learn specific topics. It may be possible to create deep learning algorithms that can identify correlations in text, sound, pictures, and other types of data and carry out categorization tasks. Additionally, it automates processes like voice translation and visual description that typically require human intellect. A crucial component of data science is deep learning, which goes hand in hand with data and project modeling [15]. While human brains consist of millions of interconnected neurons that collaborate to learn new things, deep learning relies on neural networks composed of several distinct layers of collaborating software nodes. Large amounts of labeled data and neural network topologies train deep learning models. Deep learning allows a machine to learn by doing. To understand deep learning, think of

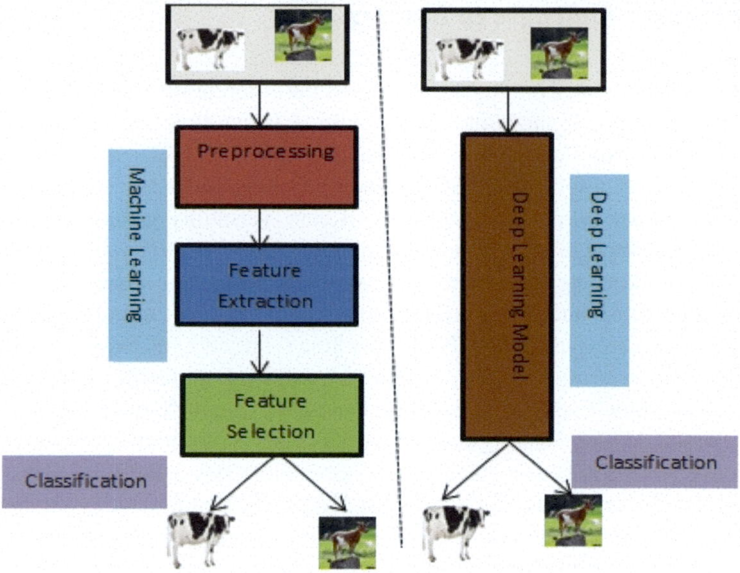

Fig. 2 Difference amongst deep learning and customary machine learning [15]

a child whose first word is "cow." By pointing to items and saying the word "cow," the child comes to understand what a dog is and is not. In response, the parent may answer, "Yes, the fact that is a cow," or "No, the fact that is not a cow." The child points to objects and becomes more aware of the traits that all dogs have in common. Without realizing it, the child is helping to understand a challenging abstraction: the concept of a cow. They are accomplishing this by building a hierarchy where each degree of abstraction is developed using the knowledge gathered from the layers of the organization above. Here, Fig. 2 shows the difference between machine learning and deep learning.

3.1 Classification of Deep Learning Techniques

The deep learning is classified into two categories which are shown in Fig. 3.

3.2 Deep Supervised Learning

This method manages labeled data. Such a strategy produces a set of inputs and outputs (xt, yt). For example, if the input is xt, the smart agent estimates $y < t = f(xt)$ and returns $t(y\hat{t}, yt)$ as the loss value. Next, to obtain a more

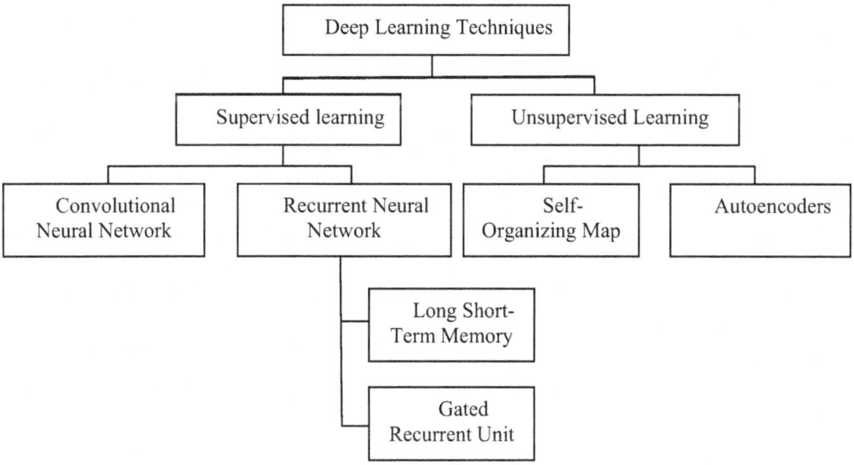

Fig. 3 Classification of deep learning techniques

accurate approximation of the intended outputs, the agent constantly adjusts the network parameters. Following a successful training outcome, the agent can discern the accurate responses to the questions posed by the environment. Deep learning (DL) supervised learning approaches include recurrent neural networks (RNNs) and convolutional neural networks (CNNs). This method's primary benefit is its capacity to gather data or produce data based on previous knowledge. This strategy has the drawback that if there aren't enough cases in the training set to fit into a class, the decision boundary may be overstretched. In general, this method of learning with remarkable results is easier to apply than other methods.

3.2.1 Convolutional Neural Networks

In the realm of deep learning, the CNN algorithm is the most popular and often utilized [16]. CNN's primary benefit over its predecessors is its autonomous, human-oversight-free ability to detect important pieces. Applications such as computer vision, facial recognition, audio processing, and others widely utilize CNNs. Similar to a conventional neural network, researchers constructed CNNs using neurons found in the brains of humans and other animals. Goodfellow et al. [17] identified the three primary benefits of CNNs: sharing of parameters, sparse connections, and comparable representations. Despite conventional fully connected (FC) networks, CNNs use 2D input-data formats, such as image signals, by leveraging shared weights as well as local connections. This strategy dramatically reduces the number of parameters, thereby increasing network performance and simplifying training. This tendency fits the visual cortex's cells, which preferentially process specific areas of a picture over the entire image, serving as local filters that identify spatial correlations in the input. Figure 4 presents a visual representation of the CNN framework for image

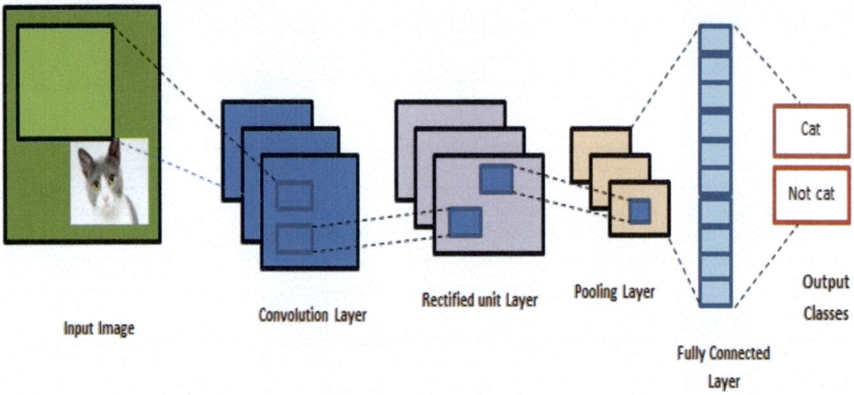

Fig. 4 Architecture of convolutional neural networks [16]

categorization. In a CNN model, each layer's input (x) is arranged in three dimensions: depth, width, and height, indicated as $m \times m \times r$, wherein width and height (m) are equal. You can also refer to the depth as the channel number; for instance, an RGB image has a depth (r) of three. The total number of kernels or filters (k) that compose each convolutional layer serves as a key function, resulting in kernels exhibiting 3-D ($n \times n \times q$), approximating the input image structure.

However, since q is either equal to m or less than r. Also, the kernels are what make local connections possible. They take in data and make k feature maps of all sizes ($m - n - 1$) by combining properties that are related (bias and weight). The convolution layer determines a dot product among its input and its weights, as indicated in Eq. (1), analogous to NLP, but with the inputs being smaller portions of the original image. We obtain this by subjecting the convolution-layer output to either non-linearity or an activation function:

$$h^k = f\left(W^k * x + b^k\right) \tag{1}$$

The subsampling layers will then all down sample their feature maps. Consequently, the training process accelerates, potentially resolving the overfitting issue by reducing the network parameters. For each feature map, the pooling function (like max or average) is applied to a surrounding region of size $p \times p$, where p is the kernel size. In the final phase, the high-level abstractions that the FC layers build through central- and low-level data mimic the final levels of a regular neural network.

The final phase uses techniques such as support vector machines, or SVMs, such as Softmax, to obtain categorized scores. Every score, for example, signifies the possibility of a given class.

3.2.2 Recurrent Neural Network

One kind of artificial neural network that is frequently utilized in voice recognition and natural language processing is the recurrent neural network. These techniques employ patterns to identify the sequential features of data and forecast the most likely course of events.

Both deep learning and expanding models that mimic neuron motion in the human brain utilize it. Their processing of a series of inputs that influence the final output through feedback loops sets them apart from other kinds of artificial neural networks [18]. Additionally, they perform particularly well in scenarios where context is necessary for result prediction. Such feedback loops allow information to remain persistent. People commonly refer to this event as a memory.

RNN use cases are typically associated with language models, where the ability to predict the next letter or word in a sentence relies on the information before it. An interesting experiment trains an RNN on Shakespeare's writings to effectively generate prose that resembles Shakespeare's writing. RNN writing is one type of computational creativity. Artificial intelligence may mimic human creativity because of its training set's understanding of semantics and syntax.

3.2.3 Long-Short-Term-Memory (LSTM)

Long short-term memory (LSTM) units, sometimes referred to as memory blocks, construct recurrent neural network (RNN) layers. People sometimes refer this network as an RNN, which is serene of LSTM units [18]. An output gate, an input gate, a forget gate, and a cell make up a typical LSTM unit. The LSTM word "memory" comes from the fact that the cell is in charge of "remembering" values across arbitrary time intervals. We can think of all three gates as "conventional" artificial neurons in a multi-layer neural network, as they use an activation function to determine the activation of a weighted sum. It is logical to regard them as controllers of the values that flow through the LSTM's connections; hence, the term "gate." There's a connection among gates and the cell. It is a type of short-term memory that can endure for an extended period. LSTMs are a great option for categorizing, analyzing, and forecasting time series when there are unknown amounts of time lag between important events. The development of LSTMs addressed the exploding and vanishing gradient problem when training conventional RNN.

3.2.4 Gated Recurrent Units

In other words, unlike the LSTM unit, the GRU unit does not require a memory unit to regulate the information flow. It has no control and can use all hidden states immediately. Due to their smaller number of parameters, GRUs may generalize with less input or train a little quicker. The modify gate selects how considerable of the prior state to retain, however the reset gate chooses how to mix fresh input with the previous memory. Whereas input and forget gates were in LSTM, update gates are in GRU. The second nonlinearity in GRU does not exist prior to computing the output. They do not possess an output gate.

Working of a Gated Recurrent Unit

1. Assume two vectors as inputs: the prior concealed state and the current input.
2. Use the procedures listed below to calculate the values of the three separate gates.
3. To get the parameterized present input and previously hidden state vectors for each gate, do element-wise multiplication (Hadamard products) among the appropriate vector and each gate's weights.
4. Apply the proper activation function element-wise for each gate on the parameterized vectors. Below is a list of gates, in addition to the activation mechanism that each gate requires.

Modify Gate: Sigmoidal Function.
Reset Gate: Sigmoidal Function.

- There is a little variation in the method used to calculate the Current Memory Gate. First, the previously concealed state vector and the Reset Gate's Hadamard product are computed. Subsequently, this vector undergoes parameterization and is appended to the already parameterized input vector.

$$\overline{h}_t = \tanh(W \odot x_t + W \odot x(r_t \odot h_{t-1})) \tag{2}$$

- To determine the current concealed condition, a vector with identical dimensions to the input must first be created. Mathematicians refer to this vector as a ones vector, symbolized by the number 1. First, find the update gate's and the previously concealed state vector's Hadamard product. To create a new vector, subtract the update gate from the ones, and then multiply the resultant vector's Hadamard product by the memory gate's current value. Lastly, sum the two vectors to obtain the concealed state vector.

$$h_t = z_t \odot h_{t-1} + (1 - z_t) \odot h_t \tag{3}$$

3.3 Deep Unsupervised Learning

This approach allows for the learning process to occur even in the absence of labeled data—that is, tickets are not required. At this point, the agent uses its internal representation to identify the critical characteristics that are required to identify the absent relationships or organizational framework in the incoming data. The domain of unsupervised learning frequently employs generating networks, reduced dimensionality, and clustering methods. New developments in the DL family, such as auto-encoders, GANs, and restricted Boltzmann machines, have done great on tasks like clustering and non-linear dimensionality reduction. Additionally, a variety of unsupervised learning applications have utilized RNNs, which include GRUs and LSTM methods. The primary downsides of unsupervised learning are its dispensation complexity and inability to distribute precise information on data sorting. Clustering is one of the most regularly used unsupervised learning techniques [18].

3.3.1 Self-organizing Map

The family of unsupervised deep learning networks uses self-organized maps to form clusters and reduce dimensionality. Teuvo Kohonen first introduced the concept in 1980. This is why we continue to use the name "Kohonen Maps" [19].

(1.a) Intuition: Firstly, we cannot use an ANN to process a dataset with a large number of features or variables because doing so will complicate our calculations. Additionally, if we must use an unsupervised deep learning framework, our only choice is to convert higher-dimensional datasets (from 3D to 2D) into lower-dimensional clusters. Second, SOM facilitates our ability to identify patterns in the data.

(1.b) Structure: In SOM, there are mainly output layers and visible input layers. The input nodes represent the characteristics of the dataset. Every input feature contributes in some way to the output features. However, we must identify patterns among the input and output layers in order to create clusters. Each synapse in the input layers carries weights. In this scenario, we consider the input values as independent weights. We'll be walking over each row in our dataset and attempting to determine which node is closest to which row.

It lacks a cross-entropy function and a cost function, in contrast to supervised deep learning frameworks. Therefore, the idea of gradient descent backpropagation is not relevant here.

(1.c) K-Means and How They're Similar to SOM: It bears a strong resemblance to the K-Means unsupervised machine learning framework in that creating clusters is its main objective. In the K-Means cluster creation procedure, the first step is to allocate k points and the number of clusters (=k). In the second phase, assign all other points to those clusters (with the closest distance to those k points). Next, we determine the centroid of each cluster. Next, we should transfer each point to the cluster's closest centroid once more. We continued this process until we reduced the error.

Steps of K-Means

The "Elbow Method" is to be used to compute the error. With this approach, all we have to do is reduce the distance amongst data points inside a group and increase the distance between clusters.

$$\underset{S}{\arg\min} \sum_{i=1}^{k} \sum_{i \in S_i} \|x - \mu_i\|^2 = \underset{S}{\arg\min} \sum_{i=1}^{k} |S_i| Var S_i \tag{4}$$

Image Source: Google Image.

where $\boldsymbol{\mu}_i$ = The centroid of the cluster.

3.3.2 Auto Encoder

The auto-encoder, a type of ANN, aims to acquire data encodings autonomously. When learning a lower-dimensional description (encoding) from higher-dimensional facts, an auto-encoder teaches the network to recognize the most important areas of the input image, typically for reduced dimensionality [20–23]. The design of auto-encoders consists of three parts:

1. Encoder: A module that creates an encoded representation of the input data from the train-validate-test set that is typically orders of magnitude less than the original data.
2. Bottleneck: The most crucial component of the network is a module holding the condensed knowledge representations.
3. Decoder: A module that aids in the knowledge representations' "decompression" and reconstruction of the data from its encoded state by the network. Then, the organization contrasts the outcome and the ground truth. Figure 5 show the engineering of the working of auto-encoder.

After study the above deep learning techniques, here we have represented the comparison between above techniques such as CNN, LSTM, GRU, SOM and autoencoder is shown in Table 1.

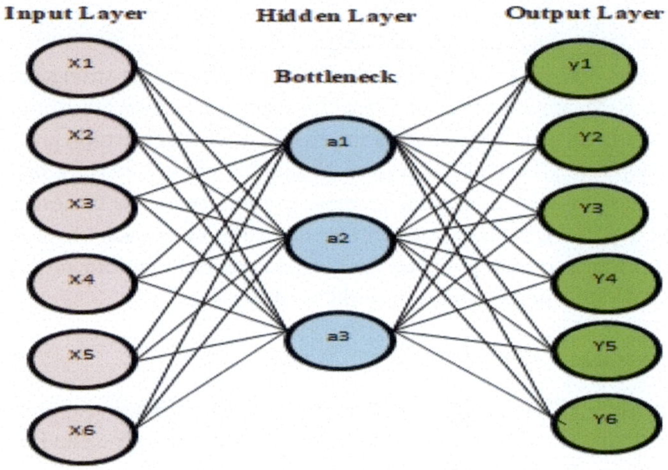

Fig. 5 Architecture of auto encoder

Table 1 Ccomparison between CNN, LSTM, GRU, SOM and autoencoder

Feature	CNN	LSTM	GRU	SOM	Autoencoder
Primary use case	Image and video processing	Grouping expectation, time series determining	Succession expectation, time series determining	Data visualization, dimensionality reduction	Data compression, feature learning
Architecture	Layers of convolutional and pooling	Recurrent units with memory cells	Recurrent units with gates for updating	Grid of neurons organizing input data	Encoder and decoder structure
Memory handling	No explicit memory handling, focuses on local spatial info	Handles long-term dependencies with memory cells	Handles dependencies with update and reset gates	No explicit memory, based on similarity	No explicit memory, uses latent space
Training complexity	High due to large number of parameters	High due to sequential nature and gates	Moderate, simpler than LSTM but effective	Low to moderate, involves unsupervised	High, involves reconstruction loss
Handling sequential data	Poor, not designed for sequences	Excellent, designed for sequential data	Excellent, designed for sequential data	Poor, not designed for sequences	Depends on application, not inherently sequential
Parameter efficiency	Efficient in spatial parameter usage	Less efficient, high number of parameters	More efficient than LSTM, fewer parameters	Efficient, fewer parameters	Varies, depends on architecture
Training time	Generally faster due to parallelism in convolution	Slower, sequential processing	Faster than LSTM, less sequential overhead	Moderate, depends on dataset size	Moderate to high, depends on complexity
Explainability	Moderate, can visualize filters and feature maps	Harder to interpret due to complex gates	Harder to interpret due to gates	High, interpretable through neuron weights	Moderate, can visualize encoded/decoded data
Use in practice	Image classification, object detection, etc	Language modeling, speech recognition	Language modeling, speech recognition	Data clustering, visualization	Data denoising, anomaly detection

(continued)

Table 1 (continued)

Feature	CNN	LSTM	GRU	SOM	Autoencoder
Advantages	Effective for image data, less overfitting	Good at handling long-term dependencies	Simpler and faster than LSTM, effective	Intuitive viz of high-dimensional data	Effective data compression, robust feature learning
Disadvantages	Not suitable for sequential data, computationally intense	High computational cost, difficult to train	Can still be computationally intense	Limited use requires careful parameter tuning	Can overfit, requires careful tuning

4 Proposed Approach

A wide variety of vehicles, each with unique features, make up Vehicular Ad Hoc Networks, or VANETs [24–27]. Typically, service providers provide connectivity to roadside units (RSUs) located alongside roadways. We are looking at the V2I correspondence framework, which comprises of a solitary RSU, otherwise called a Passage (AP), arranged in a particular region. RSUs often move and get vehicle information through the avoidance entryway, which the organization administrator initially deciphers prior to sending it to framework clients. Currently used approaches do not take into account cluster-based V2I communication. This procedure entails dividing roads into clusters and calculating characteristics like vehicle speed and traffic density, as shown in Fig. 6. It is commonly known that intensification in the number of vehicles on the road causes traffic congestion, which in turn causes average vehicle speeds to decrease. Let '*r*' signify the set of routes, and '*l*' represent the set of the present positions for every Vehicle Units (VUs), written as $l = \{r_1, r_{2,} r_2, \ldots, l_n\}$. The transition function considers the change from one phase to the next where vehicle density is exemplified as $V_{\text{d}} = \{V_{\text{d1}}, V_{\text{d2}}, V_{\text{d3}}, \ldots, V_{\text{dn}})$:

$$\Omega_{pq} : \phi \rightarrow r \quad \text{where } 1 \leq p \leq n,\ 1 \leq 1 \leq n \tag{5}$$

The function ensures that there are viable pathways on the left side of Eq. (5) using a probabilistic approach. To function effectively, the system regularly makes usage of the input factor $\varphi = (\text{l}, d, V_{\text{d}})$. We compute the car's recent position and density to govern the distance among the vehicle and the RSU. We then use this data to determine the optimal routes for traffic. We count all the cars in a given area to calculate vehicle density across every vehicle unit (VU). At any given time, all VANETs can access this information uniformly. A mathematical expression used for

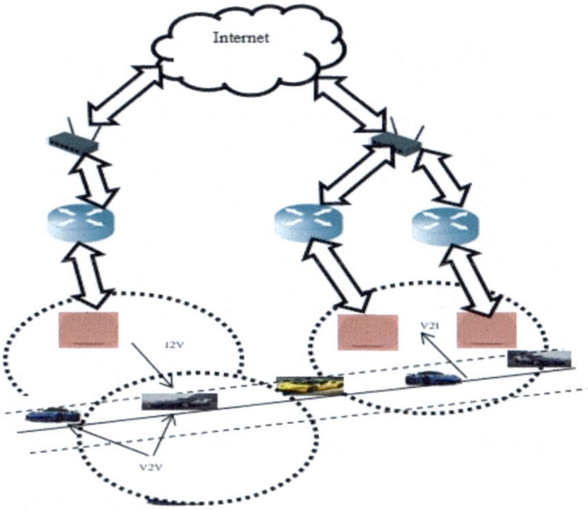

Fig. 6 Routing model of proposed approach

calculation or estimation is called an equation,

$$V_{\mathrm{d}} = \frac{\sum_{p=1}^{q} \gamma_p}{\sum_{p=1}^{q} \eta_p} \tag{6}$$

where η_p denotes both the entire vehicle time and the overall vehicle utilization in area I.

This transmission has a maximum coverage distance of 150 m and is subject to the predefined criteria. There are regional variations in the density of cars. As a result, in order to map pathways using the Matrix of Transition Probability (TPM) values for the current path, accurate mapping is necessary, as stated in Eq. (7). The TPM values can be used to identify the most effective pathways for information sharing.

$$\begin{matrix} \Omega_{11} & \Omega_{12} & \Omega_{13} & \Omega_{1n} \\ \Omega_{21} & \Omega_{22} & \Omega_{23} & \Omega_{2n} \\ \Omega_{31} & \Omega_{32} & \Omega_{33} & \Omega_{3n} \\ \Omega_{41} & \Omega_{42} & \Omega_{43} & \Omega_{4n} \end{matrix} \tag{7}$$

The suggested approach, which is constructed with minimal parametric weighting, seeks to determine the best path for this particular problem.

4.1 Intelligent Deep Route Learning Model

The proposed approach is independent of a pre-existing routing system. We use the IDRL framework to estimate traffic speeds on highways, accounting for differences in vehicle density (V_{d}). The IDRL consistently evaluates routes based on the existing channel capacity and their probability of accomplishment. You can use the car's movement and velocity in relation to nearby Roadside Units (RSUs) to assist in choosing a route. The system dynamically updates the data by comparing the exact location of the approaching unit with that of the roadside unit [28–30].

The wired communication links the roadside units, enhancing the transmitting speed. This enables surrounding RSUs to obtain information about the approaching vehicle, such as its location and speeds. GPS-based vehicle node localization is difficult due to cars' high velocity.

The route establishment phase (REP)

Neighboring VUs receive packets relayed during the REP phase. The packages comprise many components, such as the car's present location, its relative position to the surrounding region, the number of vehicles in the nearby areas, and the current route. The provided information includes the network's strengths in a specific location, latency data, and the current position of the intermediary node. By developing an intellectual system that calculates the most efficient path to accomplish the stages and reach the desired goal, we can optimize the arrangement of intermediary points.

Optimal route establishment (ORE)

The IDRL employs methodologies that utilize incentives or sanctions to ascertain the optimal path of action. The transitional function (Ω_{pq}) assists in the creation of a routing path from a starting point to a destination, making stage transitions easier. The pq values stay constant until the path threshold adjustment task is completed. The transmission of data packets over the transportation network is affected by the road capacity, which is determined by the maximum weight of monitoring that can be transported from the starting point to the destination, as well as the presence of numerous transportation routes. Two categories of weight are associated with routes: cumulative weights and route weights. These variables aid in determining the suitable course of action, which ultimately influences the decision to either reward or condemn a routing path. The reward function computes the cumulative weights, represented by the symbol μ, that vary between 0 and 1. The weights undergo uniform modification if the currently selected node is a targeted node. If the present node is not the planned destination, the transition function adjusts the cumulative weights. We can assess the reliability or effectiveness of each route using a constant function, anticipating rapid convergence towards ideal paths as it approaches 1. As the value of μ approaches zero, selecting the optimal routing path becomes more time-consuming. When transmission is not efficient, a fraction μ, which can vary between 0 and 1, is deducted from the overall weight. The Inter-National Disaster Response Law (IDRL) has the potential to mitigate or eliminate penalties in the event of a breakdown in communication.

Route assortment phase

The next step after the initial routing phase is to use a specific routing method to choose the best neighboring node within the routing table. In this process, the proximity and density of vehicles for communicating messages are considered, allowing each vehicle or node in the system to acquire the optimal phenomenology value of the nearest vehicle for communication. Vehicle density determines the vehicle intensity during sub-phases, which occur at various intervals, while the time interval remains constant. Traffic conditions cause fluctuations in the density and velocity of cars during subsequent time frames. Therefore, it is crucial to use a routing algorithm to evaluate traffic volume at a specific time to select the most efficient route.

Route selection phase using IDRL

The IDRL algorithm determines the most efficient routing paths by analyzing previous observations. The system utilizes the IDRL (Intelligent Dynamic Routing Logic) to carry out effective routing calculations by making predictions about the next Trusted Platform Module (TPM). We formulate this process as a problem involving supervised learning.

4.2 Supervised Learning

This work specifically addresses the difficulty of supervised learning using a sample space (P) as well as a labeling space (Q). The primary objective is to develop a function that, with a focus on the super-vised learning aspect of IDRL (X), connects data in P with related labels in Q. The goal is to develop a mapping function that joins the corresponding true labels (qi) of the modeled set (pi) to a Cartesian product, or $P \times Q$. Using the mapping approach, newly sampled instances from a data dispersion receive appropriate label assignments [31–33].

The proposed technique makes use of an Increased Dynamical Rule Learning (IDRL) approach to efficiently track the variations in TPM over multiple iterations, from previous to current ones. In this case, TPM forecasts the best paths for packet transfer. To develop TPM sequences, it is essential to have a thorough understanding of the system under investigation's past and present TPM. A probability distribution influences the transformation of traffic patterns, leading to apparent unpredictability and considerable fluctuation, which makes assessing protocols challenging. By utilizing the knowledge of potential transportation and organizing managers (TPMs) and a trustworthy mapping method, intelligent dynamic routing and logistics (IDRL) technology can forecast the optimal routes for future roadways. IDRLs can directly extract data from the observed TPM by applying the mapping technique. Randomized time intervals (t) separate the road networks into subsections based on both geography and time.

5 Result Analysis

We assessed the effectiveness of the proposed strategy with a network simulator (NS-2.35). We conducted the simulation in a 1000-by-1000-m region. The VanetMobiSim software precisely simulated the driving behaviors and actions of cars in urban environments. Each RSU's velocity ranged from 5 to 30 m/s when randomly distributed among several sites. We adjusted the transmission power mode to guarantee the required coverage, which included a 250-m maximum broadcast range. The source node transmitted data packets using constant bit rate (CBR) rules, with the usual packet size being 512 bytes.

This study evaluated the proposed technique by comparing it to three existing methods: I-AREOR (Increased AREOR), ARIOR (Adaptive Ranking-Based Energy-Effective Opportunistic Routing), and ARIOR (Adaptive Ranking-Based Increased Opportunistic Routing). We compare these routing systems by measuring variables like the packet delivery ratio and coverage ratio. We also took into account the RSU coverage percentage, which is a numerical representation of the amount of RSU coverage relative to the entire area. To determine the broadcast as well as endpoint location capabilities, IDRL uses RSU data.

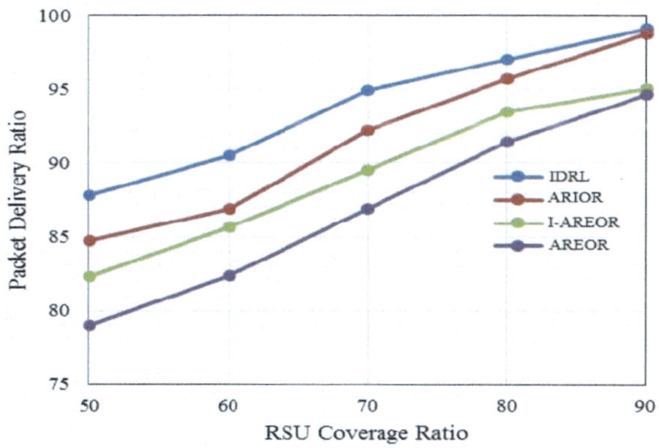

Fig. 7 Delivery ratio versus coverage ratio

Figure 7 illustrates the expansion of the transmission coverage region. The measurable information showed a steady upgrade in network execution for entirely four strategies as the quantity of street side-units (RSUs) developed. Empirical research has shown that the suggested method outperformed existing routing techniques via packet delivery ratio (PDR). The evaluated approach showed a substantial enhancement compared to the prior ARIOR, resulting in an average augmentation of 5%. The advantage stems from the use of the IDRL approach, which employs announcement messages to preserve connection information among units.

Figure 7 depicts the disparity in the packet delivery ratio (PDR) for vehicle situations with low and high densities. The proposed strategy surpasses the previously employed ARIOR methodology by a margin of 6.5%. Utilizing alternative methods that employ a distributed design leads to higher density and greater instability. The results suggest that the use of RSUs results in a significant improvement in performance, with a notable rise of 23%.

Figure 8 showcases the aftereffects of the past postpone examination. The outcomes show that the degree of RSU inclusion considerably affects the time it takes to move parcels between RSUs. By diminishing the transmission range, its results in a significant drop in relative deferral. Its results assign that the proposed strategy ensures consistency in settling RSU information, as opposed to existing directing calculations.

To make high progress rates, the IDRL approach has a great deal of potential. The projected methodology has uncovered superior accuracy in surveying other course decisions by getting information on vehicle number, speed, and direction. The dependability of information transport is essentially higher contrasted with the assortment of friend steering strategies being used today. The IDRL course helps information move productivity by bringing down dormancy and further developing framework above administration, eventually expanding the all out bundle conveyance proportion (PDR). The recreation results exhibit that IDRL steering over VANETs is greater than I-AREOR, AREOR, and ARIOR.

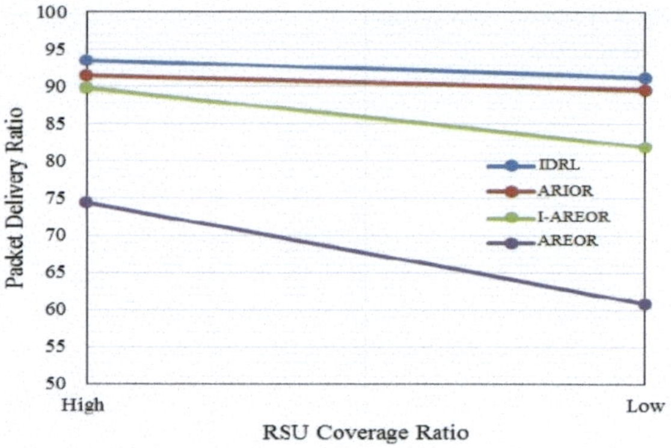

Fig. 8 Delivery ratio versus control overhead

6 Conclusion and Future Recommendation

The proposed novel Intelligent Deep Route Learning Framework for Vehicular Ad Hoc Networks (VANETs) has made an important advancement in the arena of intelligent transportation systems. This model leverages deep learning features to enhance the effectiveness, dependability, and safety of vehicle-to-vehicle contact and navigation. The advanced deep route learning framework may significantly improve VANET routing efficiency. The network's dynamic adaptation allows it to function at its best under a range of circumstances. Deep learning makes it feasible to make decisions in real time. In the event of unforeseen traffic disruptions, including accidents or road closures, automobiles can quickly process information and adjust their routes. By streamlining traffic and creating more efficient paths, the model reduces network congestion. This increases the overall effectiveness of the system and makes driving more enjoyable. We expect significant advancements in VANETs with the continued development of deep learning algorithms and the increasing accessibility of excellent traffic analysis. To summarize, our intelligent deep-route-learning model is a significant improvement for VANETs because it resolves many of the common problems with vehicular networks. Its implementation paves the way for a more secure, clever, and efficient transportation system, in keeping with the main goals of modern smart cities in addition to intelligent transportation. In future work, we will research the effect of the IDRL procedure on some other estimating boundaries and coordinate this methodology with other trial testbeds to acquire constant measurements for vehicular ad hoc network.

References

1. Zarkeshev A, Csiszár C (2020) Patients' willingness to ride on a driverless ambulance: a case study in Hungary. Transp Res Proc 44:8–14. https://doi.org/10.1016/j.trpro.2020.02.002

2. Wang Z et al (2019) Optimal control strategy design with minimum energy consumption for connected vehicle systems. In: IEEE Global Communications Conference (GLOBECOM). IEEE. https://doi.org/10.1109/Globecom38437.2019.9013378
3. Garcia-Santiago A et al (2018) Evaluation of AODV and DSDV routing protocols for a FANET: further results towards robotic vehicle networks. In: 2018 IEEE 9th Latin American Symposium on Circuits and Systems (LASCAS). IEEE. https://doi.org/10.1109/LASCAS.2018.8399972
4. Saleem MA et al (2019) Expansion of cluster head stability using fuzzy in cognitive radio CR-VANET. IEEE Access 7:173185–173195. https://doi.org/10.1109/ACCESS.2019.2956478
5. Haykin S (2005) Cognitive radio: brain-empowered wireless communications. IEEE J Sel Areas Commun 23(2):201–220. https://doi.org/10.1109/JSAC.2004.839380
6. Akan OB, Karli OB, Ergul O (2009) Cognitive radio sensor networks. IEEE Netw 23(4):34–40. https://doi.org/10.1109/MNET.2009.5191144
7. Carie A et al (2019) Cognitive radio assisted WSN with interference aware AODV routing protocol. J Ambient Intell Humaniz Comput 10(10):4033–4042. https://doi.org/10.1007/s12652-019-01282-6
8. Khan AA, Rehmani MH, Rachedi A (2017) Cognitive-radio-based internet of things: applications, architectures, spectrum related functionalities, and future research directions. IEEE Wirel Commun 24(3):17–25. https://doi.org/10.1109/MWC.2017.1600404
9. Zhang ZY, Jin CH, Liang XL, Chen Q, Peng LM (2006) Current–voltage characteristics and parameter retrieval of semiconducting nanowires. Appl Phys Lett 88(7):073102. https://doi.org/10.1063/1.2177362
10. Özgür Ü, Alivov Y, Morkoç H (2009) Microwave ferrites: Part 1: fundamental properties. J Mater Sci Mater Electron 20(9):789–834. https://doi.org/10.1007/s10854-009-9923-2
11. Kumar G et al (2018) Multidimensional security provision for secure communication in vehicular ad hoc networks using hierarchical structure and end-to-end authentication. IEEE Access 6:46558–46567. https://doi.org/10.1109/ACCESS.2018.2866759
12. Rajkumar MN, Nithya M, HemaLatha P (2016) Overview of VANET with its features and security attacks. Int Res J Eng Technol (IRJET) 3(1):137–142
13. Singh G (2016) Overview of challenges in VANET. IJIRSE 2(08):389–394
14. Tanwar R, Balamurugan S, Saini RK, Bharti V, Chithaluru P (eds) Advanced healthcare systems: empowering physicians with IoT-enabled technologies. Wiley, New York, NY
15. Chithaluru P, Kumar S, Singh A, Benslimane A, Jangir SK (2021) An energy-efficient routing scheduling based on fuzzy ranking scheme for internet of things. IEEE Internet Things J 9(10):7251–7260. https://doi.org/10.1109/JIOT.2021.3098430
16. Krizhevsky A, Sutskever I, Hinton GE (2017) Imagenet classification with deep convolutional neural networks. Commun ACM 60(6):84–90. https://doi.org/10.1145/3065386
17. Goodfellow I, Bengio Y, Courville A, Bengio Y (2016) Deep learning, vol 1. MIT Press, Cambridge. https://mitpress.mit.edu/9780262035613/deep-learning/
18. Saeed MM, Al Aghbari Z, Alsharidah M (2020) Big data clustering techniques based on spark: a literature review. PeerJ Comput Sci. 6:321
19. https://www.techtarget.com/searchenterpriseai/definition/neural-network
20. Zhang D et al (2018) Novel dynamic source routing protocol (DSR) based on genetic algorithm-bacterial foraging optimization (GA-BFO). Int J Commun Syst. https://doi.org/10.1002/dac.3824
21. Yang J et al (2019) Optimal base station antenna downtilt in downlink cellular networks. IEEE Trans Wireless Commun 18(3):1779–1791. https://doi.org/10.1109/TWC.2019.2897296
22. Liu S et al (2019) A cognitive relay network throughput optimization algorithm based on deep reinforcement learning. Wirel Commun Mobile Comput. https://doi.org/10.1155/2019/2731485
23. Bi X, Huang H, Zhang B, Wei X (2023) A hybrid routing algorithm for V2V communication in VANETs based on blocked Q-learning. IEICE Trans Commun 106(1):1–17. https://doi.org/10.1587/transcom.2021EBP3210
24. Chithaluru PK, Khan MS, Kumar M, Stephan T (2021) ETH-LEACH: an energy enhanced threshold routing protocol for WSNs. Int J Commun Syst 34(12):e4881. https://doi.org/10.1002/dac.4881

25. Chithaluru P, Tiwari R, Kumar K (2019) AREOR-adaptive ranking based energy efficient opportunistic routing scheme in wireless sensor network. Comput Netw 162:106863. https://doi.org/10.1016/j.comnet.2019.106863
26. Geng N, Bai Q, Liu C, Lan T, Aggarwal V, Yang Y, Xu M (2023) A reinforcement learning framework for vehicular network routing under peak and average constraints. IEEE Trans Veh Technol. https://doi.org/10.1109/TVT.2023.3235946
27. Mitola J, Maguire GQ (1999) Cognitive radio: making software radios more personal. IEEE Pers Commun 6(4):13–18. https://doi.org/10.1109/98.788210
28. Isik MT, Akan OB (2009) A three dimensional localization algorithm for underwater acoustic sensor networks. IEEE Trans Wireless Commun 8(9):4457–4463. https://doi.org/10.1109/TWC.2009.081628
29. Tachwali Y, Basma F, Refai HH (2010) Cognitive radio architecture for rapidly deployable heterogeneous wireless networks. IEEE Trans Consum Electron 56(3):1426–1432. https://doi.org/10.1109/TCE.2010.5606279
30. Bambos N, Chen SC, Pottie GJ (2000) Channel access algorithms with active link protection for wireless communication networks with power control. IEEE/ACM Trans Netw 8(5):583–597. https://doi.org/10.1109/90.879345
31. Li S, Da XL, Wang X (2012) Compressed sensing signal and data acquisition in wireless sensor networks and internet of things. IEEE Trans Industr Inf 9(4):2177–2186. https://doi.org/10.1109/TII.2012.2189222
32. Xie L, Jia X, Zhou K (2012) QoS multicast routing in cognitive radio ad hoc networks. Int J Commun Syst 25(1):30–46. https://doi.org/10.1002/dac.1285
33. Yuan F, Zhan Y, Wang Y (2013) Data density correlation degree clustering method for data aggregation in WSN. IEEE Sens J 14(4):1089–1098. https://doi.org/10.1109/JSEN.2013.2293093
34. Saleem MA, Shijie Z, Sharif A (2019) Data transmission using IoT in vehicular ad-hoc networks in smart city congestion. Mobile Netw Appl 24:248–258. https://doi.org/10.1007/s11036-018-1205-x

Deep Learning for Traffic Prediction in VANET

Shishir Singh Chauhan, Juhi Singh, Gauri Shanker Gupta, and Mrinal Kumar Pathak

Abstract Due to advancements in computing systems, intelligent transportation networks, and communication technologies, the development of VANETs is possible. Traffic prediction in high mobility and dynamic network topologies is highly challenging in VANETs. It is very important for high accuracy in traffic prediction as it reduces congestion, provides smooth flow of traffic, and ensures road safety. This chapter is going to discuss how deep learning technologies can be used in solving the problems above. It gives importance to different methodologies, architectures, and practical implementations. Deep learning is one of the subsets of machine learning that has gained much popularity due to its excellent ability to predict complex relationships and patterns in large datasets. It checks applicability to the problem of traffic-patterns forecasting and solving time series problems using CNN, RNN, and LSTM. This chapter refers to strategies on data preparation-data cleaning, normalization, and augmentation. Data preparation with its key demands-normalization, augmentation makes the input data correct for the deep learning model application; thus, correctness also relies on relevance. A list of relevant topics incorporates federated learning and transfer learning. Even though it doesn't solve the problem of sparse data, transfer learning helps with the adaptation and adoption of models developed from big sets of data to new and related settings. In federated learning, instead, diversity within centralised collections of data is increased because these models decentralised work to perform computations but differ by their levels of privacy. This comprehensive review of deep learning algorithms for traffic prediction in VANETs can provide much-needed insight to academics, practitioners, and stakeholders interested in intelligent transportation systems.

S. S. Chauhan (✉) · J. Singh
Manipal University Jaipur, Jaipur, India
e-mail: shishir.chauhan@jaipur.manipal.edu

G. S. Gupta · M. K. Pathak
Birla Institute of Technology Mesra, Mesra, India
e-mail: gaurishankergupta@bitmesra.ac.in

M. K. Pathak
e-mail: mrinalpathak@bitmesra.ac.in

J. Bhatia et al. (eds.), *Deep Learning Based Solutions for Vehicular Adhoc Networks*,
Studies in Computational Intelligence 1207,
https://doi.org/10.1007/978-981-96-5190-0_7

Keywords Deep learning · Traffic prediction · VANET · Automated traffic system

1 Introduction

Urbanization and increasing vehicle ownership are the leading causes of traffic congestion, which worsens transportation problems. According to Dechenaux et al. [1], the usual strategies for increasing the number of lanes may not be enough to reduce traffic congestion. The ITS aims at enhancing the efficiency of traffic and the capacity of roads with the reduction of pollution and accidents through the implementation of 5G communication and on-road sensors [2–8]. Accurate prediction of road traffic, which predicts volumes of traffic using historical data along with geographical relationships, is an integral part of Intelligent Transportation Systems (ITS) [9–12]. Such predictions underlie traffic management techniques, that include vehicular clouds (VC) and resource allocation.

Understanding how different traffic patterns interact over time is essential for predicting traffic flow [9]. Long-term trends show consistent rises in the morning and declines in the evening, which are impacted by things like daily working hours. Unpredictable variations are introduced by short-term trends that are impacted by unforeseen occurrences such as weather changes or accidents. These patterns include seasonal changes (regular cycles), random variations (chaotic changes brought on by outside sources), and trend variations (long-term directed changes) [10]. Because of these nonlinear dynamics, which are impacted by both human and environmental influences, traffic flow prediction is difficult.

In their survey of traffic prediction techniques, Nagy et al. [11] categorized predictive models into three main categories: naïve, parametric, and non-parametric. Analyzing several public datasets, their survey shows how the mobility of sensors improves predictive performance. However, Do et al. [12] specifically focused on NN models designed for traffic forecasting. The study gave a comprehensive analysis of the predictions made by neural networks in traffic forecasting, considering a wide range of neural network architectures, activation functions, and layer configurations over various prediction scenarios. All these studies together provide very important insights into the usability and practical implementation of various methods of traffic forecasting.

Recent works on DL models applied in predicting traffic flow include the examination of the structural composition of DL frameworks in NN architectures as well as the evaluation of various DL models for given applications in traffic prediction. For example, in the work that was cited as [13], the authors have considered various DL models and DL techniques applied in designing effective DL prediction systems for traffic. They carried out thorough analyses on numerous traffic status prediction tasks using a common dataset, and their comparative study on the advantages and disadvantages of each DL prediction method shed light on the effectiveness and applicability of DL models in all scenarios of traffic flow forecast. We require more such comparison analyses to gain a better understanding of practical applications of deep learning with traffic forecasting. It provides sharp prescriptions to researchers and practitioners that will be attempting to improve or implement deep learning-based approaches to achieve higher accuracy and reliability for traffic flow estimates.

The literature clearly indicates that machine learning (ML) methodologies are progressively used in traffic flow prediction jobs. Their ability to adeptly capture complex non-linear connections, handle prediction jobs with flexibility, and use large amounts of traffic pattern data is what attracts them [14]. Collectively, these components provide credibility to their increasing acceptability and efficacy in addressing traffic forecasting challenges.

This study will concentrate on the machine learning models used to traffic prediction challenges. We also consider the relevant cases that the ML model has been used for in the meantime. It is crucial that we evaluate various model types according to their correctness as well as attributes like their efficiency, hardware and data dependence, and capacity to handle certain issues. As is well known, machine learning (ML) is a vast field with a wide variety of categorization techniques for ML models based on various viewpoints. As shown in Fig. 1 in accordance with [15–18], the approaches we studied in this study are grouped by taxonomy of traffic prediction models based on ML and DL algorithms. These are further explained as follows:

Regression model: This model looks at how the dependent and independent variables are related in order to determine how to fit the dataset with a curve or line.

Example-based model: To solve the prediction problem, the example-based model first identified previous data samples that were similar to the input sequence. It then used these samples to determine what its final forecast should be.

Kernel model: The kernel technique employs a kernel function to transform the input data into a high-dimensional linear space, therefore simplifying prediction problems. The primary subjects of the inquiry are the Radial Basis Function and Support Vector Machine models.

NN model: A particular kind of computer model that emulates the information-processing capabilities of biological neurones in the brain is referred to as the neural network (NN) model. The framework has many tiers of interlinked nodes. Each node, or neurone, calculates a weighted sum of the input data before applying an activation function, which then generates an output. The particular neural network model used affects the network's design, including the amount of layers and the quantity of neurones in each layer (e.g., feedforward neural networks, convolutional neural networks, and recurrent neural networks). The input signal received by the neurone is transformed into an output signal via the activation function, which is then used to provide classifications or predictions.

Hybrid model: In forecasting, a hybrid model aggregates forecasts from many separate models to increase precision and consistency. A hybrid model strives to generate more reliable projections than any one model could by combining the advantages of many models, such as statistical, machine learning, or deep learning techniques. This method entails choosing suitable base models, figuring out how to combine their forecasts (for example, by stacking or averaging), and

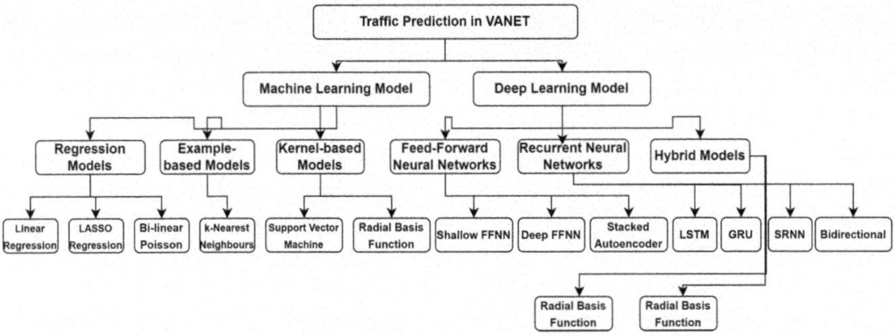

Fig. 1 Taxonomy of traffic prediction models of ML and DL

then fine-tuning the plan in light of performance assessment. Hybrid models, which combine several modelling approaches to improve predictive ability, are frequently employed in many industries where accurate forecasting is essential.

1.1 Motivation

Traffic congestion is a major transportation problem exacerbated by increasing car ownership and urbanization. It is probably that better methods, such as road expansion, are required to mitigate traffic issues. The intelligent transport system attempts to enhance the efficiency of traffic and the capacity of roads while reducing accident and pollution rates by integrating 5G communication with on-road sensors. Intelligent Transportation Systems involve the provision of accurate predictions about road traffic using historical data and geographical correlations in predicting volume hence aiding in resource distribution and management strategies.

It is true that it is only the interaction of multiple traffic patterns over time that actually defines the accurate prediction of a traffic flow. It is therefore true that while long-range patterns have regular fluctuations and are influenced owing to daily working hours, there are short-range trends introducing irregular variability, affected through unexpected events such as changes in the weather or accidents. Prediction of traffic flow is quite a challenge in view of diverse patterns including seasonal fluctuations, irregular changes, and trend reversals. These latter examples of course clearly indicate the inherently nonlinear character of traffic dynamics, induced by both human-induced and environment-induced factors.

Recent works by research studies have kept up its attention on neural network models which have been explicitly engineered with detailed evaluation of architectures, activation functions, and settings of layers. Indeed deep learning has emerged as an interdisciplinary region of machine learning for being a subarea because the subarea has made out successful data management ability in learning subtle non-linear inter-relations. Comparisons of deep models have proven that these models are robust, and helpful in predicting traffic flows.

The following chapter discusses several machine learning methods used for traffic prediction and evaluates a variety of models based on their accuracy, dependence on data and hardware, performance metrics, and ability to deal with certain challenges. It presents a complete analysis of theories behind several machine learning algorithms along with their practical implementation, giving valuable insights into how these can be applied to traffic flow forecasting within intelligent transportation systems.

1.2 Research Contribution

- This chapter considers how deep learning techniques can be exploited in traffic forecasting with detailed information about complexities introduced by the dynamic network topology and high mobility of vehicular objects.
- The chapter performs an extensive analysis of various deep learning frameworks, such as Convolutional Neural Networks (CNNs), Recurrent Neural Networks (RNNs), and Long Short-Term Memory (LSTM) networks, to prove the significance of their application in traffic prediction within VANET.
- The chapter discusses the techniques of data preparation that are important for ensuring the accuracy and reliability of input data in deep learning models, such as data cleaning, normalisation, and augmentation.
- This would bring into the use sophisticated approaches, such as federated learning, transfer learning, into the realm of VANET in demonstrating how these could offer some mitigation to the existing data privacy issues in particular areas with limited resources.
- It comprises the applications of deep learning on VANETs along with demonstrating how the use of deep learning models might help in fighting traffic congestion and optimizing the traffic flow on roads.
- It focuses on hybrid models that incorporate several deep techniques of learning to enhance both spatial and temporal traffic predictions.
- The future investigations, in conjunction with deep learning models in combination with emerging technologies on the edge computing and autonomous vehicles, will lead the way toward a more scalable, real-time traffic forecasting.

2 Fundamentals of VANETs

ITS require the establishment of VANETs. By establishing communication between vehicles, they improve driving conditions as well as traffic flow besides improving road safety. Some aspects of VANET discussed include communication protocols, architectures, and components of these systems, their characteristics besides their challenges in deployment.

2.1 Architecture and Components of VANETs

Vehicular Ad-hoc Networks are a part of an Intelligent Transportation System as this technology will provide communications between vehicles and roadside infrastructure. Structural framework: improve road safety and optimize traffic efficiency, thus improving the driving experience.

2.1.1 Architecture of VANETs

There are three key components of VANETs: vehicles, roadside units (RSUs), and the backend network or central infrastructure. To better describe that, the very basic architecture of VANETs has been represented diagrammatically in Fig. 2.

1. **Vehicles**: These are equipped with On-Board Units (OBUs) that allow them to communicate with other vehicles (V2V) and with RSUs (V2I). Modern cars also come equipped with sensors, GPS units, and other data-gathering and data-transmitting technologies.
2. **Roadside Units (RSUs)**: These units are stationary and sit beside the infrastructure of the roadway. They allow information about traffic signals, roadway conditions, and other data to be communicated using vehicles. In addition, roadside units have connectivity to the basic infrastructure.
3. **Central Infrastructure/Backend Network**: Data centres, traffic management centres, and other cloud services that gather, handle, and evaluate data from cars and RSUs fall under this category. The backend architecture facilitates sophisticated calculations, long-term data storage, and the distribution of more comprehensive traffic statistics.

2.1.2 Components of VANETs

1. **On-Board Units (OBUs)**: devices that are mounted in cars to facilitate communication. They consist of several sensors, GPS, and communication units.
2. **Roadside Units (RSUs)**: Infrastructure components that communicate with OBUs by offering roadside information services and data relay.
3. **Sensors**: gadgets that gather information on the position, direction, speed, and state of the environment of a vehicle.
4. **Communication Interfaces**: Through the use of wireless communication technologies, they enable data interchange between cars, RSUs, and the central infrastructure.

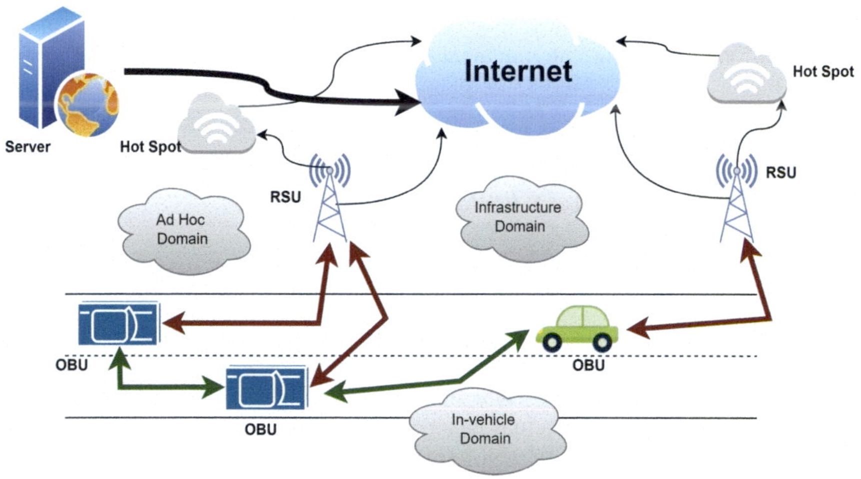

Fig. 2 Basic architecture of VANETs

2.2 Communication Protocols in VANETs

Several protocols are used in VANET communication to provide dependable and effective data sharing. These protocols are intended to meet the particular needs of communication between vehicles.

2.2.1 Types of Communication Protocols

1. **Medium Access Control (MAC) Protocols**: By controlling data transmission across wireless media, these protocols provide unhindered communication between several devices. IEEE 802.11p, which was created especially for VANETs, and DSRC (Dedicated Short-Range Communications) are two examples.
2. **Routing Protocols**: Certain routing protocols, such as AODV, DSR, and GPSR, are essential in vehicular ad hoc networks (VANETs), because cars move quickly and network circumstances change often. By starting route discovery only when required, AODV minimises overhead by creating routes on demand. DSR does not need to keep a global route database; instead, it caches routes and swiftly adjusts to dynamic network changes. GPSR forwards packets to the closest neighbour en route to the destination based on the user's geographic location. These protocols maximise the effectiveness of data transfer, which is essential for secure applications on VANETs and dependable communication.
3. **Transport Protocols**: Data transfer is dependable because to these protocols. With modifications for the high mobility and variable network circumstances of VANETs, TCP and UDP are widely utilised.

4. **Application Protocols**: These specify the precise communication needs for many applications, including entertainment services, traffic management, and safety warnings.

2.3 Characteristics and Challenges of VANETs

2.3.1 Characteristics

1. **High Mobility**: High-speed vehicle movement causes abrupt changes in network architecture.
2. **Dynamic Topology**: Robust and adaptable communication mechanisms are necessary since vehicle locations change often.
3. **Large-Scale Networks**: Vast regions may be covered with VANETs, particularly in metropolitan settings.
4. **Heterogeneous Communication**: Infrastructure components and vehicles may require and have varying communication capabilities.
5. **Real-Time Communication**: Low-latency communication is necessary for many VANET applications, including collision avoidance.

2.3.2 Challenges

1. **Scalability**: ensuring that there is no appreciable performance reduction when a high number of cars are handled by the network.
2. **Security and Privacy**: preventing unwanted access to data and guaranteeing user privacy.
3. **Interoperability**: ensuring efficient communication between various systems and devices.
4. **Reliability**: sustaining constant communication in the face of extreme environmental change and great mobility.
5. **Latency**: reducing data transmission latency to enable real-time applications.
6. **Spectrum Management**: effectively using the little amount of spectrum that is available to prevent interference and congestion.

3 Traffic Prediction Model Based on ML

This section will analyse the various models employed in traffic prediction, specifically focusing on example-based, kernel-based, and regression approaches. This presentation will encompass various forms of machine learning models, emphasising those that have emerged over the past decade. To enhance the performance of these models in practical applications, we will implement them in prediction scenarios and examine various organisational structures.

3.1 Regression Model

Traffic prediction makes use of regression models, which are frequently connected to parametric approaches. If a technique is parametric, it means that it makes assumptions about a certain distribution of traffic patterns. Because they are easy to develop and operate well for simple traffic network prediction tasks, regression models are used.

According to reference [19], the parametric method entails specifying input and output parameters inside of a mathematical model in which each parameter is directly connected to incoming data. Undefined parameters are found by analysing training data, highlighting how important factor selection and mathematical model selection are to obtaining precise predictions. Research has demonstrated that because parametric approaches have strong theoretical and mathematical underpinnings, they produce projections that are relatively accurate. When it comes to traffic forecasting, linear regression is the most used regression model. To provide accurate prediction results, researchers carefully choose weight parameters and pertinent traffic characteristics. The aforementioned predictions are obtained by linear combinations of obtainable traffic data, so highlighting the effectiveness and resilience of linear regression models in producing accurate traffic forecasts grounded in empirical data.

Rice et al. demonstrated in their work [20, 21] that a linear regression model can be effective even with a small dataset. The method proposed does take into account the fact of temporal variations in the model through the use of time-varying coefficients in adjusting parameters. Results were found to be plausible compared to the k-nearest-neighbors (k-NN) method, which involves two closest neighbors over 20 min. The study discussed that the both predictiveness and usability need to be improved rather than just focusing on the optimization of accuracy measures despite no particular accuracy statistics. This means that linear regression models work well within traffic forecasting contexts, especially within fluid environments that provide adaptations for time-related fluctuations.

Researchers conducted their study by utilizing linear regression analysis [22] to identify spatio-temporal relations that existed in a road network while predicting traffic flow patterns. They wanted to examine whether upstream traffic conditions did impact the flow of downstream traffic on a given link. The results clearly presented that upstream traffic greatly influences downstream traffic flow. The authors integrated the historical averages, upstream traffic data, and real-time traffic conditions using pre-defined weighting factors into three coherent predictive models that they presented as a solution to the problem in question. The authors also designed a method for dynamically adjusting the weight parameters based on the real-time traffic data. Validations with peak-hour traffic data proved the models sound and effective in capturing the complexities of traffic dynamics.

The model beat the historical average (HA) and basic judgement guidelines within the 15-min prediction window, according to the authors' studies. However, the simple linear regression model had a few issues that needed to be fixed. First, over intervals longer than 30 min, prediction accuracy fell short of the historical mean. Second, because these characteristics were limited to certain road segments, it was expensive to determine a set of relevant parameters that would sufficiently describe the link between traffic patterns. Finally, because the model is time-consuming when dealing with intricate road layouts and increasing traffic loads, processing time grew tremendously.

In [23], Kwon et al. coupled linear regression with a tree-based approach and a stepwise variable-selection strategy to speed up the laborious process of choosing appropriate parameters. They produced forecasts using information from occupancy, traffic flow, and historical journey times on the I-880 motorway. The model was evaluated by the authors using cross-validation (CV) over a range of prediction horizons, from 0 to 60 min, in order to solve intersection difficulties. They discovered that while historical data was still helpful for longer-term estimations, current traffic information greatly improved short-term estimates. Changing the depth of the tree significantly improved the efficiency of parameter discovery in the tree-based method. Nonetheless, the model's efficacy was significantly reliant on the initial search range, highlighting the need of a thorough comprehension of historical data unique to the road segment.

The analysed studies highlight the limitations of fundamental linear regression models in accurately capturing the nonlinear variations inherent in traffic flow dynamics. To enhance the accuracy of predicting stochastic fluctuations in traffic patterns, it is common for researchers to integrate regression with alternative methodologies. This approach acknowledges that while linear regression effectively models linear relationships, it may fall short in capturing the complex, nonlinear dynamics inherent in traffic flows. By integrating complementary methodologies, researchers aim to enhance the predictive capabilities of their models, thereby more effectively addressing the intricate and dynamic nature of traffic conditions.

The amalgamation of Granger causality theory along with the dual Lasso regression model address as a nonlinear problems in traffic data by the investigator cited as [24]. Using historical averages, they preprocessed the data to find lingering patterns and used the first Lasso regression model to remove irrelevant data. Through efficient management of both typical and unusual traffic situations, this strategy sought to enhance traffic flow forecast.

In [25], the authors developed a latent component model specifically for short-term traffic flow prediction using bi-linear Poisson regression. The accuracy and robustness of the model were enhanced by including contemporary historical data from critical route portions. In the bi-linear Poisson regression framework, a convolutive mixture was used to address temporal interdependence among different route segments. Furthermore, the authors included a stochastic variational Bayes model, enhancing the system's efficacy for real-time updating and predictive tasks, hence augmenting its applicability in dynamic traffic scenarios.

Table 1 enumerates the regression models used to predict traffic during the last

Table 1 Regression models applied in the last 10 years to forecast traffic

Papers	Work	Regression model	Area	I/O	Comment (if any)
[25]	Flow	Bi-linear	Urban	Multi-multi	Traffic prediction; Bayes method optimisation using stochastic model variations
[24]	Flow	LASSO	Free-way	One-one	Granger causality theory

decade. The table indicates that, in recent years, there have been few models for traffic forecasting using regression analysis. The fundamental nature of the regression model complicates the precise characterisation of the temporal dimensions of continuous traffic data, albeit its straightforward implementation. Regression modelling remains a viable alternative for less computationally intensive traffic prediction tasks in small, simply organised traffic networks.

Nonetheless, it is evident that the regression model has an intrinsic flaw. Regression modelling may be used to create traffic prediction models with a limited amount of historical data. It is suitable for applications requiring fast execution and those with little historical data. Conversely, it reduces the model's accuracy potential due to its vulnerability to overfitting and its inadequacy in delivering a comprehensive analysis of the dataset, including seasonal variations.

3.2 *KNN Model*

As previously outlined in Sect. 1, our methodology for traffic prediction emphasises example-based techniques, namely the KNN model. This model is preferred for its capacity to delineate spatial linkages among road segments in traffic networks. The k-nearest-neighbor (KNN) model is used for prediction and data filtration, guaranteeing dependable forecasting [26]. KNN functions as a non-parametric regression technique based on data insights, in contrast to formal prediction models. It forecasts future values via the K closest neighbours with analogous variable values, highlighting its data-driven methodology for prediction.

This approach collects the data required for forecasting by using historical data. The historical database encompasses a wide range of trends in addition to normal traffic circumstances, representing all possible patterns of traffic history. The quality of this historical information is a major determinant of prediction accuracy, especially in terms of how comprehensively it captures all potential future traffic situations. The programme depends on the relationships between every traffic system component found in the historical data to guarantee accurate projections.

The conventional KNN model is mostly used for classification jobs; however, modifications may be applied to improve its effectiveness in regression scenarios, as outlined in [27]. The integration of non-parametric regression with K-means clustering forms a predictive approach. K-means clustering organises data points into groups based on closeness, highlighting common attributes within each cluster using distance metrics to evaluate similarity. The first part of the K-means clustering technique involves determining the number of clusters (K). Thereafter, sample sites are randomly chosen to determine the first cluster centres. Subsequent data points are assigned to the next cluster centre according to proximity. Cluster centres are iteratively refined by calculating the mean of the data points allocated to each cluster until convergence is attained. Data points in proximity are chosen for the prediction phase according to their proximity to the current state vector. The neighbouring sites within each cluster are used to predict the subsequent time step. This method utilises interrelated data points inside each cluster for prediction, using clustered data structures to improve forecasting precision.

The authors anticipate traffic flow at 15, 30, 45, and 60-min intervals using KNN [28]. They specifically used previous data collected on Fridays and Saturdays for a particular segment of the route to construct their dataset. The research meticulously investigated the impact of dataset length and neighbour count on prediction accuracy, using various weekdays and prediction intervals.

To improve the process of picking K candidates in the KNN model, the authors of [29, 30] proposed the correlation coefficient distance as a measure for identifying neighbours who are more closely related. The goal of this strategy was to highlight how crucial temporal correlation data is to the prediction model. In addition to the data recordings themselves, the time interval from the present time-which was constrained to a predefined threshold, usually not exceeding one day-also had an impact on the distance between neighbours. The authors substituted a local minimum method for the traditional selection process when choosing prediction candidates in order to address the problem of dimensional reduction of K as a result of overlapping occurrences. The last forecast was refined using a Linearly Sewing Principal Component approach (LSPC), which basically turned it into a minimization issue. Their model fared better than other current models in comparison experiments conducted over different route segments.

The temporal and geographical correlations datasets are important while dealing with traffic flow predictions. For this, the MapReduce-based KNN model was developed by [31]. Unlike all previous KNN techniques, this framework introduces geographical variables. Both the upstream and downstream components constitute the base for distance metric computation. In addition, the authors introduced the Distance Weighted Voting, and this resulted in a decrease in the value of **K** in the KNN framework. This method assigns relevance to the contributions of each neighboring entity according to their proximity toward the target location, taking into consideration the various levels within the predictive framework.

For using the spatial-temporal features of their data in traffic forecasting, the authors in [32] adopted a unique approach. In contrast, in the paper in [31], the considered similar lengths were influenced by the grid distance between selected

Table 2 Progress in KNN applications for traffic forecasting over a decade

Papers	Work	State vector	Distance metric	Method of prediction
[28]	Flow	Historical records target detectors	Euclidean distance	Weighted average
[33]	Flow	Historical records target detectors	Euclidean distance	Weighted average and non weighted average
[29]	Flow	Historical records target detectors	Correlation coefficient distance	LSPC
[31]	Flow	Upstream and downstream historical record and target road	Spatial temporal traffic	Weighted average including trend adjustment
[32]	Speed	The data on grid distance Historical speed records	Gaussian weighted euclidean distance	Gaussian weighted average

road segment and others but the interest in [32] was primarily on segments that lie upstream and downstream. A weighted vector was used to assess the influence of each segment on the concerned road stretch, which enabled a significant reduction in the number of possible predictive options. [32] used a comparison of spatial correlation ratios along with a set threshold to filter out road segments that lacked sufficient spatial connectivity, thereby improving the quality of their selection process. Those candidates for the final forecast were identified by subsequently computing Euclidean distances using a Gaussian methodology. Ultimately, employed a Gaussian weighted average approach to make predictions based on the obtained distances. This approach affords traffic predictions a holistic perspective by gathering critical data across the whole road network, rather than focusing primarily on neighboring road segments.

Table 2 lists the several KNN models that have been applied to traffic condition estimation in the last ten years. These models use traffic data to extract relationships between road segments using the KNN approach. Distances in continuous traffic records or geographic distances are used to characterize interactions, depending on the sort of state vector that is being employed. Adding more historical data to the state vector may result in higher maintenance and computing expenses. Researchers usually use the Euclidean distance metric for state vectors, including single data kinds, such as historical traffic flow records [28, 33]. The weighted distance function would be one of the most effective ways to balance the effects of different kinds of data. It has been observed that the calculation time of KNN models increases with dataset size. While other studies reduce the number of possibilities in the search space by choosing them based on spatial distance, computing demands are still substantial [32]. Although spatial-temporal estimation has advanced, most current forecasts concentrate on individual road segments, which present resource issues when applied to intricate road networks.

3.3 Kernel-Based Model

Vapnik [34] developed Support Vector Machines (SVM), a fundamental component of kernel-based machine learning methodologies. Initially conceived as a versatile statistical learning framework, Support Vector Machines (SVMs) may be used for both regression and classification tasks. Support Vector Machines prioritise structural risk minimisation and are based on statistical theory, distinguishing them from several other nonlinear models and reducing the probability of encountering local optima. Support Vector Regression (SVR) denotes the use of Support Vector Machines (SVM) in regression contexts. Support vector machines (SVR) aim to get maximum generalisation performance by constructing a decision surface, or hyperplane, that maximises the margin between the hyperplane and the nearest training samples while simultaneously minimising the total deviation of training samples from the hyperplane.

Regression problems in traffic prediction frequently call for using the traditional SVM model, which is a non-linear regression technique. In order to map data into a high-dimensional space where a linear hyperplane is constructed, SVM uses kernel functions. This transformation offers a strong foundation to effectively represent complicated interactions in traffic data, enabling researchers to approach traffic prediction challenges similarly to high-dimensional linear regression assignments.

Support Vector Regression (SVR) was initially used in [35] to forecast traffic flow at a particular intersection. With an average inaccuracy rate of 6.03%, predictions were generated for the next hour using data from the five time steps prior. The test data was gathered between 6 and 10 p.m., while the training data was taken between 8 a.m. and 4 p.m. In a different research, cited as [36], SVR was used to estimate aggregate traffic speed during brief periods of time in comparison to a Multi-Layer Feedforward Neural Network (MLFNN) with a Radial Basis Function (RBF) kernel. The past 10 min worth of speed readings were used as input data to forecast speeds for the next 2, 4, and 6 min. When trained on data from the previous two days, SVR performed better than MLFNN, even though MLFNN had greater total accuracy. the gradual increment in the size of datasets signifies the worst SVR's performance and increase in the Mean Absolute Percentage Error however tested in the smaller datasets. A novel parameter, v or v-SVR, was developed by researchers in [37] to improve SVR optimisation for motorway traffic volume prediction. In comparison to MLFNN, our updated SVR model performed better in terms of MAPE (5.5% vs. 8.8%) and Root Mean Square Error (RMSE) (44.7% vs. 64.5%). These studies show how SVR and its variations have evolved and how useful they are for traffic prediction tasks.

A kernel-based model known as Support Vector Machine (SVM) is significantly influenced by the chosen kernel function. The Radial Basis Function (RBF) kernel has been extensively used in research [38, 39]. Research has examined the enhancement of SVM models by including seasonal variations into the kernel function. To include seasonal data from traffic records into SVM predictions, the authors in [40] used seasonal RBF and seasonal linear kernel functions. These algorithms implemented seasonal interval modifications to temporal data intervals. Comparative trials demonstrated that the seasonal SVM model regularly outperformed non-seasonal kernel SVM models, as well as other models like Artificial Neural Networks (ANN) and ARIMA with Kalman filter. The seasonal SVM model exhibited competitive performance across several time intervals, except the morning peak. It surpassed SARIMA when combined with a Kalman filter. The authors' comparison research demonstrated the seasonal SVM model's competitiveness in both training and prediction phases, despite SARIMA sometimes exhibiting marginally superior accuracy in certain situations and requiring double the training duration.

After choosing a suitable kernel function, parameter optimisation is an important step in SVM models. Low computation time is desired, although convergence to less-than-ideal local minima is to be avoided. The authors of [41] utilised Continuous Ant Colony Optimisation (CACO) as a means of effectively optimising SVM parameters. CACO is more successful than standard Ant Colony Optimisation (ACO) [42] because it works in a continuous search space. The study used SVM with a Gaussian Radial Basis Function (RBF) kernel to anticipate traffic flow during morning (6 am to 10 am) and evening (4 pm to 8 pm) peak hours using data from Taiwanese cities. Based on comparative studies, CACO regularly outperformed SARIMA in traffic flow prediction tasks by facilitating effective parameter adjustment over a wide range of input data features.

Unlike conventional SVM techniques, the model developed by researchers in [43] was designed to handle multi-step prediction problems. By including historical records and upstream data into the input structure, their method improves forecast accuracy. When compared to conventional SVM approaches, the model's overall prediction performance is improved due to this integration, which enables the model to more accurately capture dependencies and trends over numerous time steps. The experiment showed that prediction accuracy was greatly increased by combining upstream data with history records, particularly when prediction intervals were small, like 1. This improvement is ascribed to the significant influence that upstream traffic flow had on the particular road stretch in a little amount of time. All models became less accurate as the forecast interval increased from 1 to 3. The availability of additional historical data was beneficial for the prediction job as the prediction interval grew. The authors also pointed out that, in terms of computing efficiency, using input data that was limited to previous and historical knowledge on the particular road stretch that was being considered was the best course of action. This approach maximised prediction performance while minimising computational resources.

4 Neural Network Models

The distinguishing feature of non-parametric models, particularly those based on neural networks, lies in their ability to learn parameters directly from historical data rather than relying on pre-established values.cite41. Non-parametric models differ from traditional models with fixed parameters, offering significant advantages in today's data-rich environment due to their ability to adjust to the patterns present within the data. By utilising prior data to enhance prediction accuracy, these models establish a direct connection between predictive objectives and historical data.

Previous research indicates that non-parametric techniques, such as Support Vector Machines (SVM), significantly enhance algorithmic accuracy compared to traditional linear regression models by effectively capturing the non-linear characteristics of traffic patterns. Nonetheless, non-parametric methods are associated with certain disadvantages [44]. A substantial amount of dependable historical data is required, which can prove challenging to locate and maintain current. Moreover, in comparison to alternative methodologies, the training of non-parametric models may incur higher costs.

The 1940s saw the introduction of the neural network (NN) model by Warren S. McCulloch and Walter Pitts [45, 46], which is highly effective at handling complicated linear systems. As a black-box learning model, NN learns and adapts the input-output mapping on its own from large datasets, in contrast to standard empirical formulae. For traffic prediction jobs, its durability, associative memory, and continuous real-time parameter updates make it the perfect choice. To provide precise forecasts, the NN model incorporates important information from the study road segment and historical records with factors like weather, road state, accidents, and the effect of development on the traffic network.

A large quantity of data is needed to train a neural network, and not enough data might result in predictions that are off. The neural network is less adaptive to shifting traffic and road conditions since its applicability is usually restricted to the particular road section it was trained on. Furthermore, choosing the right amount of neurons for each buried layer is difficult yet essential. While too few neurons may reduce prediction accuracy, too many might lead to a huge network topology and longer calculation times. These elements add to the neural network's possible drawbacks in real-world traffic prediction situations.

4.1 FFNN

Neural networks are fundamentally composed of artificial neurons, each of which has a characteristic structure seen in Fig. 2. The input vector to a neuron may be described in the Eq. 1 as $\mathbf{z} = [z_1, z_2, \ldots z_d]$ if we suppose that the neuron receives d inputs $z_1, z_2 \ldots, z_d$. The neuron's output will be:

$$a = f\left(\mathbf{k}^T \mathbf{z} + b\right) \tag{1}$$

where, $\mathbf{k} = [k_1, k_2, \ldots, k_d]$ is a weight vector of dimension d. b is the bias variable and $f()$ is denoted as the activation function. The most commonly used activation functions are described in Table 3.

Table 3 Activation function

Name	Activation function
Logistic	$\sigma(X) = \frac{1}{1+\exp(-X)}$
tanh	$\tanh(X) = \frac{\exp(X)-\exp(-X)}{\exp(X)+\exp(-X)}$
ReLu	$\text{ReLU}(X) = \max(0, X)$

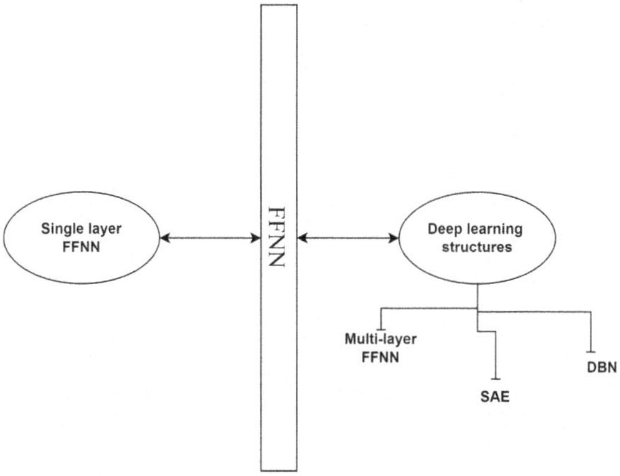

Fig. 3 Advancements in feed-forward neural networks (FFNN): state of the art

Analogous to the brain's layered information processing, the FFNN arranges neurones into consecutive layers for information processing. Each layer utilises the output of the previous layer as its input and then transmits its output to the subsequent layer. Feedforward neural networks (FFNNs) operate with unidirectional information flow, but recurrent neural networks (RNNs) have feedback loops that provide reverse information circulation. Comprehensive calculations demonstrating the internal mechanisms of FFNNs are provided in [16].

Several feed-forward neural network (FFNN) based traffic prediction methods are shown in this section. Additionally, we will look into a number of deep learning frameworks that can greatly raise FFNN's accuracy. A thorough taxonomy of the evaluated state-of-the-art FFNN literature is shown in Fig. 3.

A straightforward and obvious topology for a feed-forward neural network (FFNN) consists of a single hidden layer. In a study analogous to [47], researchers used a feedforward neural network (FFNN) to predict traffic flow with a single hidden layer architecture. The primary objective was to forecast traffic flow both at the present moment and within a specified time range that begins and concludes at a future time T. The MTL FFNN sought to enhance the accuracy of the primary prediction at time T by including supplementary tasks associated with adjacent time periods, leveraging correlations across various time intervals. This enhancement illustrates how the appropriate integration of many interconnected systems may enhance the precision of traffic flow predictions.

In the paper [48], a study has been conducted on short- and long-term traffic forecasting ability of a FFNN. A FFNN has been trained to predict the traffic flow with the help of 20% of the dataset using the input from weekday, month, and previous time step traffic volume data. It is seen that FFNNs are intuitive and fail to predict traffic flow during severe congestion. The failure to identify complex nonlinear patterns may explain this. The need for increasing the model complexity and addressing heterogeneity in the dataset arise in using FFNNs to solve real-world problems such as traffic prediction issues.

In [49], shallow FFNN was used in predicting the traffic flow on a route accommodating different kinds of vehicles. The inputs the FFNN used were numerical data on several vehicle categories, temporal information on the day and time, and speed metrics for each vehicle. Notable conclusions were that sigmoid activation is better than the tanh function, and Levenberg optimizes better than momentum optimization in training the network. Furthermore, the predictive precision enhanced by the number of hidden neurons inside the feedforward neural network, which in turn indicates that the architecture of a network plays an important role in modeling the complex interrelations of the data properly. The findings established that the addition of selected data, being the records of vehicle speed improved overall performance of the model and indicated that full contextual information was vital for efficient modeling of traffic. Findings show how diverse sources of data need to be combined, and parts of the model fine-tuned to facilitate enhancement of FFNN-based systems for traffic in better management and planning of traffic.

4.1.1 FFNN: Deep-Learning Structure

In order to forecast traffic velocity based on observations from spatially coupled sensors, researchers utilized a multilayer Feed-Forward Neural Network (FFNN) as described in [50]. The initial trend of the input was preconditioned using an l_1 trend filter in an effort to reduce noise effects, then the LASSO model was applied for identifying road segments with geometric correlations. The FFNN showed reliable performance in predicting traffic speed in both interrupted and normal conditions due to the efficient application of spatial-temporal filters.

Nevertheless, the model's efficacy decreased as the network's depth increased. In order to get around this, researchers in [51] trained a multi-layer FFNN using an unsupervised training strategy using Wasserstein Generative Adversarial Network (WGAN) techniques [52]. Following the detrending of data using historical averages, the model was created to forecast in both common and uncommon circumstances. The model does not outperform LSTM in common scenarios or SVR and ARMA in exceptional ones, even though it functions satisfactorily in normal conditions. Nonetheless, the writers point out that one of its noteworthy strengths is its adaptability to different settings without depending on a single approach. Despite its complexity, the model's training time is also noticeably less than that of other deep learning techniques.

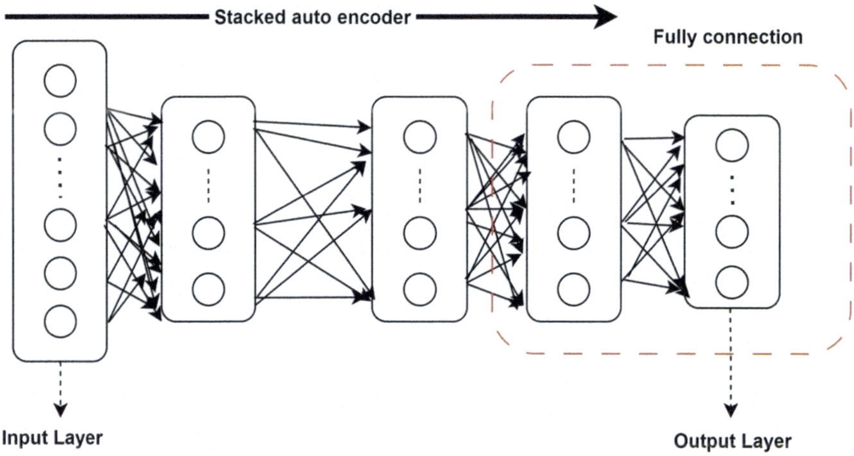

Fig. 4 Structure of SAE

One popular deep learning architecture for traffic prediction problems is the Stacked Autoencoder (SAE). As seen in Fig. 4, it comprises of several layers of autoencoders (AE). In order to extract higher-level characteristics that improve prediction accuracy, each AE layer gradually decreases the dimensionality of the input data. SAE is well-known for its capacity to minimise computing costs and effectively extract significant features from training data [53].

According to [53], autoencoders have three layers: input, hidden, and output. The buried layer receives data like an encoding technique that discards input vector attributes. The autoencoder then copies the input vector from the hidden layer to the output layer to decode these properties. The goal of unsupervised autoencoder training is to minimise the difference between the input and output vectors. This variant shows how AE extracts functional features. Let's analyse the hidden layer's output, Y, and the input vector, $Z = [z_1, z_2, \ldots, z_N]$. Equations 2 and 3 of the Autoencoder (AE) are relevant:

$$M = f\left(G_e Z + b_e\right) \tag{2}$$

$$L = f\left(G_d D + b_d\right), \tag{3}$$

Afterwards, by minimising the cost function, which is defined as follows, we may optimise the AE in Eq. 4 as

$$P(Z, L) = \frac{1}{2} \sum_{i=1}^{N} \|x_i - z_i\|^2 \tag{4}$$

The initial stage in employing the Stacked Autoencoder (SAE) model involves pre-training from the bottom up. Upon receiving the input vector, each layer of the autoencoder (AE) systematically learns to reconstruct it. The subsequent AE layer receives its input from the output generated by the hidden layer of each AE. The weights and biases of the fully linked prediction layer are initialised subsequent to the training of each autoencoder layer. Finally, a top-down backpropagation (BP) method is employed to optimise the entire network.

In [53], SVM, FFNN, and RBFNN were studied to assess the Stacked Autoencoder (SAE) model's ability to estimate traffic flow at 15, 30, 45, and 60 min intervals. The SAE model outperformed other prediction models with a mean relative error of 6.50%, according to the research. The SAE model's accuracy remains mostly intact when the forecast interval is extended. DBN is a deep-learning architecture like SAE for FFNN. The system consists on restricted Boltzmann machines (RBM), which are energy-based undirected graphical models [54, 55]. DBN, unlike SAE, includes non-interconnected units at every tier, providing the image of a completely connected network. DBN and SAE are comparable because RBM layer hidden units are inputs to the next layer. The SAE and DBN training methods are similar.

Huang et al. presented Deep Belief Networks (DBN) in [56], outperforming FFNN, SVR, and ARIMA models with over 87% accuracy over a range of prediction intervals. By fine-tuning DBN's parameters using the FireFly Algorithm (FFA), a different research [57] showed increased prediction accuracy and lower processing costs when compared to previous approaches.

4.2 *Recurrent Neural Network (RNN)*

Neurons in the same layer of a feed-forward neural network (FFNN) solely forward information from input to output; they do not exchange messages. Because of their construction, FFNNs can only produce output by using the current as their input. Nonetheless, outcomes in traffic prediction tasks frequently rely on both current and previous conditions. This restriction was addressed with the development of Recurrent Neural Networks (RNNs). Sequential relationships in traffic data are well captured by RNNs, enabling more precise modelling of how previous states affect future forecasts [58].

according to [59–61]. Because of their "short-term memory," RNNs are unique in that their neurons can process inputs from both the present layer and layers that came before it. Compared to Feed-Forward Neural Networks (FFNNs), RNNs are superior at detecting temporal correlations in traffic data because of this capacity. A thorough taxonomy of the literature on RNNs is probably depicted in Fig. 5, which also highlights the many designs, uses, and developments in the subject.

The authors of [62] conducted a thorough analysis of LSTM's effectiveness in short-term traffic flow prediction. To optimise hyperparameters such the number of hidden units in the LSTM layer and the duration of input traffic flow records, they used a single hidden layer LSTM NN model and grid search. For each of the thirty

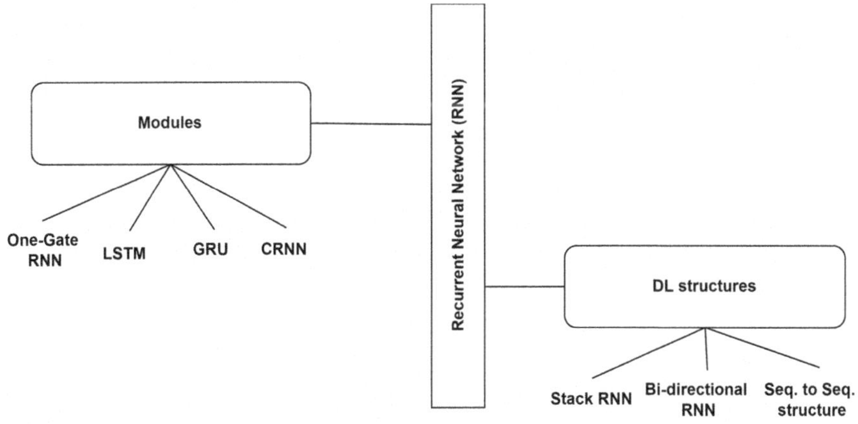

Fig. 5 RNN: state of the art

detectors, they used characteristics like MAPE and RMSE to compare LSTM against models like as Random Walk (RW), SVM, FFNN, and SAE. According to hyper-parameter analysis, the model's accuracy peaked at about 20 hidden units, and as it increased, it became less accurate. Interestingly, the LSTM NN model significantly outperformed earlier models in that it kept consistent accuracy throughout a range of input data lengths. The dataset's seasonal bias, which affected experiment stability because of noticeable seasonal fluctuations in traffic circumstances, was a weakness of the study. Test data concentrated on winter circumstances, whereas training data spanned three different seasons in 2014.

Hochreiter et al. [63] introduced LSTM in 1997 to solve long-term dependence problems in RNNs. In contrast to conventional RNNs, LSTM has input and output gates. Later, Gers et al. [64] observed that the original two-gate LSTM was occasionally ineffective with longer input time series because it oversimplified non-linear characteristics. To overcome this, Gers et al. added a forget gate to the two-gate LSTM design, leading in the creation of a three-gate LSTM module, as shown in Fig. 6.

Cho et al. (2014) created the Gated Recurrent Unit (GRU) [65], which is frequently utilised in traffic prediction because it can manage long-term dependencies with less complexity than LSTM. GRU was used in [66] to anticipate traffic flow at short intervals in both rural and urban locations. The performance of GRU was assessed using MAE and MSE metrics in comparison to RBF after the data were normalised using Min-Max normalisation. The findings demonstrated that, in both urban and rural datasets, GRU consistently outperformed RBF in terms of MSE and MAE values. With a little drop in accuracy seen exclusively in metropolitan areas, the model outperformed RBF and normal iteration approaches when employing the auto-iteration method for multi-step predictions.

In [67], traffic volume information from 30 sensors in the PeMS dataset was used to compare GRU and LSTM. Based on data from each sensor collected over the previous 30 min, both models were trained to forecast traffic flow for the next 5

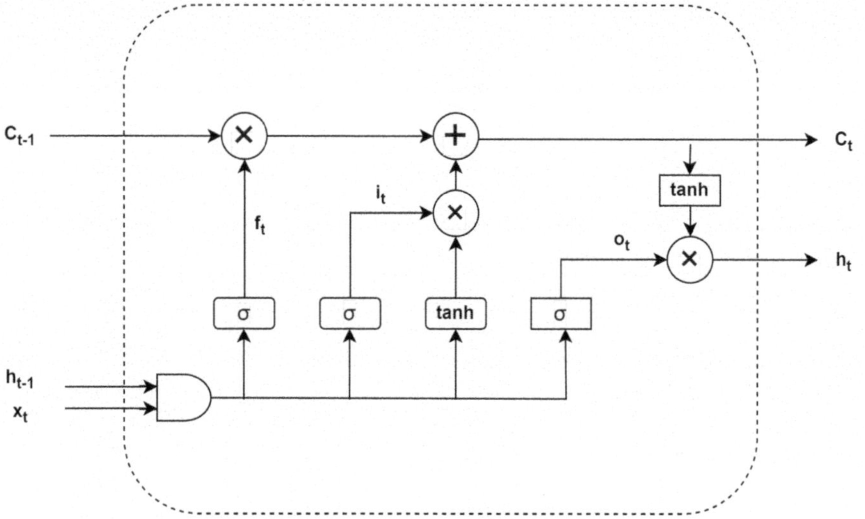

Fig. 6 Repeating module in LSTM

min. The study discovered that for predicting traffic flow, GRU and LSTM performed better than ARIMA. Although there were not many variations in overall performance between the two RNN designs, GRU demonstrated more stability in error rates across sensors when compared to LSTM.

In single-layer designs, a number of well-known RNN modules have undergone substantial testing for traffic prediction. RNNs do, however, also have a deep learning basis. In the next parts, the course will discuss many popular deep-learning RNN topologies.

4.2.1 Deep Learning Architectures for RNNs

RNNs were incorporated into deep-learning frameworks by Pascanu et al. [68] and Hihi et al. [69] in order to fully use their potential. As noted in [58], deepening the RNN structure makes it possible to extract more abstract characteristics and patterns from data. This improvement dramatically improves prediction model accuracy and performance on challenging prediction tasks.

The Stacking RNN (SRNN) is a deep-learning recurrent neural network architecture, as seen in Fig. 7. Information is sent uniformly across each of the several stacked RNN layers of an SRNN. Figure 7 illustrates that N represents the total number of layers in an SRNN, whereas L_i^j denotes the j-th RNN module at timestamp i. Assuming that h_i^j represents the output of the layer at time i and index j, we may calculate h_i^j using Eq. 5 as follows:

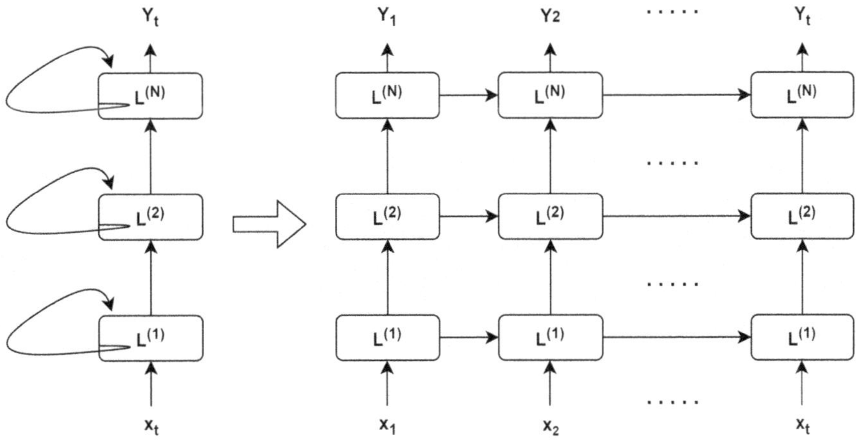

Fig. 7 SRNN structure

$$h_i^{(j)} = f\left(B^{(j)} h_{(i-1)}^{j} + G^{(j)} h_i^{(j-1)} + b^{(j)}\right) \tag{5}$$

where, B^j and G^j are weight metrics and b^j is the bias vector.

A three-layer Long Short-Term Memory Recursive Neural Network (SRNN) for forecasting Beijing traffic flow was assessed by researchers in [70]. With an emphasis on one-hour forecasting intervals, they utilised geographic data to analyse and anticipate traffic flow at each grid segment separately after dividing the region into an 8x16 grid. Predictive performance and computational economy were balanced in the design of the SRNN architecture. Promising results were obtained from their trials using a 24-step time series input, and they investigated how to enhance model convergence by optimising the learning rate.

Another deep-learning RNN architecture described in [71] is the Bidirectional Recurrent Neural Network (BRNN), as seen in Fig. 8. The main way that BRNN varies from SRNN is that each of its levels is made up of two RNN layers that pass data in opposing ways. When producing predictions, the BRNN can take into account both past and future contexts because to its bidirectional flow, which is advantageous for jobs that call for a thorough comprehension of sequence data.

The BRNN has a unique architecture that allows it to create predictions based on both the previous and following time steps. This construction consists of two RNNs delivering information in opposing ways at each layer. The BRNN, multi-layer RNN, and single-layer RNN models were contrasted in [72]. The authors also included a deep learning architecture in addition to stacked BRNN, multi-layer RNN, and a masking layer. The study concentrated on the scalability and resilience of these models in addition to accuracy. The scientists made the fascinating finding that adding additional layers to deep learning neural networks does not necessarily improve performance, highlighting the intricate trade-offs between model complexity and performance.

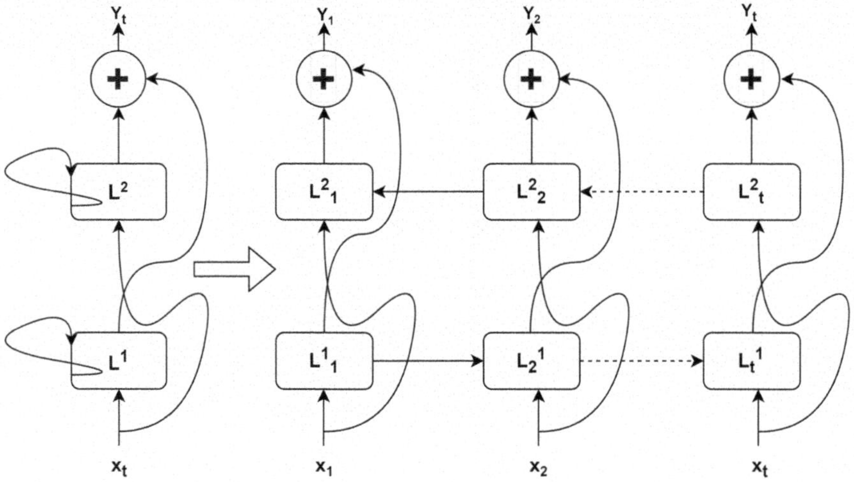

Fig. 8 Bi-RNN structure

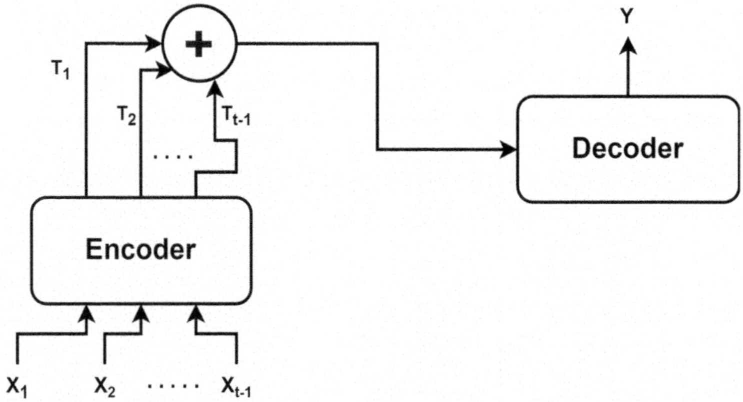

Fig. 9 Seq2Seq structure

Apart from SRNN and BRNN, another method for building RNN-based deep learning architectures is the sequence-to-sequence (seq2seq) architecture. The encoder and the decoder are the two essential parts of the seq2seq architecture, as shown in Fig. 9.

5 Case Study

This chapter references the case study titled "Mode Search Optimisation based Deep Long Short Term Memory Classifier."

Aim: In order to facilitate traffic-free travel, the research project aims to develop a deep learning-based traffic prediction and signal control model for the ITS.

Research Need: Urbanization, industrialisation, and recent ITS technological developments all contribute to increased traffic congestion, which raises fuel costs and has negative health effects. This indicates that in order to address the problems with traffic congestion that have a detrimental impact on the transportation service, a dynamic traffic management system is required.

Methods: This study develops a deep learning-based traffic prediction and control system to effectively manage peak-hour traffic and help passengers navigate the Intelligent Transportation System. The recommended mode search optimisation effectively classifies cars by specs. LSTM classifier hyperparameters are tweaked via model search to decrease training loss. Jitter, acceleration, priority, and velocity are analysed by a Mode search-based Deep LSTM classifier to forecast vehicle trajectories in the traffic signal management system.

5.1 *Comparative Analysis*

On the simulation area, the comparison study is carried out with 50, 100, and 200 automobiles, respectively. Figures 10, 11 and 12 illustrates the traffic signal controlling (TSC) technique research using 50, 100 and 200 vehicles respectively

Discussion: Table 4 provides a comparative examination of the established approach for different vehicle counts throughout the routing procedure. The implemented approach attains a peak Quality of Metric Aggregation (QMA) of 37.43 in the application involving 50 automobiles. This performance exceeds that of conventional QEOIC, ABC-based TSC, ROA-based TSC, BSO-based TSC, and CSA-based TSC methodologies by margins of 5.67%, 0.99%, 50.80%, 41.49%, and 35.40%, respectively. Moreover, the approach in question has a negligible jitter of 0.23. When applied to a sample of 100 autos, it demonstrates substantial enhancements compared to the QEOIC, ABC-based TSC, ROA-based TSC, BSO-based TSC, and CSA-based TSC techniques, with respective improvements of 83.93%, 84.87%, 3.99%, 0.29%, and 2.10%. In a relevant context, the suggested technique exhibits a minimum SDTT of 8.75, indicating enhancements of 70.25%, 55.69%, 0.83%, 77.85%, and 78.08% relative to the performance of current algorithms using 50 vehicles. The suggested approach, when implemented on a sample of 200 automobiles, yields an ideal throughput of 100.

Results and achievements: The effectiveness of the traffic prediction model is assessed through the analysis of throughput, jitter, quadratic mean of acceleration (QMA), and standard deviation of trip time (SDTT), as indicated by the experimental findings. The values for these metrics are 37.43, 0.23, 8.75, and 100, respectively. The proposed strategy demonstrates a performance improvement ranging from 5% to 42% in comparison to traditional methods.

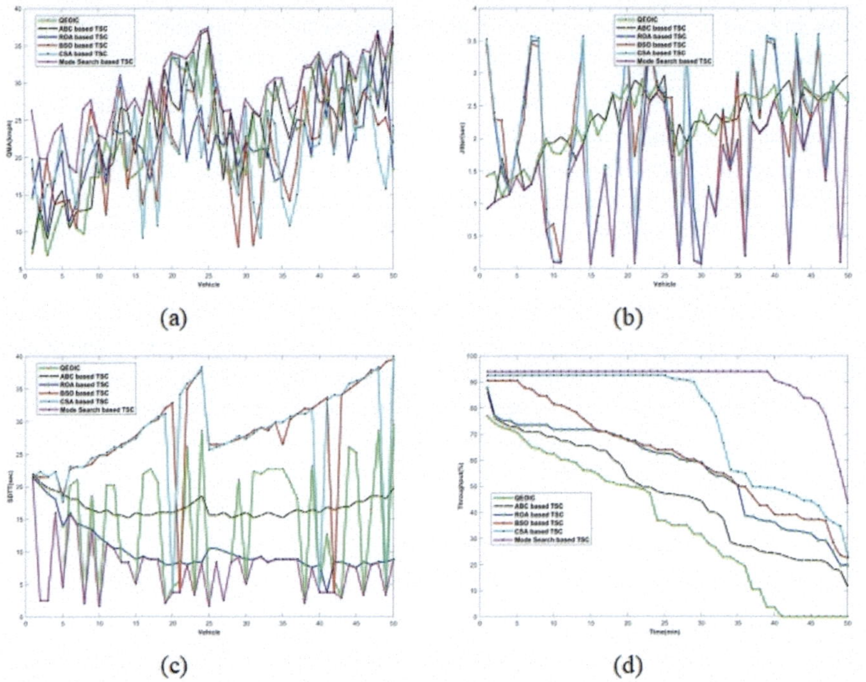

Fig. 10 Analysis with 50 vehicles with the parameters of **a** QMA, **b** Jitter, **c** SDTT, and **d** Throughput

6 Open Source Datasets

Ample data is essential to the successful development of a machine learning (ML) model. The training of the model to achieve convergence and good performance depends heavily on the data. A comprehensive dataset should include a broad range of traffic pattern information for different types of traffic prediction applications. To successfully analyse spatial-temporal interactions, more geographical information should be supplied, such as the location of data collecting locations, traffic directions, and the linkages among these places. In certain instances, the dataset ought to include pertinent data as well in order to examine the effects of important occasions, vacations, or meteorological fluctuations. The dataset's aggregation interval is another important consideration. While a larger aggregation period produces smoother time-series trends, a shorter one typically includes more random and non-linear factors, which might affect the structure of the prediction model. The development process and effectiveness of traffic prediction models are greatly influenced by each of these factors.

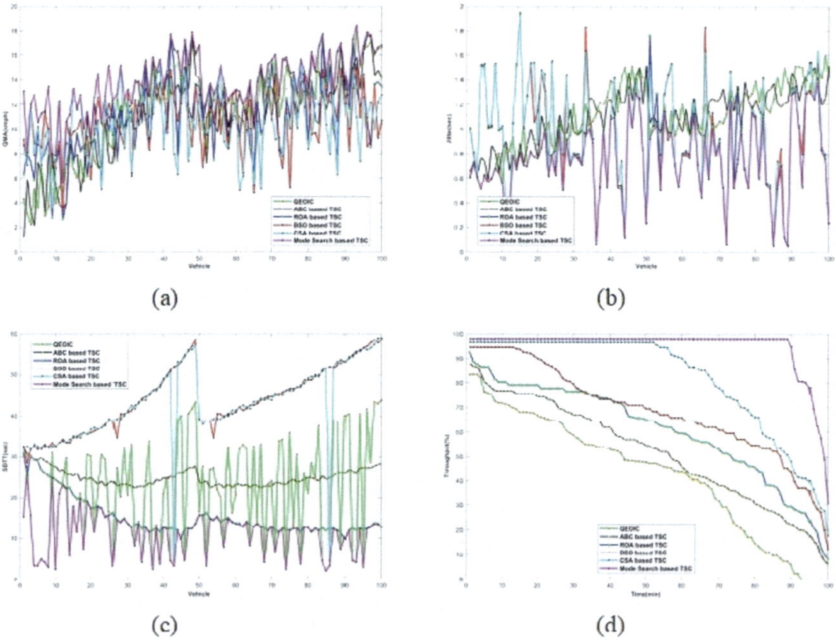

Fig. 11 Analysis with 100 vehicles with the parameter of **a** QMA, **b** Jitter, **c** SDTT, and **d** Throughput

1. **PeMS** [73]: The Performance Measurement System, or PeMS, gathers data in real time from over 39,000 distinct detectors located throughout the state of California's transportation network. Lane-specific data, including traffic flow, occupancy, speed, and other pertinent information, is provided by each detector. PeMS works effectively for short-term forecasting applications since it requires a minimum of 5 min to pass between data points. Understanding the spatial interactions between several detectors is aided by the length of road segments and the location of each detector. PeMS gathers information from a variety of sites, such as mainlines, interchanges connecting highways, on-ramps, and off-ramps.
2. **RTMC** [74]: Every 30 s, data on occupancy and traffic movement was gathered and combined into a 5-min period. The dataset offers information from 2011 through 2017. In the meantime, location data is also included in the dataset.
3. **CityPlus Smart City datasets** [75]: With the exception of July, traffic data from Denmark's urban and rural areas is gathered every year between February and November. Every 5 min, data on the number of cars and average speed are recorded. The dataset also contains each detector's physical location as well as information on pollution, the weather, and special events. This information is highly helpful for building machine learning models that employ hybrid data as input.

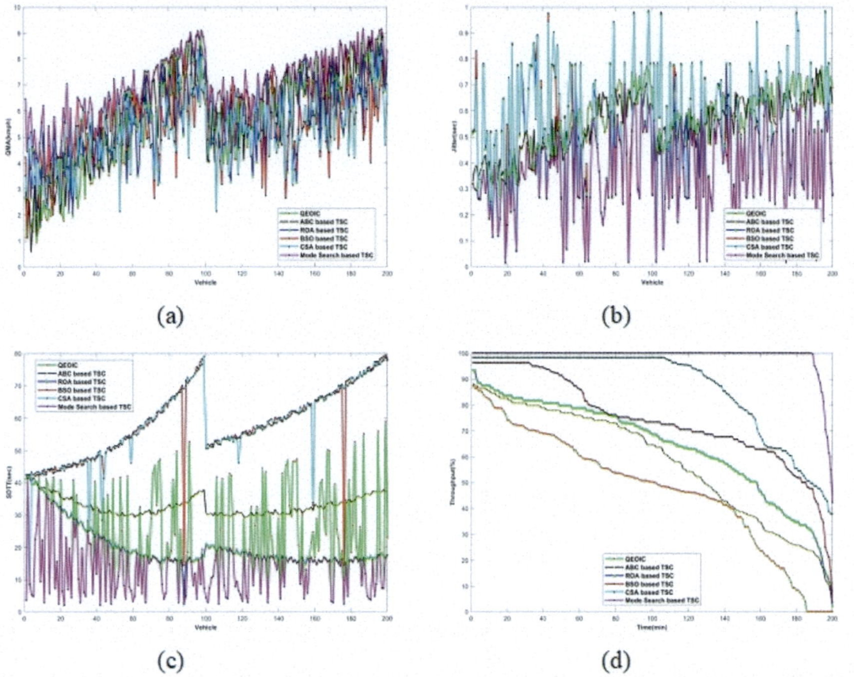

Fig. 12 Analysis with 200 vehicles with the parameters of **a** QMA, **b** Jitter, **c** SDTT, and **d** Throughput

7 Traffic Network Simulators

Models learned with empirical datasets play a critical role for applications like traffic prediction. Still, empirical datasets mostly fail to fulfill the requirements of experiments or applications all the time, and this happens because of reasons like scale, detail, or a lack of characteristic driving patterns or environmental conditions. For that reason, traffic simulation systems are equally important to enrich and supplement datasets and thus overcome problems. These models include full models of driver behavior, weather conditions, and special events that sometimes occur rarely in real operation to allow the simulation of the full variety of traffic phenomena at different scales and involving different road networks. Simulations allow researchers to generate more diversity in data than what is available from experiments, correct deficiencies in empirical datasets, and test, with great care, machine learning models against a broad variety of situations and events. This would make the predictive models more accurate as well as reliable, and also allow for complete assessment and improvements that are essential for effective traffic oversight and strategic development.

Table 4 Comparative evaluation

Methods/metrics	QEOIC	ABC-based TSC	ROA-based TSC	BSO-based TSC	CSA-based TSC	Mode search-based TSC
	Vehicles count: 50					
QMA	33.43	36.06	28.31	23.96	34.28	36.33
Jitter	3.67	3.86	3.55	3.22	3.99	3.57
SDTT	59.31	29.65	9.92	29.54	49.72	7.70
Throughput	87.24	56.61	77.85	96.72	82.47	97.06
	Vehicles count: 100					
QMA	26.61	24.61	23.63	12.76	42.88	25.89
Jitter	3.63	3.42	1.64	1.43	1.83	1.33
SDTT	42.62	38.22	32.87	55.42	56.83	22.97
Throughput	76.30	88.99	82.98	84.77	83.33	89.00
	Vehicles count: 200					
QMA	8.25	6.86	6.36	5.94	6.22	6.33
Jitter	1.89	1.82	1.88	1.48	1.48	1.47
SDTT	33.15	47.55	27.76	88.69	67.88	27.45
Throughput	88.58	78.46	94.45	95.38	97.36	100.00

Traffic simulation systems aim at realizing actual time traffic systems based on computational modeling. Generally, simulation models tend to fall into either the macroscopic or the microscopic model categories. The macroscopic models basically focus on aggregate characteristic modeling of traffic flow properties over large or whole highway systems. Normally, these tend to rely on fluid models describing traffic and include terms for density and speed-flow variables and may not have detailed records concerning all vehicles. On the other hand, microscopic simulation models focus on differences in the behavior of both the drivers and their vehicles. At a detailed level, as well, models simulate dynamic movements of traffic which include changes in lanes, acceleration as well as braking. Effects involving interactions between vehicles and how it affects traffic devices and other related infrastructures are also provided. Both the macroscopic and microscopic simulation models are crucial within the scopes of traffic engineering, urban planning, and transportation research in predicting the impacts of variations in real traffic conditions, design of infrastructure, assessment of strategies for traffic management, and provide useful information about the nature of traffic [76].

Macroscopic traffic simulation models include TransCAD [77] and EMME [78], which simulate flow as a continuous and compressible fluid of moving vehicles. Unlike microscopic models, these models focus on aggregate measures such as total traffic volume, density, and speed, which makes them faster to compute but requires less computing resource.

Macroscopic traffic simulators, as previously mentioned, do not depict the interactions within the traffic flow. Conversely, tiny traffic simulators concentrate on each vehicle's position and speed inside the traffic flow [79]. By simulating each vehicle's activity, the system mimics the flow of traffic. The user may specify the actions of the automobile, such as following, changing lanes, and destination details, in a system that simulates traffic at a tiny scale. With the help of these capabilities, users may simulate a more complex and realistic traffic scenario. The following are a few well-known miniature traffic simulators: SUMO [80], MITSimLab [81], TransModeler [82], Paramics [83], VISSIM [84], SimTraffic [85], CORSIM [86].

8 Data Pre-processing

Effective data preparation is essential before feeding data into a model. This helps to expedite convergence, focus the model's attention on the exact patterns we want to learn, and keep extraneous data from interfering with the model's performance. It might be difficult to choose the best pre-processing method, though, because it has to fit the demands of the model being utilised as well as the task at hand. An improper pre-processing method, for example, might result in poor preparation, noisy data, or outliers that could cause the model to be misled. On the other hand, very extensive pre-processing may remove significant variability from the dataset, which would make it more challenging for the model to accurately identify relevant time-series patterns. Therefore, it requires a careful approach and testing of different pre-processing strategies in an effort to preserve the properties of interest in the data while improving the model's ability to learn and generalize.

Normalization This is perhaps the most vital technique applied while preprocessing data for any kind of NN or even regression model, though a simple task it very often enhances the performance and boosts up the model to higher levels of convergence. This therefore shows that in neural network models, selection of the activation function depends on the requirement for normalizing. Most activation functions have a particular output range; for example, sigmoid has outputs ranging between 0 and 1 and tanh between -1 and 1. The activation function may not operate as well as it should if the input data is more than these ranges, which would impede model convergence. Additionally, within the same dataset, normalisation aids in standardising the scales of various data kinds. For example, normalisation guarantees that the differences in maximum speeds-which usually don't go above 200-do not adversely impact model training when traffic flow and speed data are combined. This allows for the best possible convergence rates.

In order to distinguish seasonal patterns and residual changes from the underlying data, detrending is essential in time-series research. Removing the historical average is a popular technique that effectively gets rid of long-term tendencies [87]. A different method that helps distinguish periodic trends from the original signal is the Discrete Fourier Transform (DFT), which divides time-series data into periodic components [87]. Conversely, Empirical Mode Decomposition (EMD), which has been applied in studies like [88, 89], offers a versatile technique to clearly highlight both low-frequency and high-frequency patterns. EMD is very useful for datasets that have a wide range of intricate characteristics that are present at different frequencies over the course of the time series.

9 Model Design

The main goal of any job, whether it road network prediction, multi-step prediction, or multi-task prediction, should direct the model's design. Improving forecast accuracy without appreciably lengthening processing time is the difficult part. To do this, the model should also make efficient use of spatial-temporal correlations and the connections between different input sources. This method makes sure the model uses pertinent data linkages and patterns to provide predictions that are both accurate and efficient.

In traffic forecasting, integrating a multi-task layer is a popular method for multi-task and multi-step forecasts. Scholars in [56, 90] have successfully applied this tactic. A different approach was described in [66], which included reintroducing recent prediction outcomes into the input sequence for step predictions that came after. Furthermore, to find highly correlated detectors across datasets and expose spatial-temporal correlations, K-Nearest Neighbours (KNN) was utilised. By taking use of the interdependencies between traffic data points, these techniques improve forecast accuracy. Nevertheless, KNN's computational burden is a disadvantage, particularly when dealing with big detector sets that are restricted to particular areas, which lessens its efficacy. Recent research has suggested models that combine conventional Convolutional Neural Networks (CNNs) with Recurrent Neural Networks (RNNs) to solve these issues [91]. Graph Convolutional Networks (GCNs) are used instead of original CNNs since they can handle graph data better (see [57]). GCNs employ adjacency matrices to represent spatial data (see [92, 93]). Furthermore, attention layers have been investigated as a means of capturing distinct temporal features [93–95]. Nevertheless, daily fluctuations in detector associations may constrain the use of adjacency matrices for spatial correlations [96, 97]. Yu et al. [98] suggested a temporal adjacency matrix method as a means of mitigating this.

A crucial aspect of post-model architecture configuration is the selection of optimal hyperparameters, including learning rates, layers, and hidden units in neural networks (NN), kernel function parameters in support vector machines (SVM), and the number of neighbours in K-Nearest Neighbours (KNN). Grid search is a widely used method for hyperparameter optimisation; yet, it is time-consuming due to the extensive number of parameter possibilities involved. Alternative techniques for efficiently navigating the hyperparameter space include Alternating Least Squares (ALS) [56], Firefly Algorithm (FFA) [99], Artificial Bee Colony (ABC) [100], among others. These strategies aim to uncover optimal configurations that enhance model performance across several settings, including traffic prediction.

10 Conclusion and Future Scope

This chapter delves deeper into unsolved challenges within the scope of traffic forecasting and analyzes how techniques based on machine learning techniques can be applied within such a scope. That this sophistication of algorithms towards being at a higher level for realizing elevated-level features increases in prospects for accuracy and reliability about forecasts of traffic goes directly to the increasing application of attention- and sequence-based models. Another significant improvement has been through the incorporation of CNN, or Conv-RNN into RNN. Promising results are reported on hybrids, integrating different methodologies so that it becomes more efficiently able to extract complex information rather than making use of something that could be simpler as a framework with shorter processing duration.

The problems related to the forecasting of traffic in large road networks are increasingly significant. Scientists seek systems that can handle complications of spatial and temporal interconnections in large road infrastructures. This trend speaks to the requirements of scalability and robustness in the future methodologies of traffic forecasting. Real-time predictions and large data processing are the elementary requirements of modern intelligent transportation systems.

In conclusion, hybrid models and large-scale forecasting systems are at the forefront of current research, indicating promising advances in machine learning models for traffic prediction. These advancements show how traffic control and congestion reduction might be greatly enhanced, leading to safer and more effective transport systems in the long run. In order to meet the constantly expanding needs of urban transport networks, this area has to continue its research and innovation.

The potential for deep learning applications in traffic prediction within vehicular ad hoc networks (VANETs) is extensive and holds significant promise for future advancements. A significant area of progress is found in the integration of edge computing with vehicular ad hoc networks (VANETs), facilitating real-time, low-latency traffic predictions through the processing of data nearer to its origin. A noteworthy avenue of exploration involves the application of federated learning to improve privacy-preserving traffic models, all the while ensuring the retention of prediction accuracy within decentralised networks. Furthermore, subsequent research may investigate the utilisation of Graph Neural Networks (GNNs) to enhance the understanding of spatial-temporal dependencies within traffic networks, along with the implementation of transfer learning to modify models for areas characterised by sparse traffic data. With the increasing prevalence of autonomous vehicles, it is essential to integrate deep learning models that consider their distinct traffic behaviours. Ultimately, concentrating on energy-efficient AI models will guarantee that these sophisticated systems sustain their viability in extensive, real-time traffic management scenarios.

References

1. Collins M, Etzioni S, Ben-Elia E (2024) Travel behavior and system dynamics in a simple gamified automated multimodal network. Transp Res Part A Policy Pract 183:104060
2. Samuel CN, Verdier F, Glock S, Guitton-Ouhamou P (2024) A fair crowd-sourced automotive data monetization approach using substrate hybrid consensus blockchain. Future Internet 16(5):156
3. George AS (2024) The role of fog computing in enabling real-time iot applications. Partners Univ Int Innov J 2(2):39–54
4. Moura DL, Aquino AL, Loureiro AA (2019) Towards data VSN offloading in vanets integrated into the cellular network. In: Proceedings of the 22nd international ACM conference on modeling, analysis and simulation of wireless and mobile systems, pp 235–239
5. Nya N, Baynat B (2017) Performance model for 4g/5g heterogeneous networks with different classes of users. In: Proceedings of the 20th ACM international conference on modelling, analysis and simulation of wireless and mobile systems, pp 171–178
6. Olivieri B, Endler M (2017) Dadca: an efficient distributed algorithm for aerial datacollection from wireless sensors networks by UAVs. In: Proceedings of the 20th ACM international conference on modelling, analysis and simulation of wireless and mobile systems, pp 129–136
7. Aljeri N, Boukerche A (2019) Movement prediction models for vehicular networks: an empirical analysis. Wirel Netw 25:1505–1518
8. Aljeri N, Boukerche A (2019) A probabilistic neural network-based road side unit prediction scheme for autonomous driving. In: ICC 2019-2019 IEEE international conference on communications (ICC). IEEE, pp 1–6
9. Sun P, AlJeri N, Boukerche A (2018) A fast vehicular traffic flow prediction scheme based on fourier and wavelet analysis. In: IEEE Global communications conference (GLOBECOM). IEEE 2018, pp 1–6
10. Barnett V (2005) Environmental statistics: methods and applications. Wiley
11. Nagy AM, Simon V (2018) Survey on traffic prediction in smart cities. Pervas Mobile Comput 50:148–163
12. Do LN, Taherifar N, Vu HL (2019) Survey of neural network-based models for short-term traffic state prediction. Wiley Interdisc Rev Data Mining Knowl Discov 9(1):e1285
13. Katib I, Altowaijri S (2019) Smarter traffic prediction using big data, in-memory computing, deep learning and gpus. Sensors 19(9):2206
14. Arrieta AB, Díaz-Rodríguez N, Del Ser J, Bennetot A, Tabik S, Barbado A, García S, Gil-López S, Molina D, Benjamins R et al (2020) Explainable artificial intelligence (xai): concepts, taxonomies, opportunities and challenges toward responsible AI. Inform Fusion 58:82–115
15. Goodfellow I, Bengio Y, Courville A (2016) Deep learning. MIT press
16. Shalev-Shwartz S, Ben-David S (2014) Understanding machine learning: from theory to algorithms. Cambridge University Press
17. Wang J, Boukerche A (2020) The scalability analysis of machine learning based models in road traffic flow prediction. In: ICC 2020–2020 IEEE international conference on communications (ICC). IEEE, pp 1–6
18. Sun P, Aljeri N, Boukerche A (2020) Machine learning-based models for real-time traffic flow prediction in vehicular networks. IEEE Netw 34(3):178–185
19. Friedman J (2009) The elements of statistical learning: data mining, inference, and prediction
20. Rice J, Van Zwet E (2004) A simple and effective method for predicting travel times on freeways. IEEE Trans Intell Transp Syst 5(3):200–207
21. Zhang X, Rice JA (2003) Short-term travel time prediction. Transp Res Part C Emerg Technol 11(3–4):187–210
22. Hobeika AG, Kim CK (1994) Traffic-flow-prediction systems based on upstream traffic. In: Proceedings of VNIS'94-1994 vehicle navigation and information systems conference. IEEE, pp 345–350
23. Kwon J, Coifman B, Bickel P (2000) Day-to-day travel-time trends and travel-time prediction from loop-detector data. Transp Res Rec 1717(1):120–129

24. Li L, Su X, Wang Y, Lin Y, Li Z, Li Y (2015) Robust causal dependence mining in big data network and its application to traffic flow predictions. Transp Res Part C Emerg Technol 58:292–307
25. Okawa M, Kim H, Toda H (2017) Online traffic flow prediction using convolved bilinear poisson regression. In: 2017 18th IEEE international conference on mobile data management (MDM). IEEE, pp 134–143
26. Peterson LE (2009) K-nearest neighbor. Scholarpedia 4(2):1883
27. Smith BL, Williams BM, Oswald RK (2002) Comparison of parametric and nonparametric models for traffic flow forecasting. Transp Res Part C Emerg Technol 10(4):303–321
28. Chang H, Lee Y, Yoon B, Baek S (2012) Dynamic near-term traffic flow prediction: system-oriented approach based on past experiences. IET Intell Transp Syst 6(3):292–305
29. Zheng Z, Su D (2014) Short-term traffic volume forecasting: a k-nearest neighbor approach enhanced by constrained linearly sewing principle component algorithm. Transp Res Part C Emerg Technol 43:143–157
30. Liu G, Zhao H, Fan F, Liu G, Xu Q, Nazir S (2022) An enhanced intrusion detection model based on improved KNN in WSNs. Sensors 22(4):1407
31. Xia D, Li H, Wang B, Li Y, Zhang Z (2016) A map reduce-based nearest neighbor approach for big-data-driven traffic flow prediction. IEEE Access 4:2920–2934
32. Cai P, Wang Y, Lu G, Chen P, Ding C, Sun J (2016) A spatiotemporal correlative k-nearest neighbor model for short-term traffic multistep forecasting. Transp Res Part C Emerg Technol 62:21–34
33. Zhang L, Liu Q, Yang W, Wei N, Dong D (2013) An improved k-nearest neighbor model for short-term traffic flow prediction. Proc Social Behav Sci 96:653–662
34. Vapnik V (2013) The nature of statistical learning theory. Springer Science & Business Media
35. Ding A, Zhao X, Jiao L (2002) Traffic flow time series prediction based on statistics learning theory. In: Proceedings. The IEEE 5th international conference on intelligent transportation systems. IEEE, pp 727–730
36. Vanajakshi L, Rilett LR (2004) A comparison of the performance of artificial neural networks and support vector machines for the prediction of traffic speed. In IEEE intelligent vehicles symposium. IEEE, pp 194–199
37. Zhang Y (2024) Xie Y (2007) Forecasting of short-term freeway volume with v-support vector machines. Transp Res Rec 1:92–99
38. Lin G, Lin A, Gu D (2022) Using support vector regression and k-nearest neighbors for short-term traffic flow prediction based on maximal information coefficient. Inform Sci 608:517–531
39. Almeida A, Brás S, Oliveira I, Sargento S (2022) Vehicular traffic flow prediction using deployed traffic counters in a city. Future Gener Comput Syst 128:429–442
40. Lippi M, Bertini M, Frasconi P (2013) Short-term traffic flow forecasting: an experimental comparison of time-series analysis and supervised learning. IEEE Trans Intell Transp Syst 14(2):871–882
41. Hong W-C, Dong Y, Zheng F, Lai C-Y (2011) Forecasting urban traffic flow by svr with continuous aco. Appl Math Modell 35(3):1282–1291
42. Dorigo M, Maniezzo V, Colorni A (1996) Ant system: optimization by a colony of cooperating agents. In: IEEE transactions on systems, man, and cybernetics, part B (cybernetics), vol 26, no 1, pp 29–41
43. Mingheng Z, Yaobao Z, Ganglong H (2013) Gang C (2013) Accurate multisteps traffic flow prediction based on SVM. Math Probl Eng 1:418303
44. Jabot F (2015) Why preferring parametric forecasting to nonparametric methods? J Theor Biol 372:205–210
45. Taye MM (2023) Theoretical understanding of convolutional neural network: concepts, architectures, applications, future directions. Computation 11(3):52
46. Krichen M (2023) Convolutional neural networks: a survey. Computers 12(8):151
47. Jin F, Sun S (2008) Neural network multitask learning for traffic flow forecasting. In: 2008 IEEE international joint conference on neural networks (IEEE world congress on computational intelligence). IEEE, pp 1897–1901

48. Çetiner BG, Sari M, Borat O (2010) A neural network based traffic-flow prediction model. Math Comput Appl 15(2):269–278
49. Kumar K, Parida M, Katiyar V (2013) Short term traffic flow prediction for a non urban highway using artificial neural network. Proc Social Behav Sci 104:755–764
50. Polson NG, Sokolov VO (2017) Deep learning for short-term traffic flow prediction. Transp Res Part C Emerg Technol 79:1–17
51. Lin Y, Dai X, Li L, Wang F-Y (2018) Pattern sensitive prediction of traffic flow based on generative adversarial framework. IEEE Trans Intell Transp Syst 20(6):2395–2400
52. Arjovsky M, Chintala S, Bottou L (2017) Wasserstein generative adversarial networks. In: International conference on machine learning. PMLR, pp 214–223
53. Lv Y, Duan Y, Kang W, Li Z, Wang F-Y (2014) Traffic flow prediction with big data: a deep learning approach. iEEE Trans Intell Transp Syst 16(2):865–873
54. Deng L, Yu D et al (2014) Deep learning: methods and applications. Found Trends Signal Process 7(3–4):197–387
55. Ngiam J, Khosla A, Kim M, Nam J, Lee H, Ng AY (2011) Multimodal deep learning. In: Proceedings of the 28th international conference on machine learning (ICML-11), pp 689–696
56. Huang W, Song G, Hong H, Xie K (2014) Deep architecture for traffic flow prediction: deep belief networks with multitask learning. IEEE Trans Intell Transp Syst 15(5):2191–2201
57. Goudarzi S, Kama MN, Anisi MH, Soleymani SA, Doctor F (2018) Self-organizing traffic flow prediction with an optimized deep belief network for internet of vehicles. Sensors 18(10):3459
58. Hermans M, Schrauwen B (2013) Training and analysing deep recurrent neural networks. Adv Neural Inform Process Syst 26
59. Smolensky P, Mozer MC, Rumelhart DE (2013) Mathematical perspectives on neural networks. Psychology Press
60. Pearlmutter (1989) Learning state space trajectories in recurrent neural networks. In: International 1989 joint conference on neural networks. IEEE, pp 365–372
61. Cleeremans A, Servan-Schreiber D, McClelland JL (1989) Finite state automata and simple recurrent networks. Neural Comput 1(3):372–381
62. Tian Y, Pan L (2015) Predicting short-term traffic flow by long short-term memory recurrent neural network. In: 2015 IEEE international conference on smart city/SocialCom/SustainCom (SmartCity). IEEE , pp 153–158
63. Hochreiter S, Schmidhuber J (1997) Long short-term memory. Neural Comput 9(8):1735–1780
64. Gers FA, Schmidhuber J, Cummins F (2000) Learning to forget: continual prediction with LSTM. Neural Comput 12(10):2451–2471
65. Cho K, Van Merriënboer B, Gulcehre C, Bahdanau D, Bougares F, Schwenk H, Bengio Y (2014) Learning phrase representations using rnn encoder-decoder for statistical machine translation. arXiv preprint arXiv:1406.1078
66. Guo J, Wang Z, Chen H (2017) On-line multi-step prediction of short term traffic flow based on GRU neural network. In: Proceedings of the 2nd international conference on intelligent information processing, pp 1–6
67. Fu R, Zhang Z, Li L (2016) Using LSTM and GRU neural network methods for traffic flow prediction. In: 2016 31st Youth academic annual conference of Chinese association of automation (YAC). IEEE, pp 324–328
68. Pascanu R, Gulcehre C, Cho K, Bengio Y (2013) How to construct deep recurrent neural networks. arXiv preprint arXiv:1312.6026
69. Hihi S, Bengio Y (1995) Hierarchical recurrent neural networks for long-term dependencies. Adv Neural Inform Process Syst 8
70. Zou Z, Gao P, Yao C (2018) City-level traffic flow prediction via LSTM networks. In: Proceedings of the 2nd international conference on advances in image processing, pp 149–153
71. Schuster M, Paliwal KK (1997) Bidirectional recurrent neural networks. IEEE Trans Sig Process 45(11):2673–2681
72. Cui Z, Ke R, Pu Z, Wang Y (2018) Deep bidirectional and unidirectional LSTM recurrent neural network for network-wide traffic speed prediction. arXiv preprint arXiv:1801.02143

73. Molan A, Murugesan N, Shams A, Tortora C, Rahman F, Loh J, Pande A (2020) Evaluation of coordinated ramp metering (CRM) implemented by caltrans
74. Kwon TM Annual report transportation data research laboratory
75. Ali MI, Gao F, Mileo A (2015) Citybench: a configurable benchmark to evaluate RSP engines using smart city datasets. In: International semantic web conference. Springer, pp 374–389
76. Abdelgawad K, Henning S, Biemelt P, Gausemeier S, Trächtler A (2016) Advanced traffic simulation framework for networked driving simulators. IFAC-PapersOnLine 49(11):101–108
77. Siddiqui C, Dennis M (2019) Developing a guideline for setting targets for national performance management measures to assess performance of the national highway system in south carolina. Transp Res Rec 2673(9):266–276
78. Jian S, Rey D, Dixit V (2019) An integrated supply-demand approach to solving optimal relocations in station-based carsharing systems. Netw Spatial Econ 19:611–632
79. Maciejewski M (2010) A comparison of microscopic traffic flow simulation systems for an urban area. Transp Probl 5:27–38
80. Mallissery S, Pai MM, Mehbadi M, Pai RM, Wu Y-S (2019) Online and offline communication architecture for vehicular ad-hoc networks using ns3 and sumo simulators. J High Speed Netw 25(3):253–271
81. ITSLAB M (2019) Mitsimlab
82. Ullah MR, Khattak KS, Khan ZH, Khan MA, Minallah N, Khan AN (2021) Vehicular traffic simulation software: a systematic comparative analysis. Pak J Eng Technol 4(1):66–78
83. Aguirre HRP, Lopez MQ, Salas SL, Moreno VML (2020) Virtual simulation of road traffic based on multi-agent systems. In: Virtual reality designs. CRC Press, pp 188–217
84. Kučera T, Chocholáč J (2021) Design of the city logistics simulation model using ptv vissim software. Transp Res Proc 53:258–265
85. Jin (2021) A simulation-as-a-service (SIMaaS) cloud computing platform for traffic management. Ph.D. dissertation, The University of Wisconsin-Milwaukee
86. Zhang X, Chang G-L (2014) A mixed-flow simulation model for congested intersections with high pedestrian-vehicle traffic flows. Simulation 90(5):570–590
87. Luo X, Li D (2019) Zhang S (2019) Traffic flow prediction during the holidays based on DFT and SVR. J Sensors 1:6461450
88. Duo M, Qi Y, Lina G, Xu E (2017) A short-term traffic flow prediction model based on EMD and GPSO-SVM. In: 2017 IEEE 2nd advanced information technology, electronic and automation control conference (IAEAC). IEEE, pp 2554–2558
89. Ke J, Zheng H, Yang H, Chen XM (2017) Short-term forecasting of passenger demand under on-demand ride services: a spatio-temporal deep learning approach. Transp Res Part C Emerg Technol 85:591–608
90. Yao H, Tang X, Wei H, Zheng G, Li Z (2019) Revisiting spatial-temporal similarity: a deep learning framework for traffic prediction. Proc AAAI Conf Artif Intell 33(01):5668–5675
91. Zhang J, Zheng Y, Qi D (2017) Deep spatio-temporal residual networks for citywide crowd flows prediction. In: Proceedings of the AAAI conference on artificial intelligence, vol 31, no 1
92. Yu B, Yin H, Zhu Z (2017) Spatio-temporal graph convolutional networks: a deep learning framework for traffic forecasting. arXiv preprint arXiv:1709.04875
93. Guo S, Lin Y, Feng N, Song C, Wan H (2019) Attention based spatial-temporal graph convolutional networks for traffic flow forecasting. Proc AAAI Conf Artif Intell 33(01):922–929
94. He Z, Chow C-Y, Zhang J-D (2018) Stann: a spatio-temporal attentive neural network for traffic prediction. IEEE Access 7:4795–4806
95. Do LN, Vu HL, Vo BQ, Liu Z, Phung D (2019) An effective spatial-temporal attention based neural network for traffic flow prediction. Transp Res Part C Emerg Technol 108:12–28
96. Tao Y, Sun P, Boukerche A (2019) A hybrid stacked traffic volume prediction approach for a sparse road network. In: 2019 IEEE symposium on computers and communications (ISCC). IEEE, pp 1–6

97. Tao Y, Sun P, Boukerche A (2019) A novel travel-delay aware short-term vehicular traffic flow prediction scheme for vanet. In: 2019 IEEE wireless communications and networking conference (WCNC). IEEE, pp 1–6
98. Yu B, Li M, Zhang J, Zhu Z (2019) 3d graph convolutional networks with temporal graphs: a spatial information free framework for traffic forecasting. *arXiv preprint* arXiv:1903.00919
99. Yang X-S (2009) Firefly algorithms for multimodal optimization. In: International symposium on stochastic algorithms. Springer, pp 169–178
100. Chen D (2017) Research on traffic flow prediction in the big data environment based on the improved rbf neural network. IEEE Trans Indus Inf 13(4):2000–2008

Traffic Prediction and Modeling in Vehicular Ad Hoc Networks

Nagendra Singh, Payal Suhane, Harsh Pratap Singh, Pinky Rane, and Dipti Shukla

Abstract Vehicular Ad Hoc Networks are a prominent method employed for enhancing road safety, predicting traffic, and improving the efficiency of road management. In this technique, vehicles get communication from roadside infrastructure as well as with each other. Due to such communications, drivers get information about real-time traffic conditions, road conditions, and safety hazards. One critical aspect of VANETs is traffic prediction and modeling, which play a crucial role in optimizing traffic flow, routing decisions, and resource allocation. This chapter provides a study of various modeling techniques in VANETs like static statistical methods, machine learning techniques, deep learning techniques, and hybrid models. The proposed chapter evaluates the strengths and limitations of these methods in predicting various traffic parameters, such as traffic volume, congestion levels, and travel times.

Keywords Traffic prediction · Vehicular Ad Hoc Networks (VANETs) · Traffic modeling · Mobility patterns · Communication protocols

N. Singh (✉)
Department of Electrical Engineering, Trinity College of Engineering and Technology, Karimnagar, Telangana, India
e-mail: nsingh007@rediffmail.com

P. Suhane
Department Electrical and Electronics Engineering, Vidhyapeeth Institute of Science and Technology, Bhopal, India

H. P. Singh
Department of Computer Science Engineering, Shri Vaishnav Vidyapeeth Vishwavidyalaya, Indore, MP, India

P. Rane
Department of Computer Science and Engineering, Medi-Caps University, Indore, India

D. Shukla
Department of Physics, O. P. Jindal University, Raigarh, CG, India
e-mail: dipti.shukla@opju.ac.in

J. Bhatia et al. (eds.), *Deep Learning Based Solutions for Vehicular Adhoc Networks*, Studies in Computational Intelligence 1207,
https://doi.org/10.1007/978-981-96-5190-0_8

1 Introduction

Road transportation is the most dangerous due to heavy traffic flow and rash driving. So many people were injured and lost their lives due to many accidents, and they also lost their valuable wealth in medical charges [1]. Such road transportation issues need to be resolved using AI. If we can develop an autonomous incident detection system, such automation systems will alert people to road accidents as well as road traffic conditions in real time [2]. With the help of traffic flow data, an intelligent transportation system can provide an accurate forecast of traffic flow in real time in a specific area. For the improvement of road safety and smooth traffic flow, it is required to establish communication between cars and roadside units. For the safety of road transportation systems, traditionally used network systems face many problems due to the slow and intermittent connectivity of the network. For improving road safety network management, we required some recent, efficient, and reliable systems [3].

For improving road safety, providing accurate traffic flow, and increasing driving comfort, the latest network is called VANETs. VANET is a recent development of AI used in transportation systems that provides reliable networks for vehicles. The road transportation system mostly depends on the condition of the road surface. If the road conditions are good, it is possible to avoid unnecessary jam conditions. The VANET system also checks the road condition and alerts the vehicle drivers so they can drive their vehicles accordingly. For example, if the road condition is not good, traffic is definitely higher, and you need to drive a vehicle at a slow speed. VALNET shows the condition of road traffic and provides real-time scenarios [4].

The VANET system provides an accurate prediction of traffic congestion but cannot control the traffic by itself. It helps the driver to give information about the road traffic at that particular time and area. Based on historical data, road traffic prediction in real-time is the most challenging task for the VANET network system [5]. Based on the AI and IoT systems, various people suggested smart road transport traffic flow. The traffic data used by AI helps to provide valuable information to vehicle drivers, like accidental zones, highway traffic areas, etc. Such data is also used to provide awareness to avoid dangerous circumstances and enhance client convenience with safety and security [6].

As per road conditions, the speed of vehicles changes; in rural areas, speed is low, but on the highway, speed is very high, and on the highway, most accidental cases arise. Vehicle density is not the same in the whole area of the road, and hence the density of nodes in a VANET network is random. Compared to rural areas, metropolitan areas have a higher vehicle density because most of the population stays in urban areas, and density variation depends on time. Network fragmentation sometimes happens when there is a low or irregular vehicle density. In the case of network fragmentation, vehicles move in disconnected, isolated clusters, and hence it is very difficult to manage end-to-end communication [7].

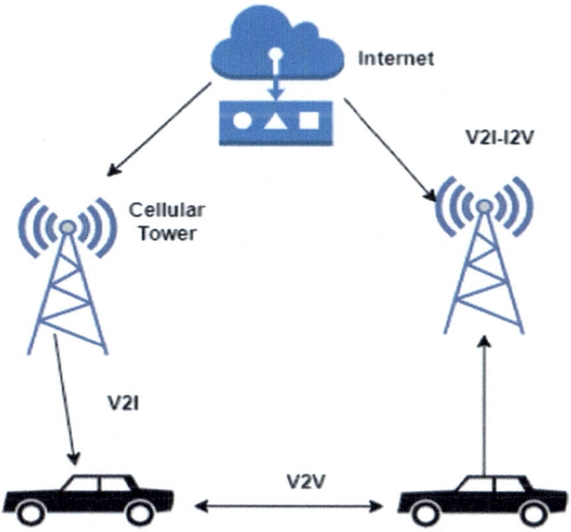

Fig. 1 VANET network structure for road transportation

1.1 VANET Network System

The VANET network has one of the important techniques named clustering, which gives optimum results from the simply obtained historical data. It helps to optimize the connected network and enhance safety services. The vehicles are connected to each other using these small clusters of nodes for communication called vehicle-to-vehicle communication. Such division of nodes into small clusters improved the performance of the vehicles, improved road safety, and provided comfort while driving [8].

VANET's network provides flow traffic forecasting, the accurate position of vehicles, and emergency alerts to vehicle drivers. VANETs also help to prevent collisions between vehicles. VANET network services mostly depend on the communication network and the exchange of data between the network tower and vehicles. VANET plays an important role in maintaining communication protocols, helping to format the data, and maintaining road transport standards [9]. Figure 1 shows the VANET network operation for the road transportation system.

1.2 The Importance of Traffic Prediction and Modeling in VANETs

1. Enhanced Traffic Management: Reducing congestion, streamlining traffic signals, and enhancing overall flow of traffic are all made possible by accurate traffic prediction.

2. Safety Applications: By anticipating possible traffic bottlenecks or collisions, predictive models enable prompt warnings and preventative actions.
3. Effective Resource Allocation: Understanding future traffic patterns in VANETs facilitates more effective resource allocation, which raises QoS.
4. Route optimization: By providing drivers with real-time updates on the most efficient routes, drivers can save both time and fuel.

1.3 Motivation and Contribution

This chapter's main motivation is to find out the best suitable method that gives an accurate prediction of the traffic flow. Using such a model, people get correct real-time information and accordingly drive the vehicles. The main contribution of this chapter is to provide a deep study of the various methods used in VANET system modeling and prediction of traffic flow. This chapter also discusses the various advantages and limitations of the VANET modeling methods. Section 2 discusses the traffic prediction techniques [10, 11]. This section gives information about the statistical methods, machine learning techniques, deep learning, and hybrid models. Section 3 shows the traffic modeling techniques. Section 4 gives the information about the challenges in traffic prediction and modeling. The future direction of the VANET system is given in Sect. 5, whereas the conclusion is shown in Sect. 6.

2 Traffic Prediction Techniques

In this section, discuss the VANET traffic prediction technique. An overview of statistical methods like time series analysis, Kalman filters, and Markov chains is discussed.

VANET performance increases more frequently due to better routing protocol contributions. Today, traffic conditions are rapidly changing due to the increasing use of vehicles and populations. Since the use of private vehicles is increasing and road conditions are also not good everywhere, the use of conventional techniques to manage traffic flow is inefficient [7].

Traffic has many bad impacts on environmental pollution and individual efficiency. For control of traffic flow and saving accidents on the road today, the VANET networking system is the most prominent and efficient transport technology. Prior to utilizing VANET technology, we must eliminate any potential incursions because the network is susceptible to various vulnerabilities [12].

In the VANET system, traffic prediction depends on historical traffic data and real-time condition information. Traffic prediction techniques are classified into statistical methods, machine learning, and hybrid models, as given below:

2.1 Statistical Methods

Statistical road safety requires preparing a model using the statistical data of traffic flow. Statistical data consists of historical data on accidents, accidental zones, heavy traffic areas, and road segments. An equation presenting the outcome of statistical road safety modeling has the function of features on the right side and an estimate of expected accident frequency on the left [8].

Statistical road safety modeling has the following objectives.

- To calculate, using their characteristics, the estimated accident frequency caused by infrastructure pieces.
- To calculate how changing certain infrastructure components may affect the predicted frequency of accidents.

Statistical techniques depend on historical traffic data and accordingly predict future trends. The following methods are included in the static method.

2.1.1 Time Series Analysis

One of the key elements of intelligent transportation systems is that they try to put plans and regulations into road transportation networks. This is an advanced travel and management information system (ATMIS). This road management system provided accurate, real-time traffic information for the vehicle drivers. So that driver is alert during the trip [7].

In order to monitor and assess the current situation and subsequently provide advisory information to the general traveling public, traffic control centers require this data. Accurate and real-time traffic information can now be obtained by management centers and travelers alike, thanks to sophisticated data collection techniques and technology [13].

The journey times and delays on brief road segments could be used to display this information. The majority of travelers consider travel time to be an oblique indicator of travel expenses. In essence, timely and spatially distributed congestion should be mitigated by effectively disseminating anticipated trip information [14]. Time Series Analysis in Vehicular Ad Hoc Networks used many models, for example, ARIMA (Autoregressive Integrated Moving Average), SARIMA (Seasonal ARIMA), and LSTM (Exponential Smoothing Methods), etc.

2.1.2 Kalman Filters

It is a state-space model that can be used to simulate multi-input and multi-output systems. It is applicable for stationary and dynamic systems. The Kalman filter is applicable for traffic state modeling because of this property [8]. State variables are updated by the Kalman filter based on the preceding data observation.

The Kalman filter has the following features:

(1) The storage of historical data does not require any additional storage space.
(2) The Kamman filter computes the historical data efficiently.

Since the Kalman filter theory is based on linear Gaussian problems, it is not applicable to linear real-world problems. So it is required to use an algorithm that is developed in such a way to convert the application of the Kalman filter for both linear and nonlinear applications so that it can easily be applied to real-world problems like traffic prediction in real time. For non-linear models, the extended Kalman filter (EKF) is an effective technique because it is based on the Gaussian law. This technique used a linear model of the problem and simulated it using the Taylor series for the optimum solution [14].

For finding the best solution to the nonlinear model of the real-world problem, the Unscented Kalman Filter (UKF), a more advanced version of the Kalman filter than the EKF, is one of the recent innovations. Unscented Kalman Filter (UKF) is the enhanced version of KF used for the solution of nonlinear problems. Such an advanced version of KF classified the parameters, evaluated the mean and covariance of the data, and hence easily handled the nonlinear set of data.

UKF is not required to linearize the errors as compared to EKF; hence, its performance is better than EKF. For linear problems, the quality of UKF estimates is almost the same as standard Kalman filters. In various tracking solutions, standard Kalman filters are used. The VANET system uses the Kalman filter for the prediction of the transmitted messages. Under this system, every vehicle uses its past location to estimate and record its location at various intervals, then sends out the real-time position. In this method, first the prediction information is taken and checked for errors; after that, it is transmitted.

2.1.3 Markov Chains

Markov chaining is a mathematical tool used in traffic prediction. It does not have any memory storage parts suitable for modeling various random processes, including traffic flow. One kind of stochastic process that works with systems that change from one state space to another is the Markov chain [9]. For the prediction of traffic flow, Markov chains used the following steps:

Step 1: Using the different traffic conditions like speed limits, population density, and flow rate of vehicles first defines the state of the problem.
Step 2: For the simulation of the Markov model, it required historical traffic data flow from different areas in different time frames.
Step 3: With the help of the collected data, prepare a model.
Step 4: Using a transition matrix, simulate the collected data.
Step 5: With the help of a simulation model, find out the best results, which means accurate traffic flow.

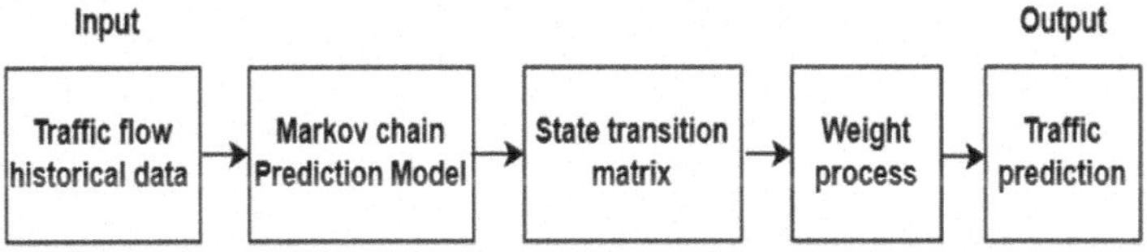

Fig. 2 Markov chain process for traffic prediction

For the prediction of road traffic flow, Markov chains are a simple and powerful technique. With the help of historical data, this technique developed the simulation model and helped the vehicle drivers achieve the best control of traffic and accurate traffic conditions, as shown in Fig. 2. It also helps people with the planning and management of traffic control in real-time.

This technique is combined with the other adaptive models for the estimation of more accurate and dynamic predictions of traffic systems. These probabilistic models are helpful for simulating traffic light cycles and vehicle movements, as they forecast future states based on the probability of the current state.

2.2 *Machine Learning Techniques (ML)*

ML techniques can learn complex patterns from large datasets and give effective predictions of traffic flow. The following ML technique is used for the prediction of traffic flow:

2.2.1 Artificial Neural Networks

It is based on the function of the human brain's neurons and has emerged as a powerful tool in traffic prediction because it can model complex and non-linear data. Computational models known as artificial neural networks are composed of layers of neurons, which are interconnected processing units [15]. Three different types of layers were used in the ANN model shown in Fig. 3.

A weight is assigned to each neuronal connection, and this weight is changed during training in order to reduce prediction error. ANNS are used in traffic prediction to forecast future traffic conditions, including speed, flow, and congestion levels, by learning from past traffic data.

The ANN system, when applied in the VANET system, helps to identify various traffic challenges and different kinds of security issues of vehicles. ANN learns from the traffic data and suggests the most suitable VANET system.

Following steps are used by ANNs for the prediction of traffic flow:-

- Data Collection: It collects historical traffic data like traffic flow, speed of vehicles, weather conditions, vehicle density in different time zones, etc.

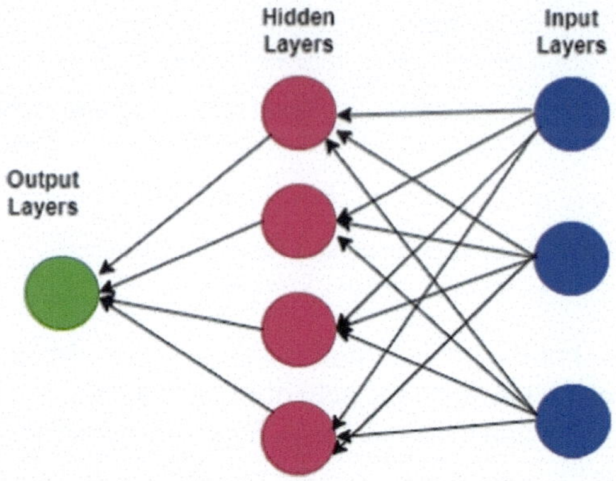

Fig. 3 The different ANN layers used for the prediction of traffic flow

- Data preprocessing: It involves handling missing values, cleaning the gathered data, normalizing the features, and dividing the data into training and testing sets.
- Network Architecture Design: For the design of the neural network architecture, It decides the required layers, the requirements of neurons in each layer, and the type of activation function.
- Training the Network: In the ANN architecture, send the processed data for training to adjust the weights and biases through back propagation and optimization techniques.
- Validation and Testing: For the validation of the model, we have to test it using a set of traffic historical data.
- Prediction: Based on the traffic data model predict the accurate traffic flow.

ANN predicts the traffic conditions based on the data collected on different time and days of the month. Data are collected on particular days within a time frame, along with weather conditions. The following layers are included:

1. Input Layer: This layer required historical data on vehicle speed, traffic conditions, and weather conditions in different time zones on different days of the month.
2. Hidden Layers: Multiple layers with neurons to learn the complex patterns in the data.
3. Output Layer: Produces the predicted traffic speed.

Advantages and Limitations of ANN method

Advantages

- Non-linearity: Capable of modeling complex, non-linear relationships in traffic data.
- Adaptability**:** Can adapt to different types of traffic data and prediction tasks.
- Generalization: Effective in generalizing from historical data to make accurate predictions.

Limitations

- Data Requirements: Requires a large amount of high-quality data for training.
- Computational Cost: Since training neural networks is a very long process and requires a large network system, we need to find out the cost of computation.
- Black-box Nature: Difficult to interpret the internal workings and decision-making process.

Because ANNs can recognize and simulate intricate patterns in traffic data, they are very useful for predicting traffic patterns. ANN provides high accuracy in the test data. It also helps in planning and controlling the traffic flow accurately. Due to its versatility and high skill in predicting traffic flow with the help of the ANN model, it is a very beneficial technique.

2.2.2 Support Vector Machines (SVM)

Support VANETs primarily use vector machines to classify traffic data, identify anomalies, provide security, and predict traffic flow. SVM manages the data and creates the routing matrix. This matrix enhances the performance of the routing system. For the identification of malicious behavior in the VANET, SVM used the IDS detection system. Based on the various data like speed of the vehicle, density of the traffic, and position of the vehicle, the SVM-based VANET model predicts the collision of the vehicles.

SVM is one of the most reliable ML techniques recently used for traffic prediction. SVM techniques use supervised learning models for the simulation of historical data. This technique first divides the traffic data into various classes so that it can easily make accurate decisions [16].

Traffic historical data like vehicle speed, density of population, and road conditions are taken into account for the design of the simulation model. After that, the model is tested on various sets of data to find the best solution.

The following steps are employed for traffic flow prediction using the SVM technique.

1. For the preparation of the simulation model, it is required to collect historical data on traffic flow.
2. The collected data needs to be analyzed and divided into various classes.
3. Now design the best model using the historical traffic data.
4. As per the collected data and its nature, we have to choose an appropriate kernel function.
5. Train the SVM model using the training data to find the optimal hyper plane or regression line that fits the data.
6. For the validation of the suggested model, it is required to test it using various sets of data and find out its best performance.
7. The trained model is used to predict the actual traffic conditions.

Advantages and Limitations of SVM

Advantages

- With the help of kernel functions, SVM can handle nonlinear data easily.
- Robustness: They are robust to over-fitting, especially in high-dimensional spaces.
- Versatility: SVMs are versatile techniques that easily handle classification and regression tasks.

Limitations

- Computational Complexity: For large data sets, computation is very intensive in the SVM technique.
- Parameter Selection: Choosing the right kernel and tuning parameters is a challenging task that requires lots of experimentation.

2.2.3 Random Forests

Random forests based on the ensemble learning method are highly effective for traffic prediction. This method easily handles complex and large datasets. It can provide accurate traffic predictions by averaging multiple decision trees. During training, Random Forests build several decision trees and produce the mean forecast for each tree or the mode of the classes [17]. This technique increases accuracy and robustness by utilizing the power of numerous models.

The Following Steps are used by Random Forests for prediction of the Traffic flow.

- Data Collection: Collect the comprehensive historical traffic data, including features such as traffic volume, speed, density, weather conditions, and time of day.
- Data Preprocessing: In the processing of data, first analyze the missing values, and then divide them into training and testing sets. After that, normalize the features of the data sets.
- Feature Engineering: Create new features to improve model performance.
- Model Training: By using traffic data, we need to train the Random Forest model. This involves building multiple decision trees with different subsets of data and features.
- Model Validation and Testing: Using cross-validation techniques we have to validate the model and test its performance using a traffic data set.
- Prediction: Use the trained Random Forest model to predict future traffic conditions.

Advantages and Limitations of Random forest method

Advantages

- Accuracy: Random Forests often provide high predictive accuracy by averaging multiple decision trees.
- Robustness: They are robust to over fitting due to the combination of bagging and random feature selection.
- Handling Large Datasets: Can handle large datasets and high-dimensional data effectively.
- Interpretability: Feature importance can be extracted, providing insights into the most influential factors.

Limitations

- Computational Complexity: For the large number of trees, training the model and prediction of traffic computation is very intensive.
- Storage: Requires more memory to store multiple decision trees.
- Black-Box Nature: Despite being more interpretable than some other models, the ensemble nature can still obscure individual decision paths.

2.2.4 Deep Learning

A subset of ML technology is deep learning methodology. It has transformed numerous fields by using multi-layered neural networks to simulate intricate patterns in data. Deep learning techniques are strong tools for traffic prediction because they can capture complex temporal and spatial correlations in traffic data and produce forecasts that are incredibly precise [18].

Neural networks with several layers that have the ability to learn hierarchical data representations are known as deep learning models. Large-scale data sets and intricate patterns are the hallmarks of these models' effectiveness in tasks including picture recognition, natural language processing, and traffic prediction.

Deep learning is applied in the VANET system to identify the complexity level of the vehicle communication system. The advancement of DL in the VANET system helps to identify the exact traffic situation, improve the security of the vehicles, and make a robust autonomous system.

Figure 4 shows the various deep learning architectures used for traffic predictions. DL follows the various steps shown in Fig. 5 to identify the traffic flow.

Advantages and limitations of the deep learning technique

Advantages: The deep learning method has the following advantages.

- Accuracy: Deep learning models can achieve high accuracy by capturing complex patterns in traffic data.
- Flexibility: capable of handling various types of data (e.g., temporal, spatial) and prediction tasks (e.g., speed, volume).
- Scalability: it can be scaled to handle large datasets and complex models.

Fig. 4 Various deep learning architectures

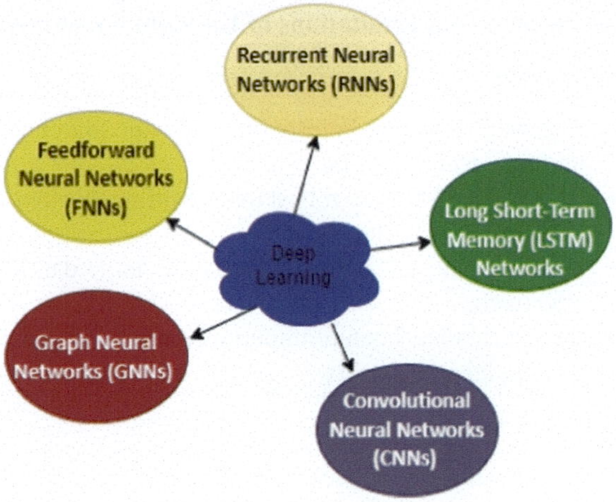

Fig. 5 Various steps used in DL for prediction of traffic flow

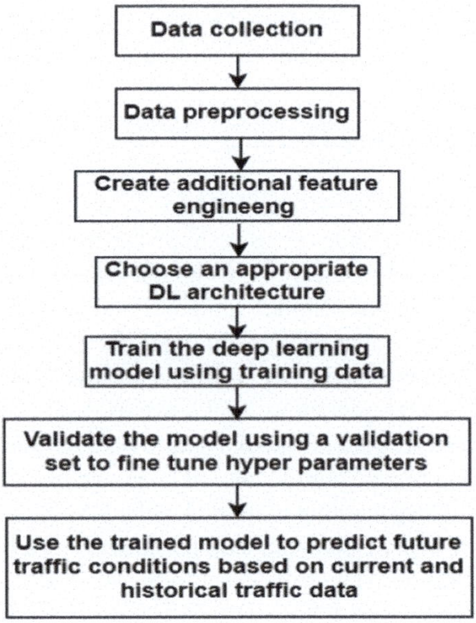

Limitations: The DL technique has the following limitations.

- Data Requirements: For training of the model requires high-quality large data set.
- Computational Cost: It re large hardware for training the model and hence cost increases.
- Complexity**:** Models can be complex and difficult to interpret, often considered "black boxes."

2.3 *Hybrid Model of Traffic Prediction*

Utilizing the advantages of several machine learning and statistical techniques, hybrid models for traffic prediction increase the precision and resilience of traffic forecasting. The following are a few popular hybrid models for traffic forecasting:

2.3.1 ARIMA and Neural Networks

In this approach, traffic prediction is achieved by combining the auto-regressive integrated moving average model with neural network. Neural networks (NN) [19] are used to represent non-linear patterns and complex interactions in the residuals from the ARIMA model after linear temporal relationships in traffic data are captured by ARIMA (Auto regressive integrated moving average) models.

2.3.2 Wavelet Transform and Machine Learning Models

This model finds the optimal traffic flow prediction solution by combining the wavelet transform with machine learning methods. In order to extract pertinent characteristics and separate noise from the traffic time series data, wavelets split the data into distinct frequency components. These features are fed into ML models such as SVM, ANN, XGBoost, etc. to make predictions [20].

2.3.3 Spatiotemporal Models

It is an amalgam of LSTM and graph convolution networks (GCN). Whereas LSTMs manage the temporal dynamics in the traffic network, GCNs are employed to capture spatial dependencies.

2.3.4 Deep Learning with Traditional Methods

This model combines Long Short-Term Memory (LSTM) networks with a Kalman filter. Kalman filters are employed for real-time correction and refining, while long-term dependencies are captured by LSTMs [8]. The following steps required to implement the hybrid model for the prediction of traffic is shown in Fig. 6.

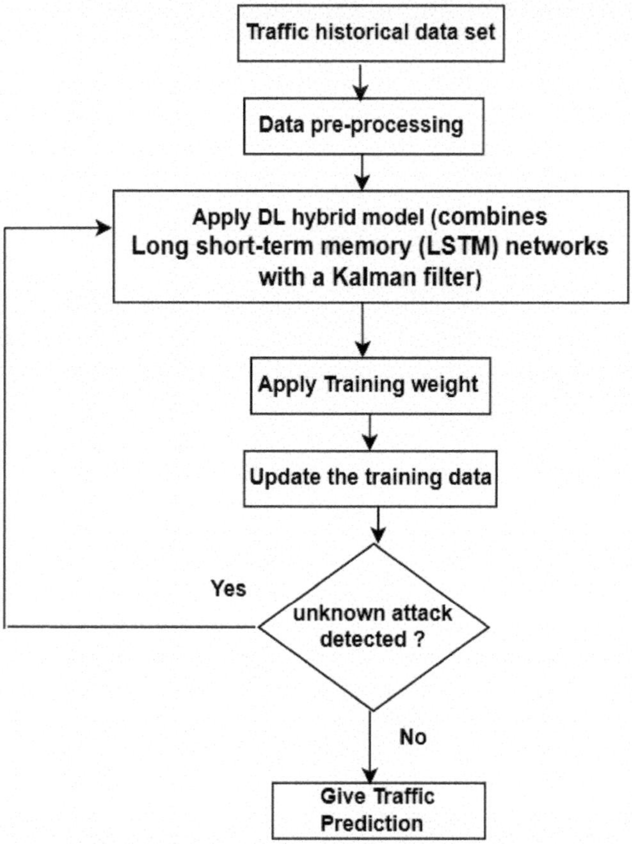

Fig. 6 Implementation of a hybrid model

2.4 Advantages of Hybrid Models

- Enhanced Accuracy: By integrating many models, hybrid techniques can frequently perform better than single models by identifying more intricate patterns in the data.
- Robustness: A greater range of data properties, such as noise, non-linearity, and missing values, can be handled by hybrid models.
- The hybrid model can simulate the data of various traffic flows and gives the results accordingly.

3 Traffic Modeling Techniques

In this section, we discussed the different modeling techniques used for the VANET system. Vehicle data such as vehicle speed, road conditions, and weather conditions are required for the modeling of the simulation model in the VANET network. The following categories are used in the traffic modeling techniques:

3.1 Microscopic Models

This modeling system includes information about the movement of individual vehicles and interacts with the VANET network. Such techniques include the following models:

- Car-Following Models: In this technique, data is collected on driver behaviors, which means how the driver follows the next vehicles and how he adjusts the speed so that he maintains the proper gap between the vehicles.
- Lane-Changing Models: In this technique, data is collected on how cars change their lines on the road. This technique also collects data on car speed, driver behavior, and the density of the population.
- Agent-Based Models: This model collects data on traffic dynamics. In such cases, they consider cars to be autonomous agents.

3.2 Mesoscopic Models

Mesoscopic models group the vehicles into clusters to balance detail and computational efficiency. These models follow the traffic flow characteristics. They cannot be considered the behavior of individual vehicles [21]. These modeling techniques include the following:

- Cellular Automata: These models create cells on the road, each of which can have a car or be vacant, and they update the status according to pre-established guidelines.
- Queueing Models: These models are helpful for studying traffic congestion because they represent traffic as a sequence of queues at crossings and road segments.

3.3 Macroscopic Models

Macroscopic models treat traffic as a continuous fluid and consider the aggregate behavior of traffic flow.

- **Lighthill–Whitham–Richards (LWR) Model**: This model describes traffic flow, capturing phenomena, and congestion using partial differential equations.
- **Gas-Kinetic Models**: These models use equations to represent the speed and density distribution of cars, drawing comparisons between gas dynamics and traffic flow.

4 Challenges in Traffic Prediction and Modeling

This section discusses the various challenges that arise in modeling the VANET system and predicting accurate situations. Many advanced techniques are used in VANET's networking system for traffic prediction, but still, it has so many challenges, as follows:

Data Heterogeneity: The collection of data from various sources like sensors, GPS, and cameras is difficult because of differences in data formats and quality.
Real-Time Processing: For processing data and generating forecasts in real time, traffic prediction models require efficient algorithms and robust computing infrastructure.
Scalability: Traffic models should be flexible so that they can scale the data if vehicle numbers increase and there are more extensive road networks.
Adaptability: We are required to model quickly if traffic patterns change due to accidents, road works, or weather conditions.
Privacy and Security: We have to ensure the privacy and security of vehicular data unless it is critical, as breaches can have severe consequences.

5 Future Directions

This section discusses the different advanced techniques that will be used in the future for modeling the VANET system. In the future, traffic prediction and modeling in Vehicular Ad Hoc Networks will involve the following areas:

- Integration of AI and IoT: AI methods used with IoT devices can improve data gathering, processing, and forecast precision.
- Advanced Deep Learning Models: We need to apply more advanced DL techniques that can handle complex data easily.
- Collaborative Learning: With the collaboration of other techniques, we can train the testing model for better solutions.
- Autonomous Vehicles: we have to model the vehicles on the current data set to find accurate solutions.
- Advanced Machine Learning Techniques**:** This is the time to concentrate on moving towards more sophisticated, advanced techniques that can effectively handle large nonlinear data.

- Data Fusion and Integration: For the best and most accurate prediction of traffic, it is required to collect the data from various sources, like traffic cameras, IoT-based devices, and GPS-enabled vehicles.
- Real-Time and Dynamic Modeling: The traffic conditions are very dynamic and hence affect the accuracy of the prediction of the correct situation in real time. So to obtain better results, it is required to develop an adaptive model that supports the dynamic conditions and gives accurate results in real-time.
- Edge Computing: Roadside infrastructure can be used to implement new edge computing technologies with IoT systems to collect real-time traffic data.

6 Conclusion

Road traffic is increasing day by day. It directly influences the life cycle of people. In the metropolis, it takes 1–2 h to reach the destination due to heavy traffic, even if the distance is not very far. If people get information about the traffic situation, they may be able to start their ride accordingly, follow the low-traffic area, and save time. Many classical methods were used to find the traffic flow conditions, but they were not accurate. Traffic data is very complex because it contains a lot of information. In that case, traditional methods are unable to predict the accurate traffic flow.

The VANET system is one of the ideas that is used to find out the traffic flow in real time. but the VANT system is also facing many problems due to high mobility and dynamic changes in traffic. So VANET systems require specific traffic forecasting and special features. The handling of complex traffic data requires some advanced algorithms. In this paper, we discuss the many advanced techniques, like ML and DL, hybrid methods, etc., that are able to handle complex and nonlinear traffic data and give the best traffic prediction in VANETs. However, such techniques have limitations in that they require large historical data and a long computation system.

References

1. Oussama S, Saad H, Zibouda A (2020) Survey on vehicular ad hoc networks clustering algorithms: overview, taxonomy, challenges, and open research issues. Int J Commun Syst 1–20
2. Rohini G, Gnana Kousalya C, Singh N, Ratansing PV (2023) Autonomous forecasting of traffic in cellular networks based on long-short term memory recurrent neural network. Cybern Syst 1–13
3. Singh HP, Singh N, Mishra A, Sen SK, Swarnkar M, Pandey D (2024) Logistic regression based sentiment analysis system: rectify. In: 2024 IEEE international conference on big data & machine learning (ICBDML), Bhopal, India, pp 186–191. https://doi.org/10.1109/ICBDML60909.2024.10577296
4. Xia Z (2021) A comprehensive survey of the key technologies and challenges surrounding vehicular ad hoc networks. ACM Trans Intell Syst Technol 12(4):1–37

5. Singh N, Sharma AK, Sarkar I, Prabhu S, Chadaga K (2024) IoT-based greenhouse technologies for enhanced crop production: a comprehensive study of monitoring, control, and communication techniques. Syst Sci Control Eng 12(1)
6. Leandro N (2018) Balico: localization prediction in vehicular ad hoc networks. IEEE Commun Surv Tutorials 20(4):2784–2803
7. Veena RS, Khambra G, Sayal R, Singh N et al (2024) An energy efficient ipv6 packet delivery outline for industrial IoT OVER G.9959 protocol based wireless sensor network (WSN). SN Comput Sci 5:813. https://doi.org/10.1007/s42979-024-03195-4
8. Singh N, Chandel P, Shukla S, Soni A, Sen D, Khan S (2024) Design of microstrip antenna for multipurpose wireless communication, pp 12–18. eBook ISBN9781003422440. https://doi.org/10.1201/9781003422440
9. Avleen KM, Batra S, Husanbir SP (2019) Security of vehicular ad-hoc networks: a comprehensive survey. Comput Sec 1–14
10. Singh N, Singh HP, Mishra A, Khare A, Swarnkar M, Almas SK (2024) Blockchain cloud computing: comparative study on DDoS, MITM and SQL injection attack. In: 2024 IEEE international conference on big data & machine learning (ICBDML), Bhopal, India, pp 73–78. https://doi.org/10.1109/ICBDML60909.2024.10577412
11. Sondhiya V, Dwivedi K, Kulsum Almas S, Singh N (2023) Energy management of nanogrid through flair of deep learning from IoT environments, artificial intelligence techniques in power systems operations and analysis. Taylor& Francis. ISBN. 9781003301820
12. Guo J, Zhang Y, Chen X et al (2018) Spatial stochastic vehicle traffic modeling for VANETs. IEEE Trans Intell Transp Syst 19:416–425
13. Rohini G, Gnana Kousalya C, Singh N, Ratansing PV (2023) Autonomous forecasting of traffic in cellular networks based on long-short term memory recurrent neural network. Cybern Syst 1–13
14. Bhatia J, Dave R, Bhayani H, Tanwar S, Nayyar A (2020) SDN-based real-time urban traffic analysis in VANET environment. Comput Commun 149:162–175
15. Banoth R (2023) Stochastic modeling and performance analysis in balancing load and traffic for vehicular ad hoc networks: a review. Int J Netw Manage 35(3):234–247
16. Singh N, Chakrabarti T, Chakrabarti P, Panchenko V, Budnikov D, Yudaev I, Bolshev V (2023) Analysis of heuristic optimization technique solutions for combined heat-power economic load dispatch. Appl Sci 13(18):10380. https://doi.org/10.3390/app131810380
17. Choubey H, Arya V, Singh J, Choudhary N, Sharma A, Singh N (2023) Efficient model based on deep learning for the classification of dementia. Int J Recent Innov Trends Comput Commun 11(10):1056–1061
18. Rohini G, Gnana Kousalya C, Singh N, Patil VR (2023) Autonomous forecasting of traffic in cellular networks based on long-short term memory recurrent neural network. Cybern Syst 1–13. https://doi.org/10.1080/01969722.2023.2166245
19. Kavitha AK (2023) Deep learning model for traffic flow prediction in wireless network. Automatika 64(4):848–857
20. Abir M (2021) Survey on artificial intelligence (AI) techniques for vehicular ad-hoc networks (VANETs), pp 1–16
21. Qimei CUI (2022) Vehicular mobility patterns and their applications to internet-of-vehicles: a comprehensive survey. Science China Inf Sci 65:1–42

Intelligent Data Dissemination in Vehicular Networks: Leveraging Reinforcement Learning

Jitendra Bhatia, Maanit Shah, Rushi Prajapati, Khush Shah, Premal Shah, Harshal Trivedi, and Dhaval Joshi

Abstract Data dissemination in Vehicular Ad Hoc Networks (VANETs) is vital for the development and operation of intelligent transportation systems, as it enables the rapid and reliable exchange of critical information among vehicles and infrastructure. However, the dynamic nature of VANETs, characterised by high node mobility and frequently changing network topologies, poses significant challenges for conventional routing protocols. The major challenges of traditional routing protocols struggle with scalability, Quality of Service (QoS), and efficient data dissemination. Machine learning (ML) based traditional routing algorithms that typically rely on predefined datasets for training and can struggle to adapt to the dynamic and unpredictable nature of VANET environments. In contrast, reinforcement learning (RL) excels by learning from interactions with the environment in real-time. RL-based routing algorithms can adaptively optimize routing decisions based on the constantly changing network conditions, such as vehicle density, mobility patterns, and communication link quality. This chapter explores the potential of Reinforcement Learning (RL) to address these challenges by enabling adaptive routing protocols that dynamically adjust to network conditions. We provide a comprehensive overview of the fundamentals of RL and examine how these concepts can be applied to develop RL-based routing strategies in VANETs. Through detailed analysis and discussion, the

J. Bhatia (✉)
School of Technology, Nirma University, Ahmedabad, India
e-mail: jitendra.bhatia@nirmauni.ac.in

M. Shah
School of Arts and Sciences, Ahmedabad University, Ahmedabad, India
e-mail: maanit.s@ahduni.edu.in

R. Prajapati · H. Trivedi · D. Joshi
Sahana Systems Ltd., Ahmedabad, India
e-mail: rushi.prajapati@sahanasystem.com

K. Shah
Indian Institute of Technology, Gandhinagar, Palaj, India
e-mail: khush.shah@iitgn.ac.in

P. Shah
Adani Institute of Infrastructure Engineering, Adalaj, India

J. Bhatia et al. (eds.), *Deep Learning Based Solutions for Vehicular Adhoc Networks*, Studies in Computational Intelligence 1207,
https://doi.org/10.1007/978-981-96-5190-0_9

chapter demonstrates the ability of RL to enhance the scalability, QoS, and overall performance of data dissemination in VANETs, paving the way for more robust and efficient vehicular communications in future ITS deployments.

Keywords Vehicular ad hoc networks · Quality of service · Reinforcement learning · Machine learning · Intelligent transport system · Software defined networking · Deep learning

1 Introduction

VANET architecture comprises several key components that work in conjunction to establish a reliable communication network. At the core of this system are the vehicles, each equipped with On-Board Units (OBUs) containing necessary hardware and software to enable vehicle-to-vehicle (V2V) and vehicle-to-infrastructure (V2I) communications [1]. OBUs often include GPS receivers, a variety of sensors and short-range wireless interfaces which allow vehicles to exchange data on position, speed and surrounding environment [2]. Figure 1 illustrates a simple VANET configuration with V2V and V2I communication. In addition to the OBUs, fixed infrastructure elements called Roadside Units (RSUs) expand the network coverage and provide supplementary services [3]. To achieve this, RSUs are positioned strategically along roadways. RSUs work as a gateway in order to facilitate the exchange of data between vehicles and the broader network infrastructure [1]. RSUs also enable localised services such as traffic signal control, parking information, and internet access.

Various channels are used for communication in VANETs. V2V communication enables direct information sharing between neighbouring vehicles, which is crucial to applications like cooperative driving and collision avoidance [4]. On the other hand, V2I communications enable vehicles to interact with stationary infrastructure, facilitating services such as electronic toll collection and traffic signal prioritisation [5]. Additionally, vehicle-to-everything (V2X) communication extends the

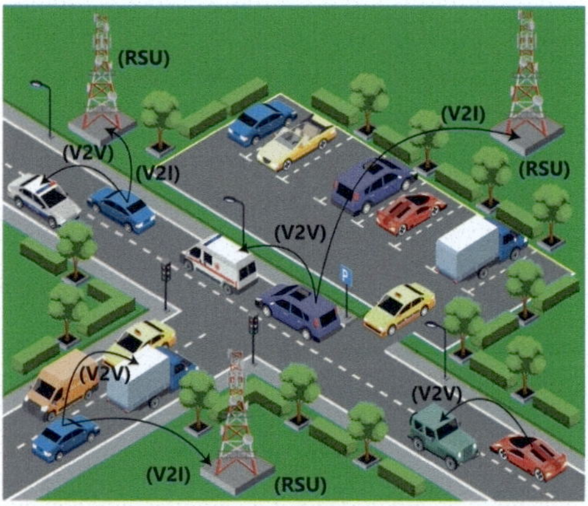

Fig. 1 A simple VANET configuration

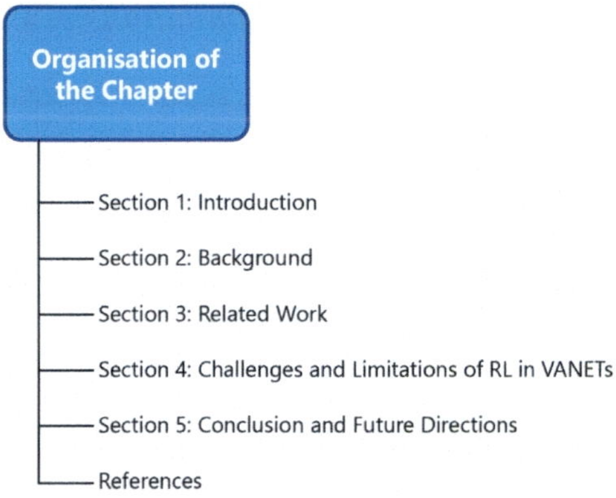

Fig. 2 Organisation of the chapter

network's capabilities to interact with pedestrians and other elements of the larger transportation ecosystem [6]. The organisation of the chapter is shown in Fig. 2.

VANETs play a vital role in modern Intelligent Transportation Systems (ITS), which facilitate V2V and V2I communication with the aid of multi-hop for end-to-end data delivery [7]. The primary aim of VANETs is to improve road safety and traffic efficiency, with applications in real-time navigation and infotainment delivery as well. However, these networks tend to be highly dynamic due to the mobility of vehicles [8]. Hence, this poses significant challenges towards achieving consistent and efficient data dissemination [9]. In these dynamic environments, routing protocols that employ reinforcement learning have been proposed as a potential solution to overcome these challenges and achieve stable data dissemination compared to other ML-based traditional algorithms.

1.1 Importance of Data Dissemination

Data dissemination in VANETs is paramount to the effective functioning of various applications aimed at enhancing road safety, traffic efficiency, and the overall driving experience [10]. By facilitating the real-time exchange of critical information—such as accident alerts, traffic conditions, and emergency services—data dissemination ensures that vehicles can react promptly to potential hazards, thereby preventing accidents and improving safety. Additionally, it supports dynamic traffic management by optimising route planning and reducing congestion, while also enhancing driver convenience through the distribution of up-to-date navigation and service information. As VANETs scale, efficient data dissemination is essential to managing network load, reducing redundancy, and ensuring secure, reliable communication [10]. Ultimately, robust data dissemination is the backbone of VANETs, enabling them to fulfil their potential in transforming road transport systems into safer, smarter, and more responsive networks.

1.2 Challenges and Limitations of Conventional Routing Protocols

VANETs tend to be highly dynamic due to the mobility of vehicles, which leads to frequent changes in the network topology [8]. This presents significant challenges when it comes to achieving and maintaining stable communication and effective routing [9]. These challenges may be partially attributed to the inability of conventional routing protocols to adapt to frequent topological changes, as they are typically designed for more static or less dynamic networks. As a result, the overall performance of VANETs can be compromised. These challenges can be further categorised into four primary areas (as shown in Fig. 3).

1. **Scalability**: Conventional routing protocols struggle with scalability in VANETs due to the highly dynamic nature of the network [11]. The number of vehicles in a given area can vary significantly, leading to challenges in maintaining consistent communication.
2. **Quality of Service**: Maintaining quality of service (QoS) is a significant challenge as well. This is because conventional protocols often find it difficult to deliver consistent QoS due to unpredictable vehicle mobility and varying network density in VANETs. Increased rate of route disruptions caused by high mobility result in frequent packet loss and delay that eventually degrades overall network performance [11].
3. **Limited Bandwidth Availability**: The exchange of control packets in a VANET to establish and maintain routes eats up a significant portion of the available bandwidth, which reduces the bandwidth available for actual data transmission, and this, in turn, causes conventional routing protocols to perform poorly as their routing overheads increase [12].
4. **Efficient Broadcasting**: Efficient broadcasting is essential for disseminating information in VANETs. However, conventional protocols often face challenges in managing broadcast storms, where the simultaneous transmission of messages by multiple nodes leads to network congestion and packet collisions [13]. This issue is exacerbated by the high mobility and dynamic topology of VANETs, making it difficult to ensure reliable message delivery [12].

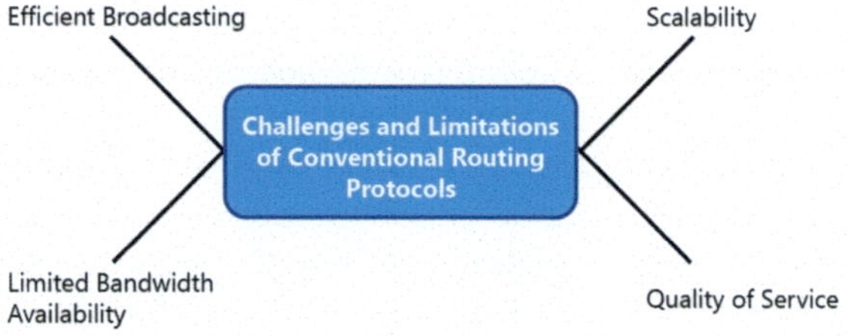

Fig. 3 Challenges and limitations of conventional routing protocols

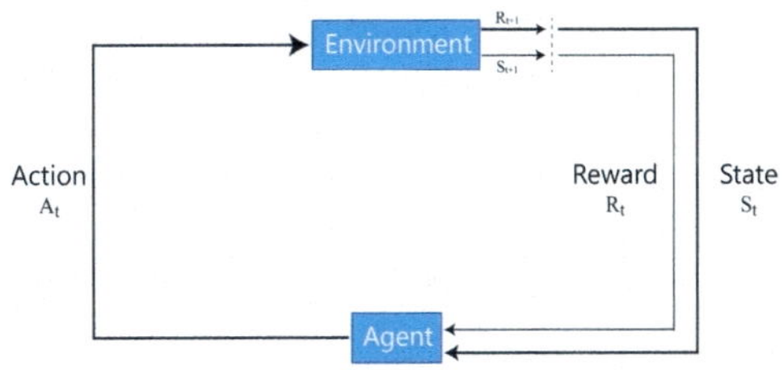

Fig. 4 The agent-environment interaction [15]

To overcome these challenges, emerging routing approaches leverage advanced techniques to adapt to the dynamic and complex nature of VANETs. By integrating intelligent mechanisms that can dynamically adjust to changing network conditions with the aim of enhancing scalability, QoS, bandwidth efficiency, and overall network performance [14].

2 Background

A promising technique to overcome the challenges and limitations that the dynamic nature of VANETs brings is the application of machine learning, more specifically, reinforcement learning. Before delving into specific RL-based VANET routing protocols, it's essential to understand the fundamentals of reinforcement learning and why it may provide a solution to many of the challenges and limitations of VANETs.

2.1 Fundamentals of Reinforcement Learning

Reinforcement learning combines the ideas of machine learning and optimal control to model intelligent agents. In reinforcement learning, a learner called an agent takes some action in its surroundings called the environment. According to the agent's action, the environment updates its state and rewards the agent [15].

The environment refers to the aspects of the problem that the agent cannot directly control or manipulate. In contrast, the actions that the agent takes are the decisions to maximise some reward signal [15]. These actions are part of the agent's strategy to interact with the environment in a manner that leads to the highest cumulative reward. The state of the environment includes all relevant information that could assist the agent in choosing its actions. Importantly, it is assumed that the agent has partial, if not complete, knowledge of the environment. For example, the agent may have some knowledge about how its actions and the resulting states lead to the calculation of rewards. However, even though the agent understands this reward function, it is still considered part of the environment because the agent cannot arbitrarily change it. The agent observes the state of the environment, then takes actions guided by a specific policy, and obtains rewards or penalties accordingly. This ongoing cycle of interaction, learning from the consequences of actions, and adjusting future behaviour is central to reinforcement learning. An overall view of the interactions between the agent and the environment is shown in Fig. 4. This distinction between what the agent knows and can control is critical in reinforcement learning. An agent may have a thorough understanding of the problem domain, much like a human who knows the rules of solving a Rubik's cube but still struggles actually to find the optimal solution [15]. The agent-environment relationship, therefore, represents the fundamental limitations on the agent's control rather than just the limitations on its knowledge.

A policy, denoted as π, is a strategy used to determine how a learning agent behaves at any moment. It is a probability distribution over all the possible actions in a state given that specific state. Essentially, a policy defines the mapping from the observed states of the environment to the actions that should be taken in those states [15].

$$\pi(a|s) = P\left[A_t = a|S_t = s\right] \tag{1}$$

Having established a foundational understanding of the core components of reinforcement learning, we can now delve deeper into the nature of tasks that an agent may encounter. Here, the structure of tasks plays a major role in defining how agents learn and adapt.

An agent's interactions with its environment can be divided into a number of subsequences referred to as episodes. Every episode ends with a distinct terminal state, following which the system either reverts to a specified starting state or moves to a state drawn from a standard distribution of initial states. These types of tasks are referred to as episodic tasks [15]. However, this is not always the case. In many cases, the interaction between the agent and the environment does not inherently divide into distinct episodes but continues indefinitely without interruption. These are known as continuing tasks [15].

2.1.1 Markov Decision Processes

The Markov Property states, "The future is independent of the past given the present." Ideally, a state signal should compactly summarise past actions while retaining all pertinent information. This usually involves more than just the most recent actions but not more than the entire history of past sensations. A state signal that preserves all pertinent information is considered Markov or is said to exhibit the Markov Property. Mathematically, the Markov Property [15] can be expressed as,

$$P[S_{t+1}|S_t] = P[S_{t+1}|S_1, \dots, S_t] \tag{2}$$

A Markov Process consists of a sequence of random states $S_1, S_2, \dots$ that adhere to the Markov Property. This process can be characterised by the tuple $\langle S, P \rangle$, where

S denotes a finite collection of states, and P represents the matrix of state transition probabilities [15]. For a Markov state s and its subsequent state $\bar{s}$, the probability of transitioning between states is defined by,

$$P_{s\bar{s}} = P[S_{t+1} = \bar{s} | S_t = s] \tag{3}$$

The state transition matrix P defines transition probabilities from all states to all successor states [15].

$$P = \begin{bmatrix} P_{11} & P_{12} & \cdots & P_{1n} \\ P_{21} & P_{22} & \cdots & P_{2n} \\ \vdots & \vdots & \ddots & \vdots \\ P_{n1} & P_{n2} & \cdots & P_{nn} \end{bmatrix} \tag{4}$$

By leveraging the ideas of the Markov Property and Markov Process, we can use a mathematical framework called the Markov Decision Process (MDP) to formalise the interaction between the agent and the environment. A Markov Decision Process can be represented as a tuple $\langle S, A, P, R, \gamma \rangle$ where,

1. S is a finite set of states.
2. A is a finite set of actions.
3. P represents a state transition matrix, where $P^a_{s\bar{s}} = P\left[S_{t+1} = \bar{s} | S_t = s, A_t = a\right]$.
4. R denotes a reward function, defined as $R^a_s = E\left[R_{t+1} | S_t = s, A_t = a\right]$.
5. γ is a discount rate, $\gamma \in [0, 1]$.

Having established a solid understanding of MDPs and their role in formalising reinforcement learning problems, we now turn our attention to value functions.

2.1.2 Value Functions

Reinforcement learning algorithms rely heavily on value functions as they quantify the expected long-term rewards associated with different states and actions. Value functions provide us with the notion of "how goo" it is for an agent to be in a specific state or take a certain action from that state by taking into account the cumulative future rewards [15]. There are two primary types of value functions,

1. **State-Value Function**: The state-value function, denoted as $V_\pi(s)$, represents the expected reward return when the agents start from a state s and follow a policy π [15]. It essentially measures the long-term value of being in a specific state under a given policy.

$$V(s) = E[G_t | S_t = s] = E_\pi \left[\sum_{k=0}^{\infty} \gamma^k R_{t+k+1} \,\middle|\, S_t = s \right] \tag{5}$$

2. **Action-Value Function**: The action-value function, denoted as $Q(s, a)$, represents the expected return when the agent starts from state s and performs the

action a while following a policy π [15]. It provides a more granular view by evaluating the long-term value of choosing a specific action in a given state.

$$Q(s, a) = E_\pi [G_t \mid S_t = s, A_t = a] = E_\pi \left[\sum_{k=0}^{\infty} \gamma^k R_{t+k+1} \middle| S_t = s, A_t = a \right] \quad (6)$$

Value functions serve as the foundation for various RL algorithms. Accurately estimating value functions allows agents to make better decisions about choosing their next action in a way that maximises the cumulative reward. The relationship between the value functions, the agent's policy, and the environment's dynamics within an MDP form the basis for understanding reinforcement learning techniques. The main objective of reinforcement learning is not just evaluating a policy. It is to find the most optimal policy that maximises the cumulative reward. This brings us to the concept of optimal value functions. Optimal value functions represent the maximum expected return achievable from a specific state or state-action pair when following the optimal policy.

1. **Optimal State-Value Function**: The optimal state-value function $V_*(s)$ represents the maximum state-value function achievable across all policies. $V_*(s)$ indicates the maximum reward that can be obtained when starting in state s [15]. However, it doesn't tell us what policy to follow.

$$V_*(s) = \max_\pi V_\pi(s) \quad (7)$$

2. **Optimal Action-Value Function**: The optimal action-value function $Q_*(s, a)$ represents the maximum action-value function across all policies [15].

$$Q_*(s, a) = \max_\pi Q_\pi(s, a) \quad (8)$$

Having established a thorough understanding of value functions, we now turn our attention to practical methods for estimating these functions. One of the key challenges in reinforcement learning is having to update value estimates efficiently as the agent interacts with its environment. This challenge is effectively addressed by temporal difference learning.

2.1.3 Temporal Difference Learning

Temporal Difference Learning is a technique employed by value-based reinforcement learning algorithms to learn value functions iteratively. The algorithm begins by arbitrarily initialising a table $V(S)$, where each state in the MDP is assigned an initial value. A positive learning rate α is selected. A given policy π is constantly evaluated, with a reward r being obtained. Using this reward r, the value function is updated for the current state.

$$V(S_t) \leftarrow (1 - \alpha)V(S_t) + \alpha[R_{t+1} + \gamma V(S_{t+1})] \tag{9}$$

where,

1. S_t is the current state.
2. S_{t+1} is the next state.
3. α is the learning rate, $(0 < \alpha < 1)$.
4. γ is the discount factor.
5. $R_{t+1} + \gamma V(S_{t+1})$ is known as the TD target.
6. $R_{t+1} + \gamma V(S_{t+1}) - V(S_t)$ is the Temporal Difference Error.

Having explored temporal difference learning, we now look at value-based RL algorithms that leverage temporal difference learning to find optimal policies.

2.1.4 Q-Learning

Q-learning is a value-based reinforcement learning algorithm. The algorithm's model-free nature allows it to function in a stochastic environment [16]. In Q-learning, an agent explores its environment, aiming to maximise the cumulative reward as it navigates from its current state to a specific goal state. The algorithm uses a table, referred to as a Q-table, to store the Q-values linked to each state-action pair (as illustrated in Fig. 5). The Q-table is initialised with arbitrary fixed values. At each time step t, the agent chooses an action A_t that allows it to transition to a new state S_{t+1} from its current state S_t while receiving a reward R_{t+1} [15]. The Q-value $Q(S_t, A_t)$ is adjusted in the Q-table at every time step t:

$$Q^{\text{new}}(S_t, A_t) \leftarrow (1 - \alpha)Q(S_t, A_t) + \alpha(R_{t+1} + \gamma \max_a Q(S_{t+1}, a)) \tag{10}$$

where,

1. S_t is the current state.
2. S_{t+1} is the next state.
3. R_{t+1} is the reward received when moving from state S_t to state S_{t+1}.
4. α is the learning rate, $(0 < \alpha < 1)$.
5. γ is the discount factor.
6. $Q(S_t, A_t)$ is the current value.
7. $\max_a Q(S_{t+1}, a))$ is the estimate of optimal future value.
8. $R_{t+1} + \gamma \max_a Q(S_{t+1}, a)$ is the temporal difference target.

When the state and action spaces grow, the size and complexity of Q-tables that store learned values for state-action pairs grow exponentially [12]. Q-learning faces

Fig. 5 Q-learning schematic

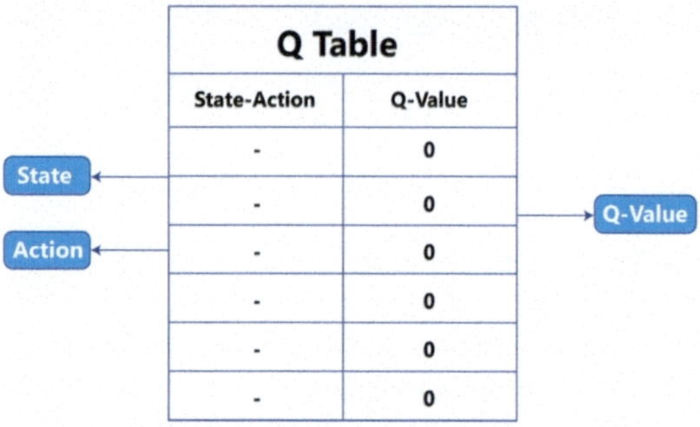

a challenge in such cases as it becomes difficult for the algorithm to maintain a Q-table, which may cause convergence to not be achieved. This is a limitation as most real-world problems involve continuous state and action spaces [16].

2.1.5 Deep Reinforcement Learning

Deep reinforcement learning (DRL) enhances traditional reinforcement learning approaches by integrating deep learning techniques [17]. At the core of DRL is the deep Q network (DQN), a deep neural network that approximates Q-values [18]. DRL processes unprocessed sensory data from the environment using deep learning and produces outputs based on approximations.

A replay buffer stores all the experiences of the agent, with each experience being represented as a tuple (S_t, A_t, R_t, S_{t+1}). A random mini-batch is selected from this replay buffer to train the DQN [19]. The DQN's weights are updated at each iteration. An additional neural network, known as the target network, is used to stabilise the learning process. This network shares the same architecture as the DQN but has fixed parameters. The target network is used to compute the target, as opposed to the DQN, which is used to adjust the network parameters. Figure 6 provides an overview of the basic workings of DRL.

While deep reinforcement learning does overcome the challenge of dealing with continuous state spaces, it still struggles to handle continuous action spaces.

2.1.6 Deep Deterministic Policy Gradient (DDPG)

Deep Deterministic Policy Gradient (DDPG) is an off-policy actor-critic algorithm. Its characteristics allow it to handle continuous action spaces in reinforcement learning problems [21]. DDPG builds upon the deterministic policy gradient (DPG) framework and combines it with an approach similar to that used in DRLs. It incorporates deep neural networks as function approximators. DDPG employs two types of neural network: an actor network and a critic network. The actor network determines the

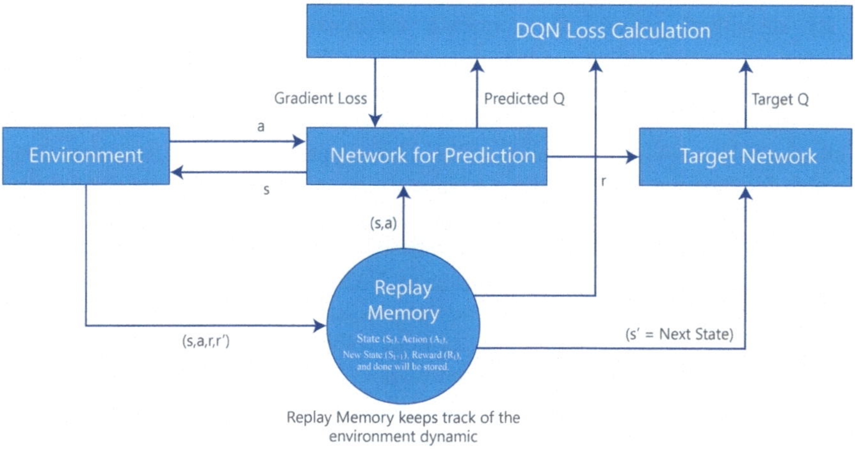

Fig. 6 Overall view of DRL [20]

Fig. 7 Benefits of using RL in VANETs

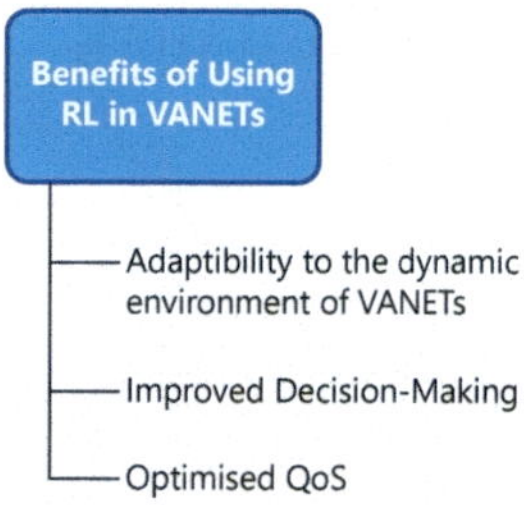

optimal action to take in a given state to maximize cumulative reward. On the other hand, the critic network estimates the Q-value for each state-action pair.

Like DQN, DDPG uses a replay buffer to store experiences (state, action, reward, next state) randomly sampled in mini-batches during training [21]. This helps break correlations in the training data, leading to improved learning stability. Additionally, DDPG utilises target networks for both actors and critics. These target networks are slowly updated versions of the main networks that provide stable learning targets, further enhancing the algorithm's stability.

2.2 Need of Reinforcement Learning in VANETs

Reinforcement learning has emerged as a promising approach to enhance routing protocols in VANETs. As discussed, conventional routing methods often face challenges due to the dynamic nature of VANETs, as high vehicle mobility and dynamic link connectivity make efficient routing difficult [22, 23]. Some benefits of using RL in VANETs have been highlighted in Fig. 7.

RL algorithms can effectively adapt to the dynamic environment of VANETs by learning optimal strategies for various tasks, such as routing, spectrum management, and resource allocation [24]. By continually interacting with their environment, RL agents improve their decision-making processes, resulting in more efficient and robust VANET operations [25]. This adaptability makes RL particularly suitable for handling the complex and constantly changing environment of vehicular networks [24]. RL-based algorithms offer adaptive solutions by continuously learning and optimizing routing decisions based on the real-time environment [12]. RL algorithms improve the performance of VANET routing by optimizing various QoS parameters such as end-to-end delay (EED), packet delivery ratio (PDR), throughput, routing overhead, number of hops (NoH), and security [26, 27]. In the following section, we will dive into the diverse landscape of RL-based data dissemination approaches for VANETS. We will explore how different reinforcement learning techniques are applied to address the unique challenges of VANETs.

3 Related Work

This section explores various RL-based adaptive routing approaches for VANETs that effectively tackle the challenges and limitations brought about by the dynamic nature of VANETs by enabling vehicles to learn optimal routing strategies based on environmental feedback. We look at various RL-based routing solutions, each employing distinct algorithms to enhance routing performance. A taxonomy of the RL-based VANET routing techniques covered in this section is shown in Fig. 8.

3.1 QGrid

QGrid is a greedy algorithm based on q-learning, proposed by Li et al. [28]. The geographical area is divided into grid regions in order to conceptualise the learning environment. The q-values of these grids are calculated based on historical data on vehicle movement patterns. The next hop that is geographically closest to the final destination is selected greedily by vehicles. A second-order Markov Chain is used to decide the next hop to forward data from the sender vehicle. The hop with the highest probability of entering the next-best zone is chosen. Figure 9 shows an example of QGrid.

3.2 Adaptive Routing Protocol Based on RL (ARPRL)

Wu et al. [29] propose a routing protocol that uses hello packets to maintain the optimal path. ARPRL maintains a routing table (or Q-table) and a neighbour table. It

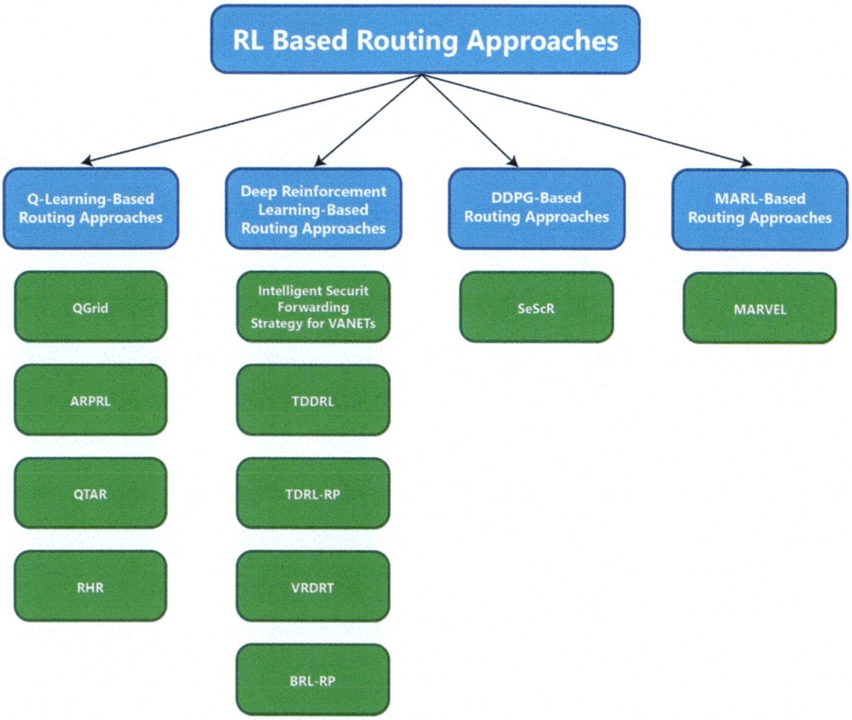

Fig. 8 Taxonomy of RL-based routing approaches

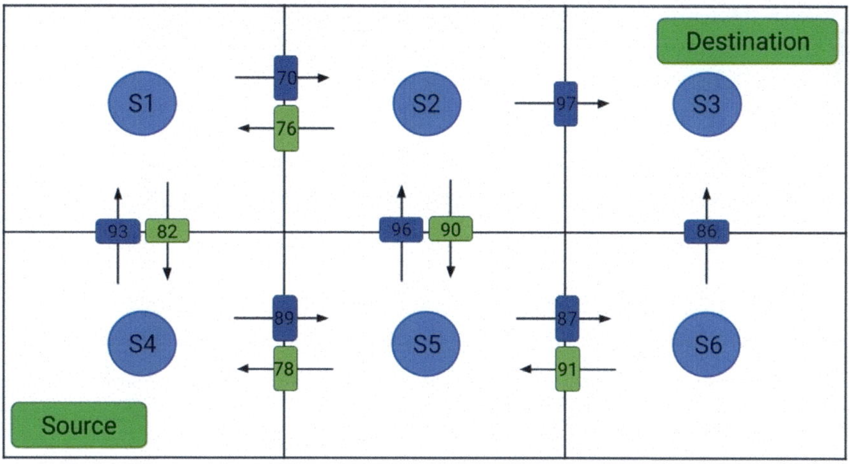

Fig. 9 A QGrid routing protocol example. It displays a destination vehicle in grid region S3 and a source vehicle in grid region S4 [12]. The q-values for entering or leaving each grid are represented by the arrows and matching values inside the rectangular boxes

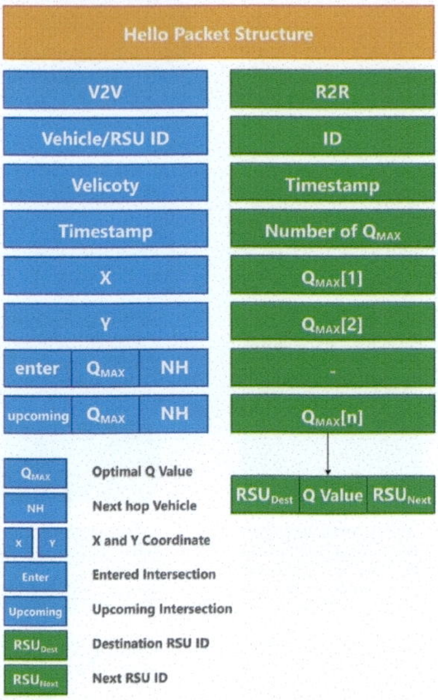

Fig. 10 V2V and R2R hello packet structure used in QTAR [12]

uses a hello timer to detect the arrival or exit of neighbour nodes. Hello packets include the vehicle's speed, direction, position and Q-table data. Vehicles first check their Q-table for the best next hop to make routing decisions. If none are found, they initiate a route probe request. ARPRL uses proactive learning for faster convergence and reduces route looping with modified hello packets that include position, timestamp, and Q-values. Information from the Q-table is broadcasted to neighbouring vehicles with every hello packet interval to keep Q-tables updated.

3.3 Q-Learning-Based Traffic-Aware Routing Protocol (QTAR)

Q-learning-based traffic-aware routing protocol (QTAR) proposed by Wu et al. [30] to enhance data transmission in urban vehicular networks. QTAR combines the principles of geographic routing with q-learning in order to optimise RSU-to-RSU (R2R) as well as vehicle-to-vehicle (V2V) communications. In V2V routing, packets act as agents, and vehicles represent states. In contrast, in R2R routing, hello packets act as agents, and the neighbouring RSUs represent states. The hello packet structures used in V2V and R2R are shown in Fig. 10. The primary goal of QTAR is to ensure high connection reliability and minimise EED.

3.4 RL-Based Hybrid Routing Algorithm (RHR)

Ji et al. [31] propose a hybrid routing system for real-time path exploration that can identify blind paths that may represent a risk of link disconnection. For dynamic VANET systems, the routing table is brought up to date to minimise sporadic link disconnections. Instead of focusing on a single path, RHR explores different possibilities. RHR uses a packet-carry-on feedback system to reward and penalise various paths.

3.5 Deep Reinforcement Learning-Based Intelligent Security Forwarding Strategy for VANETs

Liu et al. [32] propose a DRL-based adaptive approach to routing. The state space encompasses node attributes like delay, distance, maliciousness, and message characteristics, and the action space represents potential forwarding nodes. The protocol assumes that the control layer serves as the learning agent in an SDN environment within the network infrastructure. It uses a DNN (deep neural network) to learn optimal paths from source to end goal (i.e. destination). The trust value of vehicles is calculated based on factors like forwarding capability and data from control packets. The state transition probability for DRL considers the combination of the forwarding ratio of vehicles and locations in this approach. The reward function evaluates forwarding decisions based on various factors, using different equations for malicious and non-malicious nodes. By updating network parameters based on accumulated rewards, the learning process iteratively enhances the forwarding strategy. This approach aims to optimise routing performance in a dynamic VANET environment by considering various factors such as message urgency, reliability, and security concerns.

3.6 Trust-Based Deep RL-Based VANET Routing Protocol (TDDRL)

Zhang et al. [33] propose a routing algorithm designed for software-defined networking (SDN) in VANETs that employs deep reinforcement learning to find optimal routing paths within a centralised SDN architecture. In order to provide security, TDDRL uses trust scores to choose next-hop nodes. The protocol operates under the assumption that the network framework is an SDN setup, where the control layer serves as the learning agent [33, 34]. It applies DNN to gain an understanding of the optimal route to the destination node. Factors such as data from control packets and forwarding capability are used to calculate the trust value of vehicles. State transition probability in DRL formulation takes both location and forward ratio into account.

3.7 *Software-Defined Trust-Based Deep Reinforcement Learning Routing Protocol (TDRL-RP)*

Zhang et al. [35] propose a secure routing protocol for VANETs that leverages DRL. It uses an SDN controller with CNN to determine the best path that should be adopted. This approach has a trust model that takes into account neighbour vehicles' trust value, speed and direction, among other factors, before making the routing decisions. In TDRL-RP, network infrastructure forms the environment, while SDN controllers are responsible for acting as agents in the DRL mechanism. This protocol uses the delivery ratio in decision-making and the vehicle's current position. As a result, two different matrices are formed for vehicle position and forwarding rate so that the controller can choose the appropriate node for data transmission. The trust value of vehicles is updated periodically, allowing for dynamic changes in trust status. Control packets are used to determine trust values, calculated using a weighted equation considering direct trust and trust with data packets.

3.8 *Vehicle Routing with Deep Reinforcement Learning Technique (VRDRT)*

Saravanan et al. [36] propose a routing algorithm that uses DRL to predict vehicle movements in a road segment. The purpose of the algorithm is to replace the SCF (store carry forward) method by estimating the most congested road segment. Its aim is to gather and keep vehicle information about roads while using DRL to predict traffic conditions. VRDRT segments roads into a multitude of clusters based on vehicle density and utilises a transmission matrix to examine the finest available route. The VRDRT algorithm operates in two phases: the RSP (route selection phase) and the REP (route establishment phase). In the REP phase, vehicles broadcast hello messages containing their current status, including density and distance-related information, to neighbouring vehicles. RSP selects the optimal next-hop neighbour vehicles based on the decisions of the DRL agent.

3.9 *Blockchain and RL-Based VANET Routing Protocol (BRL-RP)*

Zhang et al. [37] propose a secured routing protocol for VANETs that utilises blockchain technology to implement security features and addresses the security threats VANETs face in roadside regions with limited infrastructure. The protocol employs an optimisation model based on the MDP. It uses duelling Double Deep Q-Learning to solve optimisation problems related to trust features, computational capability, and consensus nodes. The BRL-RP protocol computes trust among vehicles based on their previous interactions and data-sending behaviour. It uses a data packet structure that includes information such as neighbours' ID, vehicle position, velocity, throughput, trust value, and sequence number. Trust is calculated using a direct trust equation that considers the ratio of correctly forwarded packets to total sent packets within a time window. The trust values range from 0 to 1, with 0 indicating a malicious node and 1 representing a fully trusted node.

3.10 SDN-Enabled Spectral Clustering-Based Optimized Routing Using Deep Learning (SeScR)

Nahar and Das [38] propose a spectral clustering-based routing scheme to optimise routing in the dynamic environment of vehicular networks.

The core of SeScR's routing approach lies in its use of the DDPG technique. DDPG is particularly effective for routing decisions in large-scale networks with extensive/continuous action spaces, which is characteristic of VANETs. This technique helps determine whether a vehicle should act as a relay node in the data transmission process. The actor-network suggests actions, and the critic network evaluates them. Vehicles act as agents, using beacon message exchanges to learn and decide on packet-forwarding actions. This iterative process, which optimises routing efficiency in dynamic vehicular environments, continues to run until packets reach their destinations.

3.11 Multi-agent Reinforcement Learning for VANET Delay Minimization (MARVEL)

Lu et al. [39] propose a novel approach to addressing the routing challenges in VANETs using Multi-Agent Reinforcement Learning (MARL). MARVEL models the routing problem as a multi-agent scenario. Each individual vehicle becomes an agent that trains itself to make optimal routing decisions in a decentralised fashion, using factors such as overall latency, vehicle position, speed, direction, and local channel information to make optimal next-hop choices. The reward system is designed to enforce successful packet delivery and penalise transmission failures; this makes sure that overall network latency is minimised effectively during learning. Various RL-based VANET routing approaches discussed within this section indicate the flexibility and potential of reinforcement learning-based data dissemination in VANETs.

For a comprehensive overview of these algorithms, refer to Table 1, which contains a summary of the RL-based VANET routing protocols discussed in this section.

Table 1 Comparison of RL-based routing protocols for VANETs

Algorithm	RL approach	Key features	Limitations	Performance metrics	Specific benefits
QGrid [28]	Q-Learning	Grid-based Q-table for state-action values, link stability	Complexity increases with network size	End-to-end delay, connection reliability	Improved route stability in grid-based networks
ARPRL [29]	Q-Learning	Adapts to real-time network conditions, considers link quality	High computational overhead due to constant adaptation	End-to-end delay, connection reliability	Fast adaptation to network changes
QTAR [30]	Q-Learning	Traffic awareness, minimizes delay, maximizes delivery	Suboptimal paths due to traffic changes	Delay, reliability, quick convergence mechanism	Enhanced real-time traffic management
RHR [31]	Q-Learning	Combines Q-Learning with other metrics	High complexity in metric combination	End-to-end delay, connection reliability	Balanced performance improvement
Deep RL-based Intelligent Security Forwarding [32]	DRL	Uses deep neural networks for forwarding, focuses on secure dissemination	Vulnerable to adversarial attacks	Security feature	Robust data transmission
TDDRL [33]	DRL	Trust evaluation, improves delivery	Slow trust evaluation	End-to-end delay, connection reliability	Improved reliability through trust management
TDRL-RP [35]	DRL	SDN for centralized control, RL with trust	High SDN and trust management overhead	End-to-end delay, connection reliability	Efficient centralized routing
VRDRT [36]	DRL	Optimizes vehicle routing, minimizes time/fuel	Requires extensive data	Vehicle direction, offline learning	Cost-effective route optimization
BRL-RP [37]	DRL	Blockchain with deep RL	High blockchain overhead	Security feature	Enhanced data security
SeScR [38]	DDPG	DDPG with spectral clustering	High computational needs	Energy efficiency, performance	High efficiency in large-scale networks
MARVEL [39]	MARL	Multi-agent approach to minimise delay, considers vehicle density and link quality	Complex agent coordination	Latency minimisation, link quality	Effective delay reduction in multi-agent scenarios

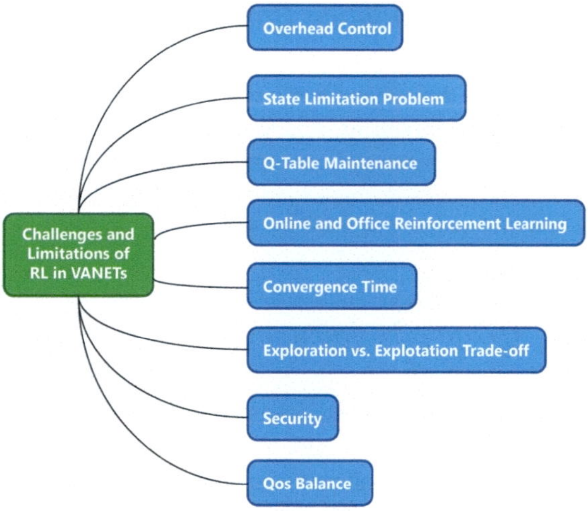

Fig. 11 Challenges and Limitations of RL in VANETs

4 Challenges and Limitations of RL in VANETs

Even though RL algorithms enhance VANET routing protocols, they still present numerous challenges and limitations [12]. Understanding these obstacles is crucial for developing more robust and effective RL-based solutions in VANETs. Figure 11 provides an overview of key challenges and limitations. Each of these factors introduces unique complexities that must be addressed in order to leverage RL in the dynamic environment of VANETs.

4.1 Overhead Control

Routing overhead is the additional load that routing systems place on wireless networks in order to establish and maintain routes. This load is caused by the protocols exchanging relatively smaller control packets [40]. RL-based routing protocols in VANETs face difficulties related to network overhead due to the limited bandwidth available in vehicular networks. This overhead restricts the channel's capacity for transmitting data packets. This results in suboptimal routing performance as the extra burden created by control packets can congest the network [41]. Even though these control packets are typically small in size, their cumulative effect can become significant, especially in dynamic networks. As nodes continuously move or join and leave the network, the frequency and volume of these control packets increase. This increase can consume a significant portion of the network's bandwidth, leading to inefficiencies and reduced overall performance, as the network spends more time and resources maintaining routes rather than transmitting actual data.

4.2 State Limitation Problem

The size of the state space increases as the number of vehicles increases. Consequently, the amount of time spent exploring to find the optimal states increases as well. The algorithms discussed in this chapter use a fixed threshold to reduce the number of states through random sampling. This may cause convergence at a local optimum. As a result, the best neighbouring states might stay hidden [42].

4.3 Q-Table Maintenance

As the state space expands, the size and complexity of Q-tables that store learned values for state-action pairs also grow [12]. This increases memory usage and processing times for updating and accessing the table. In a dynamic VANET environment, such delays can lead to outdated routing decisions that minimise the effectiveness of the RL approach.

4.4 Online and Offline Reinforcement Learning

Offline reinforcement learning involves learning from a fixed dataset without any additional interactions with the environment. This method is useful when online data collection is costly or impractical [43]. Conversely, in online reinforcement learning, real-time interactions with the environment are used to facilitate continuous learning and policy improvement. Striking a balance between online and offline learning poses a significant challenge in VANET routing.

4.5 Convergence Time

The number of iterations an algorithm needs to discover an optimal solution is called the convergence time [44]. The time these algorithms take to converge to an optimal solution can be substantial. In case of very slow convergence rates in the rapidly changing VANET environment, routing decisions could become outdated or suboptimal, leading to network performance degradation.

4.6 Exploration Versus Exploitation Trade-Off

For a reinforcement learning agent to maximise its rewards, it needs to balance two key strategies. While the agent should favour actions that have proven successful in the past, capitalising on its existing knowledge to secure rewards, it must also venture into uncharted territory by attempting new actions, as this exploration is crucial for identifying potentially better choices in future scenarios [15]. This creates a fundamental tension between leveraging known effective action to obtain reward (exploitation) and experimentation in order to make better action selections in the future (exploration). Striking the right balance between these two approaches is essential for the agent's long-term success in optimising its reward acquisition. In the context of VANETs, the highly dynamic nature of the environment can invalidate the calculation of the previous-hop result [45].

4.7 Security

The key limitation in terms of security is the vulnerability of RL systems to malicious attacks. RL algorithms rely on information from neighbouring vehicles, which can be falsified or manipulated [46]. This makes the routing system susceptible to various attacks, including message spoofing and impersonation. Ensuring the integrity and authenticity of this information without introducing significant overhead is a major challenge.

4.8 QoS Balance

Optimising multiple, often conflicting, QoS parameters simultaneously is complex [47]. RL-based algorithms need to take into account factors such as throughput, delay, and priority. However, achieving overall optimal performance involves a trade-off between these factors. Striking a balance between these factors to attain the overall optimal performance poses a complex optimisation challenge that current RL-based methods find challenging to address efficiently.

5 Conclusion and Future Directions

Major Research is deeply being focused on enhancing the effectiveness of routing algorithms in VANETs. This chapter explored the various state-of-the-art schemes for routing protocol that includes ML/DL and RL based approaches. RL-based methods stand out as they tend to enhance system performance through time and gaining experience. This contrasts with other artificial intelligence methods, which are constrained by their initial input data. RL's ability to continuously adapt and refine its strategies sets it apart in the field. This chapter examines VANET routing protocols that leverage RL techniques, exploring how this adaptive approach can be applied to address the challenges of vehicular networking. The efficiency of RL-based data dissemination in VANETs depends on adaptive, efficient and secure protocols that are capable of efficiently handling the rapidly changing nature of vehicular networks. It is important to have feedback systems that consider link quality, mobility, and network topology. The future direction advocates on enhancing the scalability and robustness of these RL models by offering a promising approach that enables decentralised decision-making among multiple agents [48]. Furthermore, the integration of RL with emerging technologies such as 5G and edge computing is another major direction for future research. 5G networks can significantly boost the performance of RL-based VANETs through faster and more reliable data transmission, as they offer low latency and high bandwidth [24]. Moreover, edge computing can help in reducing decision-making latencies of RL agents by processing data near the source [49]. This way, there is real-time decision-making and better management of network resources, which is essential for the dynamic environment of VANETs.

References

1. Yeferny T, Hamad S (2021) Vehicular ad-hoc networks: architecture, applications, and challenges. *arXiv preprint*arXiv:2101.04539
2. Katiyar A, Singh D, Yadav RS (2020) State-of-the-art approach to clustering protocols in VANET: a survey. Wirel Netw 26:5307–5336. https://doi.org/10.1007/s11276-019-02171-5
3. Xia Z, Wu J, Wu L, Chen Y, Yang J, Yu PS (2021) A comprehensive survey of the key technologies and challenges surrounding vehicular ad hoc networks. ACM Trans Intell Syst Technol (TIST) 12(4):1–30. https://doi.org/10.1145/3436285
4. Yogarayan S, Razak SFA, Azman A, Abdullah MFA (2021) A mini review of peer-to-peer (P2P) for vehicular communication. Indones J Electr Eng Inf (IJEEI) 9(1):185–197
5. He W, Li H, Zhi X, Li X, Zhang J, Hou Q, Li Y (2019) Overview of V2V and V2I wireless communication for cooperative vehicle infrastructure systems. In: 2019 IEEE 4th Advanced Information Technology, Electronic and Automation Control Conference (IAEAC). IEEE, pp 127–134
6. Farsimadan E, Palmieri F, Moradi L, Conte D, Paternoster B (2021) Vehicle-to-everything (V2X) communication scenarios for vehicular ad-hoc networking (VANET): an overview. In: International Conference on Computational Science and Its Applications. Springer International Publishing, Cham, pp 15–30
7. Yao Y, Xiao B, Wu G, Li X, Yu Z, Zhang K, Zhou X (2017) Voiceprint: a novel Sybil attack detection method based on RSSI for VANETs. In: Proceedings of the 47th Annual IEEE/IFIP International Conference on Dependable Systems and Networks (DSN). IEEE, pp 591–602. https://doi.org/10.1109/dsn.2017.10
8. Khatri S, Vachhani H, Shah S, Bhatia J, Chaturvedi M, Tanwar S, Kumar N (2021) Machine learning models and techniques for VANET based traffic management: implementation issues and challenges. Peer-to-Peer Netw Appl 14:1778–1805. https://doi.org/10.1007/s12083-020-00993-4
9. Bugarcic P, Jevtic N, Malnar M (2022) Reinforcement learning-based routing protocols in vehicular and flying ad hoc networks–a literature survey. Promet-Traff Transp 34(6):893–906. https://doi.org/10.7307/ptt.v34i6.4159
10. Bhatia J, Dave J, Bhavsar M, Tanwar S, Kumar N (2021) SDN-enabled adaptive broadcast timer for data dissemination in vehicular ad hoc networks. IEEE Trans Veh Technol 70(8):8134–8147. https://doi.org/10.1109/TVT.2021.3092065
11. Ahamed A, Vakilzadian H (2018) Issues and challenges in VANET routing protocols. In: 2018 IEEE International Conference on Electro/Information Technology (EIT). IEEE, pp 723–728

12. Nazib RA, Moh S (2021) Reinforcement learning-based routing protocols for vehicular ad hoc networks: a comparative survey. IEEE Access 9:27552–27587. https://doi.org/10.1109/ACCESS.2021.3058388
13. Yogarayan S, Razak SFA, Azman A, Abdullah MFA, Ibrahim SZ, Raman KJ (2020) A review of routing protocols for vehicular ad-hoc networks (VANETs). In: 2020 8th International Conference on Information and Communication Technology (ICoICT). IEEE, pp 1–7
14. Nama M, Nath A, Bechra N, Bhatia J, Tanwar S, Chaturvedi M, Sadoun B (2021) Machine learning-based traffic scheduling techniques for intelligent transportation system: opportunities and challenges. Int J Commun Syst 34(9):e4814
15. Sutton RS, Barto AG (2018) Reinforcement learning: an introduction. MIT Press
16. Gaskett C, Wettergreen D, Zelinsky A (1999) Q-learning in continuous state and action spaces. In: Proceedings of the Australasian Joint Conference on Artificial Intelligence. Springer, pp 417–428
17. Luong NC, Hoang DT, Gong S, Niyato D, Wang P, Liang Y-C, Kim DI (2019) Applications of deep reinforcement learning in communications and networking: a survey. IEEE Commun Surv Tutor 21(4):3133–3174
18. Rodoshi RT, Kim T, Choi W (2020) Deep reinforcement learning based dynamic resource allocation in cloud radio access networks. In: Proceedings of the International Conference on Information and Communication Technology Convergence (ICTC). IEEE, pp 618–623
19. Li Y, Hu X, Zhuang Y, Gao Z, Zhang P, El-Sheimy N (2020) Deep reinforcement learning (DRL): another perspective for unsupervised wireless localization. IEEE Internet Things J 7(7):6279–6287
20. Park JH, Farkhodov K, Lee SH, Kwon KR (2022) Deep reinforcement learning-based DQN agent algorithm for visual object tracking in a virtual environmental simulation. Appl Sci 12(7)
21. Illicrap TP, Hunt JJ, Pritzel A, Heess N, Erez T, Tassa Y, et al (2015) Continuous control with deep reinforcement learning. *arXiv preprint*arXiv:1509.02971
22. Eiza MH, Ni Q (2013) An evolving graph-based reliable routing scheme for VANETs. IEEE Trans Veh Technol 62(4):1493–1504. https://doi.org/10.1109/TVT.2013.2244625
23. Mehta K, Malik LG, Bajaj P (2013) VANET: challenges, issues and solutions. In: Proceedings of the International Conference on Emerging Trends in Engineering and Technology (ICETET), pp 78–79
24. Lansky J, Rahmani AM, Hosseinzadeh M (2022) Reinforcement learning-based routing protocols in vehicular ad hoc networks for intelligent transport system (ITS): a survey. Mathematics 10(24):4673. https://doi.org/10.3390/math10244673
25. Zhang R, Wu K (2023) A new dynamic clustering scheme for VANETs driven by deep reinforcement learning. In: Proceedings of Third International Conference on Machine Learning and Computer Application (ICMLCA 2022), vol 12636. https://doi.org/10.1117/12.2675347
26. Chettibi S, Chikhi S (2011) A survey of reinforcement learning based routing protocols for mobile ad-hoc networks. In: Özcan A, Zizka J, Nagamalai D (eds) Recent trends in wireless and mobile networks: CoNeCo WiMo 2011, vol 162, pp 94–102. Communications in Computer and Information Science. Springer. https://doi.org/10.1007/978-3-642-21937-5
27. Mammeri Z (2019) Reinforcement learning based routing in networks: review and classification of approaches. IEEE Access 7:55916–55950. https://doi.org/10.1109/ACCESS.2019.2915224
28. Li F, Song X, Chen H, Li X, Wang Y (2018) Hierarchical routing for vehicular ad hoc networks via reinforcement learning. IEEE Trans Veh Technol 68(2):1852–1865
29. Wu J, Fang M, Li X (2018) Reinforcement learning based mobility adaptive routing for vehicular ad-hoc networks. Wirel Personal Commun 101:2143–2171
30. Wu J, Fang M, Li H, Li X (2020) RSU-assisted traffic-aware routing based on reinforcement learning for urban VANETs. IEEE Access 8:5733–5748
31. Ji X, Xu W, Zhang C, Yun T, Zhang G, Wang X, et al (2019) Keep forwarding path freshest in VANET via applying reinforcement learning. In: 2019 IEEE First International Workshop on Network Meets Intelligent Computations (NMIC). IEEE, pp 13–18
32. Liu B, Xu G, Xu G, Wang C, Zuo P (2023) Deep reinforcement learning-based intelligent security forwarding strategy for VANET. Sensors 23(3):1204

33. Zhang D, Yu FR, Yang R, Tang H (2018) A deep reinforcement learning-based trust management scheme for software-defined vehicular networks. In: Proceedings of the 8th ACM Symposium on Design and Analysis of Intelligent Vehicular Networks and Applications (DIVANet). ACM, p 17
34. Bhatia J, Mehta R, Bhavsar M (2018) Variants of software defined network (SDN) based load balancing in cloud computing: a quick review. Fut Internet Technol Trends 164–173
35. Zhang D, Yu FR, Yang R (2018) A machine learning approach for software-defined vehicular ad hoc networks with trust management. In: Proceedings of the IEEE Global Communications Conference (GLOBECOM). IEEE, p 16
36. Saravanan M, Ganeshkumar P (2020) Routing using reinforcement learning in vehicular ad hoc networks. Comput Intell 36(2):682–697
37. Zhang D, Yu FR, Yang R (2019) Blockchain-based distributed software defined vehicular networks: a duelling deep Q-learning approach. IEEE Trans Cogn Commun Netw 5(4):1086–1100
38. Nahar A, Das D (2020) SeScR: SDN-enabled spectral clustering-based optimized routing using deep learning in VANET environment. In: Proceedings of the 2020 IEEE 19th International Symposium on Network Computing and Applications (NCA). IEEE, pp 1–9
39. Lu C, Wang Z, Ding W, Li G, Liu S, Cheng L (2021) MARVEL: multi-agent reinforcement learning for VANET delay minimization. China Commun 18(6):1–11
40. Abuashour A, Kadoch M (2018) Control overhead reduction in cluster based VANET routing protocol. In: Ad Hoc Networks. Springer, Cham, pp 106–115
41. Abboud K, Zhuang W (2015) Impact of microscopic vehicle mobility on cluster-based routing overhead in VANETs. IEEE Trans Veh Technol 64(12):5493–5502
42. Kalakanti AK, Verma S, Paul T, Yoshida T (2019) RL SolVeR pro: reinforcement learning for solving vehicle routing problem. In: Proceedings of the 1st International Conference on Artificial Intelligence and Data Science (AiDAS), pp 94–99
43. Levine S, Kumar A, Tucker G, Fu J (2020) Offline reinforcement learning: tutorial, review, and perspectives on open problems. arXiv preprint
44. Wei Q, Lewis FL, Sun Q, Yan P, Song R (2017) Discrete-time deterministic Q-learning: a novel convergence analysis. IEEE Trans Cybern 47(5):1224–1237. https://doi.org/10.1109/TCYB.2016.2542923
45. Jiang T, Grace D, Mitchell PD (2011) Efficient exploration in reinforcement learning-based cognitive radio spectrum sharing. IET Commun 5(10):1309–1317
46. Chen Y, Huang S, Liu F, Wang Z, Sun X (2019) Evaluation of reinforcement learning-based false data injection attack to automatic voltage control. IEEE Trans Smart Grid 10(2):2158–2169
47. Tong W, Hussain A, Bo WX, Maharjan S (2019) Artificial intelligence for vehicle-to-everything: a survey. IEEE Access 7:1082310843
48. Mianji E, Fardad M, Muntean G-M, Tal I (2024) A survey on multi-agent reinforcement learning applications in the Internet of Vehicles
49. Kong X et al (2022) Deep reinforcement learning-based energy-efficient edge computing for Internet of Vehicles. IEEE Trans Indus Inf 18(9):6308–6316. https://doi.org/10.1109/TII.2022.3155162

Smart Traffic Forecasting in Urban Junctions: A Deep Learning and VANET Approach

Yousef Abuzir and Saleh Y. Abuzir

Abstract Urban traffic congestion emerges as a major challenge for transportation systems, adversely influencing both operational efficiency and the comprehensive quality of life. This chapter presents a method to predict congestion at intersections using advanced deep learning techniques. Using data from Vehicle Ad-Hoc Networks (VANET), this study focuses on understanding how various factors shape traffic patterns. These include weather, time of day, and special events. The process begins with collecting and cleaning data, setting the foundation for analysis. Several deep learning models are then employed to improve traffic predictions. A key feature of the approach is its integration with VANET systems. This enables real-time solutions like adjusting traffic lights, offering alternative routes, and sending alerts to drivers. The outcomes go beyond technical improvements, touching on broader benefits such as easing congestion and making roads safer. The work concludes by identifying areas where research could expand to refine smart traffic management further.

Keywords Traffic prediction · Deep learning · Urban junctions · Vehicular ad hoc networks · RNN · LSTM · CNN · Traffic flow optimization

1 Introduction

Traffic is a huge problem, especially in cities. Roads are packed, gas is wasted, and the air is polluted. More and more people are moving to cities, and there are more and more cars, making the problem even worse [1]. A study for the INRIX Global Traffic Scorecard [2] shows that big cities are always jammed up, and drivers waste a lot of time stuck in traffic. This hurts the economy and the environment [3–5].

Y. Abuzir (✉)
Faculty of Technology and Applied Sciences, Al-Quds Open University, Ramallah, Palestine
e-mail: yabuzir@qou.edu

S. Y. Abuzir
Civil and Environmental Engineering, Brescia University, Brescia, Italy
e-mail: saleh.abuzir@unibs.it

J. Bhatia et al. (eds.), *Deep Learning Based Solutions for Vehicular Adhoc Networks*,
Studies in Computational Intelligence 1207,
https://doi.org/10.1007/978-981-96-5190-0_10

To fix this, we need to accurately predict traffic patterns. If we can do that, cities can adjust traffic lights or reroute cars to keep traffic flowing smoothly.

These days, multiple technologies are available to provide real data on traffic congestion, which can be used in smart traffic systems analysis and prediction. Advanced predictive models provide officials and decision-makers with accurate and timely traffic forecasts, enabling better traffic management and congestion reduction.

This chapter introduces a new approach to improve traffic prediction at city intersections. By combining powerful Deep Learning (DL) techniques with real-time data from smart cars (Vehicular Ad Hoc Networks, or VANETs), we aim to make city traffic smoother, reduce congestion, and improve the environment.. Unlike older ways, our method uses the constantly changing data from VANETs, which helps in making better and faster traffic predictions. This new combination improves how well we can predict traffic and also lets us manage traffic in real time, making traffic lights work better and helping cars find new routes.

- The chapter seeks to address the following key goals.
- Examine DL applications in urban traffic prediction: Use VANET data to achieve high-accuracy traffic forecasts at city intersections.
- Review existing traffic prediction approaches: Highlight the benefits and limitations of traditional methods compared to DL-based models.
- Outline data collection and preprocessing strategies: Focus on essential techniques for preparing data to train predictive models effectively.
- Evaluate model performance: Compare different DL algorithms using various metrics to determine their accuracy and reliability in predicting traffic.
- Explore specific DL architectures:
  - Recurrent Neural Networks (RNNs) [6]: Suitable for handling sequential data like traffic flow by considering past trends to forecast future conditions.
  - Long Short-Term Memory (LSTM) [7]: An advanced RNN variant designed to manage long-term dependencies, especially where current traffic patterns are influenced by events from much earlier.
  - Convolutional Neural Networks (CNNs) [8]: Primarily used for image and sequence data, adapted here for analyzing complex traffic datasets.
- Integrate DL with VANETs: Investigate how DL models can leverage VANET infrastructure to enable real-time traffic management [9, 10].
- Assess societal and environmental impacts: Explore how accurate traffic predictions can support sustainable urban development and better quality of life.

Overview

The chapter offers a detailed exploration of DL-based traffic prediction methods. It begins by tracing the evolution of traffic forecasting techniques, contrasting traditional statistical methods with modern DL approaches. Data preparation is a key focus, with discussions on collection, preprocessing, and pattern extraction. The effectiveness of DL architectures—RNNs, LSTMs, and CNNs—is analyzed through comparisons with actual traffic data, highlighting the most accurate models.

Additionally, the chapter examines how integrating DL models with VANETs can facilitate real-time solutions such as adaptive signal control and vehicle rerouting. The chapter also explores the wider implications of these advancements. These include their potential to ease congestion, improve road safety, and support more sustainable urban transportation systems.

The insights presented emphasize the critical role of DL techniques in driving proactive traffic management and fostering greener, more efficient cities.

2 Background and Related Work

Figuring out how much traffic there will be is a big deal these days, thanks to smart machines. Old ways like ARIMA and Kalman were used, but they couldn't keep up with the changing traffic patterns. Machine learning, with tools like SVMs and KNN, made things better, but they still had some problems. Other machine learning tricks, like SVM and KNN again, made predictions more accurate, but they weren't perfect [11, 12]. Models like RNNs and LSTMs have been really good at finding hidden patterns in traffic data. But we still need to deal with bad data and complex models to make really reliable traffic predictions [13–15].

2.1 Literature Review of Traffic Prediction Methods

Traffic prediction has been a hot topic for researchers for a long time. Like other fields, modern technology improves traffic forecasting. This technology has moved from traditional statistical methods to advanced ML methods and deep Learning (DL) strategies. Initial endeavors heavily relied on historical traffic data and statistical models such as ARIMA and Kalman filters. Despite establishing a foundation, these techniques encountered challenges in addressing the dynamic and complex patterns of traffic [11, 12].

Today, ML techniques and methodologies become very crucial in traffic congestion predication. These approaches offered improved prediction accuracy by uncovering non-linear relationships within traffic data [16]. Nevertheless, these methodologies encountered constraints, particularly when confronted with substantial amounts of real-time data and the need to adapt swiftly to evolving traffic circumstances.

In recent times, DL has emerged as a forcefully tool for traffic prediction. It has provided researcher and decision maker with an exceptional ability to analyze spatial and temporal traffic data. These models support huge datasets and knowledgeable models to deliver more precise and reliable traffic predictions [16].

Wang et al. [11] utilized DL methodologies, specifically LSTM, AdaBoost, and gradient descent (GD), to enhance the precision of traffic congestion predictions. The final model performed better because LSTM had been integrated into AdaBoost's predictions.

An extensive assessment of prediction strategies concerning traffic, including DL methods, is done in the paper [12]. They are classified according to various points of view, and the most recent approaches applied in a host of traffic prediction applications are discussed. The models capture complex traffic patterns in the data by RNNs and LSTMs—challenges: data quality, model complexity, real-time adaptability.

The enhanced Fuzzy C-Means (FCM) cluster approach works excellently in analyzing traffic state and is very effective LSTM models for urban traffic flow prediction. The FCM clustering approach does excellently about assessing the state of urban traffic [13].

Gowda et al. illustrated that DL models improve the accuracy of traffic prediction for complicate transportation datasets. Nevertheless, challenges like poor data quality and the possibility of real-time change are still prevalent [14].

Other researchers developed a model for forecasting exact data on traffic flow, which also includes the time for traffic light cycles, accidents, rallies, and even roadwork that could result in traffic bottlenecks. ML techniques, such as Support Vector Machine (SVM) and KNN, yielded higher accuracy than conventional methods. The RNNs with LSTMs are DL models that effectively capture complex patterns in the data [15].

In the article by Tepelea, the author compared the traditional and DL methods of image depth measuring and concluded that the DL method stands above others by accuracy, particularly when such images are complex and uncalibrated. Traditional techniques are restricted to small datasets; meanwhile, bottomless learning deals with vast ones. DL excellently with accuracy, especially in complex patterns [17].

More recently, Shiri et al. proofed that DL can deal with huge, high-dimensional datasets with CNNs. ML techniques, such as SVM and KNN, yielded excellent results and high accuracy compared to traditional methods. DL provides highly adaptive learning from new data continuously. They present a detailed survey of diverse DL models, (i.e., CNN, RNN, Generative ML models, Deep Reinforcement Learning, and Deep Transfer Learning), discussing the architecture, uses, benefits, advantages, applications, and limitations of every model [18].

DL is reshaping decision-making techniques applicable across various sectors, as elaborated by the authors. It utilizes multiple layers of non-linear feature transformation of input data through the process of representation learning to tackle challenges caused by big datasets. DL can model very complicated traffic patterns with high accuracy. DL requires much computational resources to do effective real-time processing [19].

DL is getting more important in our everyday lives. It has already made drastic differences in industry, such as cancer detection, healthcare, autonomous vehicles, and predictive analytics and forecasting, as well as speech recognition and other areas. Traditional techniques deal with small datasets, while DL technologies apply to a large-scale dataset. DL brilliantly captures intricate patterns with a high level of accuracy [20].

Figure 1 summarize the different techniques used for traffic predictions. Figure 2 explore the different classification of problems and DL used.

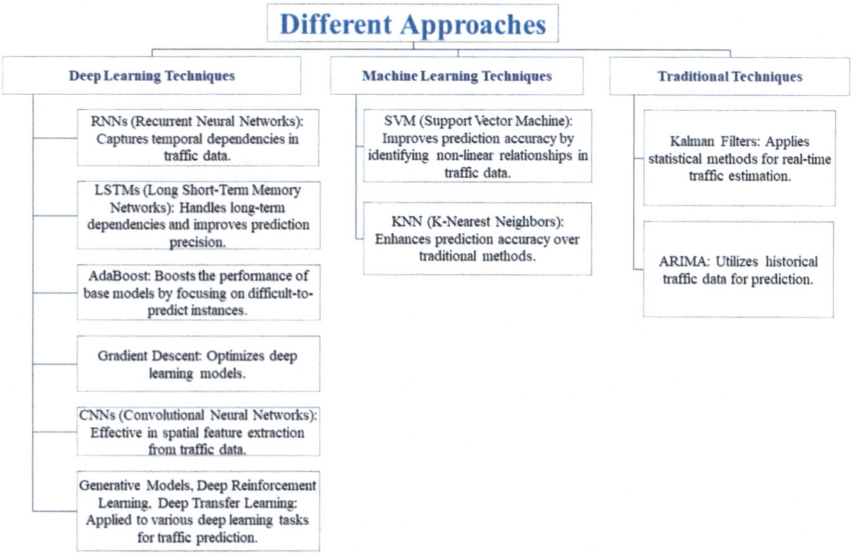

Fig. 1 A summary the different techniques commonly found in this review

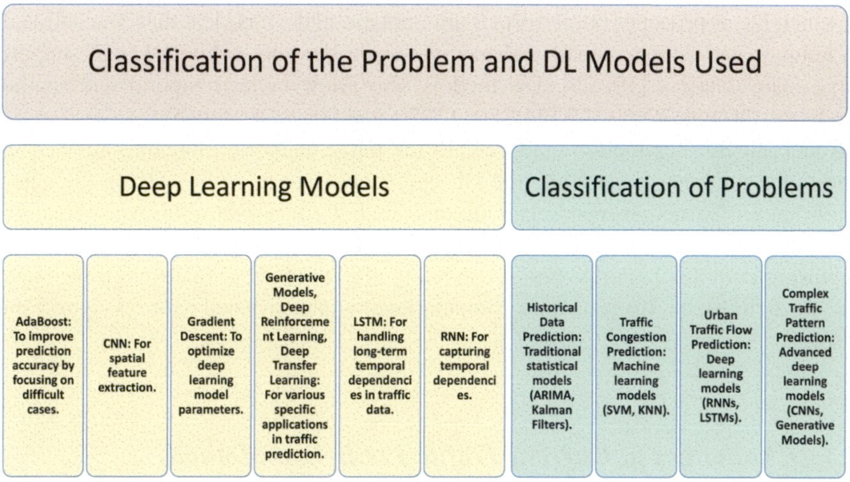

Fig. 2 Classification of the problem and DL models used

2.2 Discussion on Traditional Techniques Versus DL Approaches

Comparing traditional techniques and DL approaches sheds light on the advancements and limitations of each method in traffic prediction (Table 1).

Table 1 Traditional techniques versus DL approaches [19–22]

Aspect	Traditional techniques	DL approaches
Data handling	Limited to small and medium-sized datasets	Capable of handling large-scale, high-dimensional datasets
Temporal dependencies	Basic temporal modeling (e.g., ARIMA)	Advanced temporal modeling (e.g., RNNs, LSTMs)
Spatial dependencies	Limited spatial consideration	Effective spatial feature extraction (e.g., CNNs)
Flexibility	Less flexible, requires manual feature engineering	Highly flexible, automatic feature extraction
Adaptability	Less adaptive to new patterns	Highly adaptive, learns from new data continuously
Accuracy	Moderate accuracy, struggles with complex patterns	High accuracy, excels in capturing complex traffic patterns
Computational resources	Requires fewer computational resources	Computationally intensive, requires significant resources
Real-time processing	Limited real-time capabilities	Effective real-time processing and predictions

Table 1 compares different between traditional and DL approach and shows items in which DL approaches outperform traditional methods. It is clear that conventional techniques can support small and medium-scale datasets, while DL can support large-scale datasets [19, 20]. DL models also allow better temporal and spatial modeling through RNNs, LSTMs, and CNNs and hence are capable of much more advanced feature extraction compared to the basic temporal modeling allowed in traditional methods [21]. While the DL approaches are more flexible on the automatic feature extraction approach, traditional methods do not show this type of feature flexibility and adaptability of DL. The accuracy is very high with DL, mainly in capturing complex patterns, besides showing effectiveness in real-time processing and predictions, but the process is computationally intensive and requires significant resources.

2.3 Challenges in Current Traffic Prediction Methods

Despite significant progress, several challenges remain in traffic prediction methods [14, 21–24]:

- *Quality and Availability of Data*: High-quality, real-time traffic data is crucial for accurate predictions. Data can be insufficient, unclear, noisy, unstructured, incomplete, or unavailable.

- *Model Complexity*: DL Models are complex and require advanced and high computational resources for training, testing and running. This intricacy may hinder their deployment in resource-constrained settings.
- *Scalability: Making prediction models work in big cities with different traffic* patterns is a big challenge. Keeping them accurate across different areas with changing traffic conditions is tough.
- *Integration*: Putting advanced prediction models together with current traffic management systems, like traffic lights, is tricky and complicated.
- *Adaptability*: Traffic can change quickly due to accidents, road closures, or weather. Prediction models need to be able to adjust quickly to provide accurate real-time forecasts.
- *Privacy* and Security: Collecting and processing real-time traffic data raises privacy and security concerns. It's important to collect data legally and protect it from cyberattacks.

To overcome these challenges, researchers, city planners, and tech developers need to work together to keep pushing the boundaries of research and development. Future advancements in data collection technologies, model optimization, and infrastructure integration will be vital in overcoming these hurdles and enhancing the effectiveness of traffic prediction systems.

Table 2 compare two case studies and showcase the impact of predictive congestion management systems in two different urban settings, highlighting the importance of integrating advanced technologies with existing infrastructure to address congestion and improve overall mobility.

3 Data Collection and Preprocessing

The effectiveness of any traffic prediction model tied to the high quality and comprehensiveness of the data it uses. For urban traffic forecasting, the dataset should encompass various data types collected from multiple sources at urban junctions. These datasets typically include traffic volume counts, vehicle speeds, occupancy rates, weather conditions, and, increasingly, connected vehicle data. Table 3 summarizing the different types of data in the dataset of traffic congestions and Table 4 summarizing the different sources of data may be collected in the dataset.

The following sections explores a case study using DL techniques to predict traffic congestion.

1. It starts by loading and exploring a dataset containing traffic data from four junctions.
2. The data includes attributes like date, time, junction, and number of vehicles.
3. The case was studied through a series of steps:

    - Data exploration to identify patterns and trends.
    - Feature engineering based on existing data.

Table 2 Comparison of two case studies [11, 13, 23]

Aspect	Los Angeles, California	Singapore
Location	Los Angeles, California	Singapore
System name	LADOT predictive traffic management system	LTA traffic prediction system
Objective	Optimize traffic flow, reduce congestion	Forecast traffic conditions, proactively manage congestion
Implementation	Implemented by LADOT	Developed by Singapore's Land Transport Authority (LTA)
Approach	Leveraged advanced ML techniques and real-time traffic data	Combined traditional statistical models with DL techniques
Integration	Integrated predictive models with existing traffic infrastructure (signals, cameras, sensors)	Integrated DL models with traditional prediction methods
Data sources	Real-time traffic data, weather patterns, social media feeds	Historical traffic data, weather patterns, social media feeds
Impact	Reduced travel times, improved air quality, enhanced mobility	Proactive congestion management, improved traffic flow
Success factors	Effective integration of predictive models with traffic infrastructure	Combination of DL and traditional methods for accurate predictions

Table 3 A summary of the different types of data in the traffic dataset [24, 25]

Data type	Description
Traffic volume counts	Number of vehicles passing a junction (vehicles per hour)
Vehicle speeds	Average speed of vehicles at different times of day
Occupancy rates	Proportion of time a sensor detects a vehicle
Weather conditions	Temperature, precipitation, visibility
Connected vehicle data	Vehicle positions, speeds, travel directions
Incident reports	Accidents, roadworks, other traffic flow disruptions

- Data preprocessing for modeling readiness, and the construction.
- Building and Comparing DL Models: We'll create and compare different Deep Learning models like RNNs, LSTMs, CNNs, and Gradient Descent.
- Training and Testing: These models will be trained and tested using our dataset.
- Analysis and Applications: We'll analyze the results and explore how this approach can be used to improve traffic management and urban planning.

This case study aims to provide valuable information about traffic congestion, helping decision-makers make better choices.

Table 4 A summary of the different sources of data in the traffic dataset [26]

Data source	Description
Inductive loop sensors	Embedded in road surface, detect metal objects (vehicles)
Traffic cameras	Capture real-time video feeds, extract vehicle data
Radar and Lidar sensors	Use radio waves/laser light to detect vehicles, measure speeds
Global positioning system (GPS)	Provides live vehicle location and speed information
Mobile phone data	Aggregated location data indicates traffic density and movement
VANETs	Vehicles share data on speed, location, travel direction (real-time)
Weather stations	Provide local weather data impacting traffic flow
Crowdsourced data	User-reported data on traffic, incidents, road hazards (complements sensor data)

In the following sections, we'll dive deeper into a case study that uses various DL techniques for traffic prediction. We'll start by loading and exploring a dataset with information about traffic at four different intersections. By carefully analyzing the data and building models, we aim to create a predictive framework that can accurately forecast traffic congestion levels.

3.1 Description of the Dataset Collected from Urban Junctions

The main goal is to create a model that predicts traffic congestion at four different intersections. This model will provide valuable information about traffic patterns to improve traffic management and infrastructure planning. Ultimately, this could reduce traffic congestion and improve traffic flow.

This section also highlights some challenges:

- Data Sparsity: Some intersections might not have a lot of data.
- Variability: Traffic patterns can change based on the time of day, day of the week, and other factors.
- Feature Engineering: Extracting useful information from raw data is important for creating effective models.

Potential Benefits:

- Improved Traffic Management: Better traffic predictions can help plan traffic signals, road closures, and other management strategies.
- Urban Planning: Long-term data can inform infrastructure development projects to address recurring congestion issues.

Importing Libraries

```
Import necessary libraries:
Import NumPy for numerical operations.
Import Pandas for data manipulation.
Import OS for operating system interactions.
Import Matplotlib for plotting graphs.
Import Seaborn for advanced data visualization.
Import datetime for handling date and time.
Import TensorFlow for MLand DL models.
Import MinMaxScaler from scikit-learn for feature scaling.
Import mean_squared_error from scikit-learn for evaluation
metrics.
Import Sequential, Dense, LSTM, Gated Recurrent Unit (GRU),
Dropout, Conv1D, MaxPooling1D, Flatten from TensorFlow Keras
for building neural network models.
Import Adam optimizer from TensorFlow Keras.
Import warnings module to handle warning messages and suppress them.
Load the dataset:
Use Pandas to read a CSV file from a specified path.
Store the data in a variable named dataset.
Display the first few rows of the dataset:
Print the first few rows to understand the structure and content of
the data.
```

Data Description: This dataset consists of numbers of vehicles at four junctions on hourly basis. This dataset comprises 48.1 k (48,120) observations of the number of vehicles each hour. The CSV file provides four features (Table 5):

- DateTime: Timestamp when the data was recorded.
- Junction: Identifier for each of the four junctions (1, 2, 3, 4).
- Vehicles: The number of vehicles counted at that time.
- ID: A unique identifier for each record.

Explanation of data sources

The sensors at each junctions collecting traffic data across different times and periods. Some of the junctions have provided only sparse data.

Initial Observations:

- The data spans multiple timestamps with vehicle counts for four junctions.
- The Vehicles feature represents the traffic volume at each timestamp

Table 5 Data example first 5 items form traffic dataset

DateTime	Junction	Vehicles	ID
2015-11-01 00:00:00	1	15	20151101001
2015-11-01 01:00:00	1	13	20151101011
2015-11-01 02:00:00	1	10	20151101021
2015-11-01 03:00:00	1	7	20151101031
2015-11-01 04:00:00	1	9	20151101041

3.2 Data Exploration

To enhance data analysis and prepare it for DL models, we perform date conversion and feature engineering on the 'DateTime' column. First, converting it to a datetime object allows us to easily extract components like year, month, day, hour, and day of the week (Table 6). These features capture temporal patterns and trends valuable for DL models. After extracting these informative features, we can streamline the dataset by dropping the original 'DateTime' column. This focuses our analysis on the primary crucial information for model building and avoids redundancy in the data. Overall, this preprocessing step ensures the data is well-suited for further exploration and subsequent use in ML tasks.

The result in Table 7 analyzes the traffic dataset's structure, offering insights into its columns, data types, and memory usage. Key observations from the analysis:

- The DataFrame is comprised of 48,120 rows and 8 columns.
- All columns hold integer data with no missing values.
- The entire DataFrame occupies roughly 2.9 megabytes of memory.
- This analysis aids in comprehending the dataset's structure, data integrity (like missing values), and the data types stored in each column.

Figure 3 is a timeseries plot which shows traffic volume at four junctions over time. The x-axis tracks the timeline, and the y-axis measures traffic volume. Each line in the plot reflects a separate junction, and the colors differentiate between the junctions allowing for visual comparison of traffic patterns. There is a cyclical pattern to the traffic volume, with traffic volume being higher during weekdays and lower during weekends. There appears to be some variation in traffic volume between the different junctions. For example, Junction 1 seems to generally have higher traffic

Table 6 The structure and the description of the DataFrame

<class 'pandas.core.frame.DataFrame'>
RangeIndex: 48,120 entries, 0 to 48,119
Data columns (total 8 columns):

#	Column	Non-Null Count	Dtype
0	Junction	48,120 non-null	int64
1	Vehicles	48,120 non-null	int64
2	ID	48,120 non-null	int64
3	Year	48,120 non-null	int64
4	Month	48,120 non-null	int64
5	Day	48,120 non-null	int64
6	Hour	48,120 non-null	int64
7	DayOfWeek	48,120 non-null	int64

dtypes: int64(8)
memory usage: 2.9 MB

Table 7 The structure and the description of the traffic dataset after date conversion

#	Column	Non-null count	D type
0	Junction	48,120 non-null	int64
1	Vehicles	48,120 non-null	int64
2	ID	48,120 non-null	int64
3	Year	48,120 non-null	int64
4	Month	48,120 non-null	int64
5	Day	48,120 non-null	int64
6	Hour	48,120 non-null	int64
	RangeIndex	48,120 entries, 0 to 48,119	
	Data Columns (total)	8 columns	
	Memory Usage	2.9 MB	

volume than the other junctions. Examining Junction 4 (yellow line), we can see it generally experiences lower traffic volume compared to the other junctions. This observation is particularly evident during weekdays when the other junctions exhibit higher peaks. There could be several reasons behind this, such as Junction 4 being located on a less travelled route or having a different traffic light configuration that promotes smoother flow.

It is difficult to say for sure from this plot, but there may also be some seasonality to the traffic volume, with traffic volume being higher during certain times of the year. This plot provides a good overview of traffic patterns at the four junctions over time. It can be used to track trends in traffic volume and enable comparisons across various junctions.

The plot in Fig. 4 is a line chart that shows the average monthly traffic volume at four junctions. The x-axis represents the month, and the y-axis represents the average vehicle count. Each line in the chart corresponds to a different junction, and

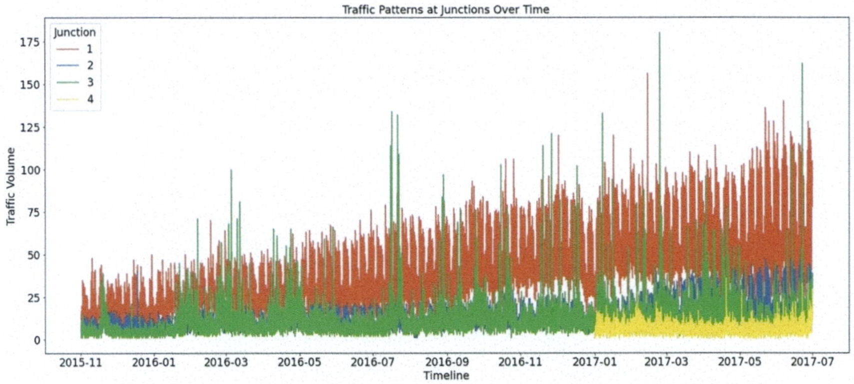

Fig. 3 Traffic volume at four junctions over time. Weekday peaks and variations between junctions are evident

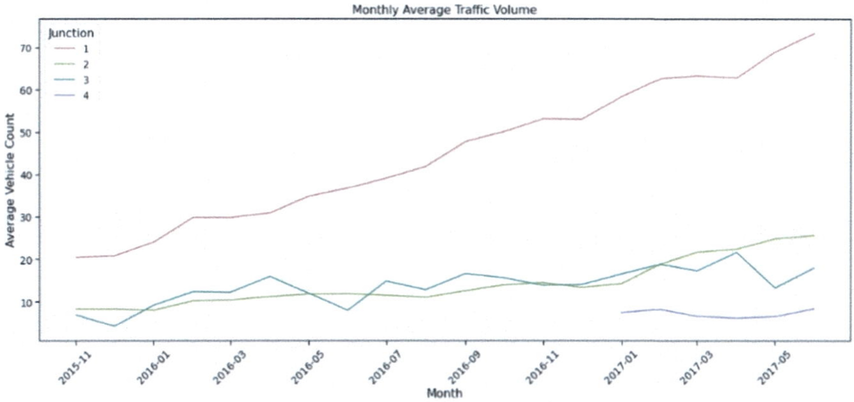

Fig. 4 Monthly average number of vehicles

the color differentiates between them. There seems to be a seasonal pattern to the traffic volume, with some months experiencing higher traffic than others. It's difficult to pinpoint specific trends without knowing the month labels, but some lines appear to show peaks in the summer or warmer months (potentially) and lower traffic in colder months.

There are variations in traffic volume between the different junctions. Some junctions appear to have consistently higher average traffic throughout the year (possibly Junction 1 or the line in red), while others experience lower average traffic (likely Junction 4 or the yellow line).

The bar chart in Fig. 5 shows the average traffic volume by day of the week for four junctions. Each bar in the chart corresponds to a specific junction, and the color differentiates between them. There is a clear distinction in traffic volume between weekdays and weekends. Traffic volume is noticeably higher on weekdays (Monday to Friday) than on weekends (Saturday and Sunday). Weekdays likely see more traffic due to people commuting to and from work or school, more commercial deliveries and business.

The line graph in Fig. 6 shows the average traffic volume by hour of the day for four junctions. There appears to be a distinct rush hour pattern during weekdays, with traffic volume increasing significantly during certain hours in the morning and evening. This suggests commuting patterns as a major influence on traffic during these times and vary between junctions. The rise in traffic during mornings and evenings likely corresponds to people commuting to and from work or school. There might also be some influence from delivery trucks or commercial vehicles contributing to traffic volume, especially during mornings or evenings.

The bar chart in Fig. 7 illustrates the average traffic volume at each junction. By analyzing the bar heights in this chart, you can identify junction 1 has the highest average traffic volume. This junction might be a priority for further investigation or congestion management strategies.

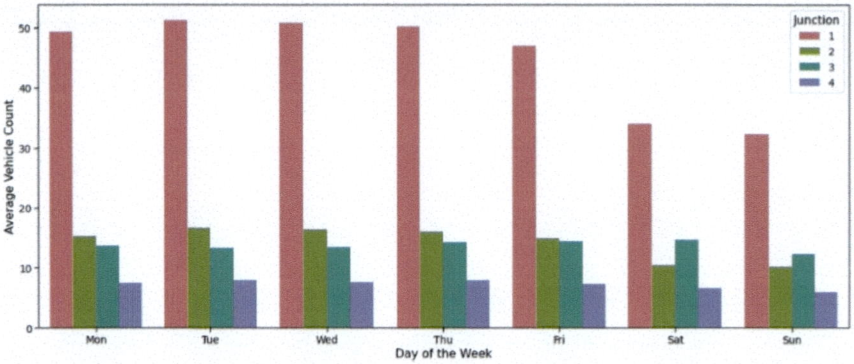

Fig. 5 Daily average traffic volume by day of the week

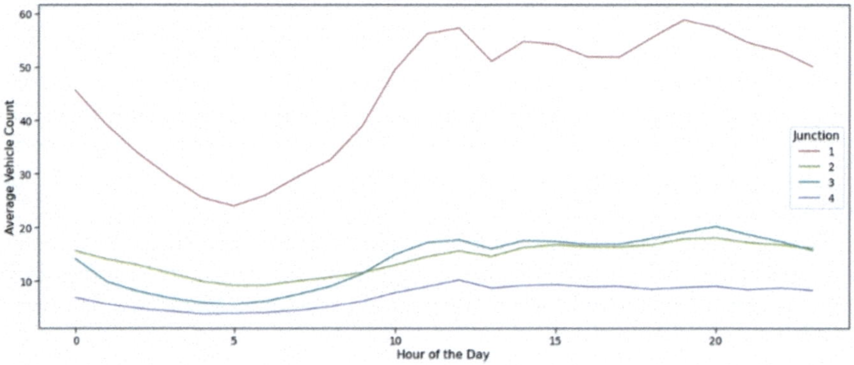

Fig. 6 Hourly average traffic volume

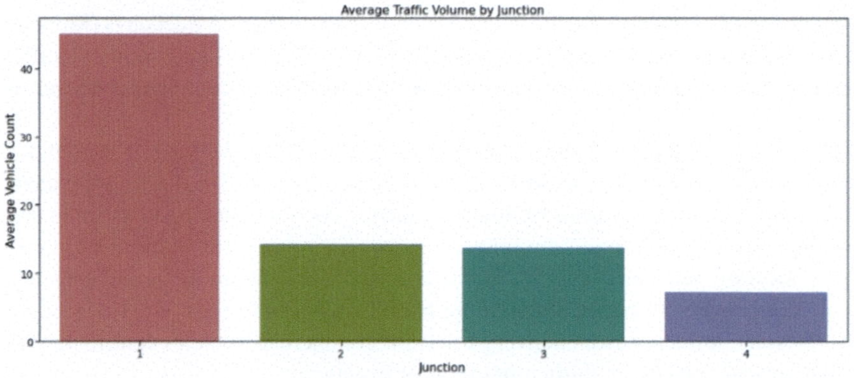

Fig. 7 Average traffic volume by junction identify junction 1 has the highest average traffic volume

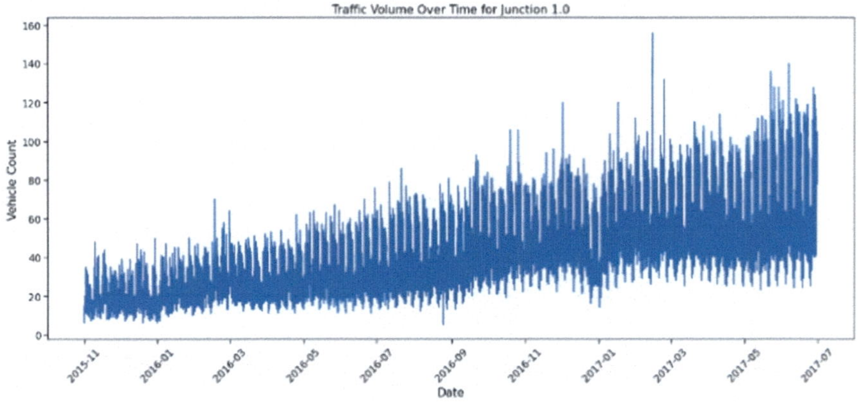

Fig. 8 Traffic volume over time for junction 1

Figure 8 is a time series plot showing the traffic volume at Junction 1 over time. There seems to be a cyclical pattern to the traffic volume, with potentially higher traffic during some parts of the year and lower traffic during others.

Correlation Matrix and Heatmap

Correlation matrix and heatmap can visualize the relationships between the different features in traffic dataset. They reveal how features move together, providing valuable insights. Strong correlations, positive or negative, suggest features might be redundant or have opposing influences. Conversely, weak correlations indicate features are likely independent. Heatmaps visually represent these correlations using color intensity, making it easier to identify patterns [27, 28].

In the realm of traffic prediction, correlation analysis plays a pivotal role in various scenarios. It serves the purpose of identifying features that exhibit a robust correlation with traffic volume (Vehicles), thereby potentially excluding those with feeble correlations. Correlation analysis can identify features that have a negative impact within the model. Of paramount importance is its ability to offer a profound comprehension of the interactions among features, thereby guiding subsequent decisions related to data preparation and model construction. Using a correlation matrix and heat map, the researcher can understand how traffic features such as year, month, and day of the week affect traffic volume. This acquired knowledge can then be utilized to cherry-pick the most pertinent features and ready the data for optimal performance in DL models.

The heatmap depicted in Fig. 9 examines the interconnections among specific attributes within a traffic forecasting dataset, such as Year, Month, Day, Hour, DayOfWeek, and Vehicles (which represents the target variable, likely denoting traffic volume). The correlation matrix and heatmap provided yield valuable insights regarding the associations between the features in your traffic prediction dataset and traffic volume. In the subsequent section, a brief overview is presented:

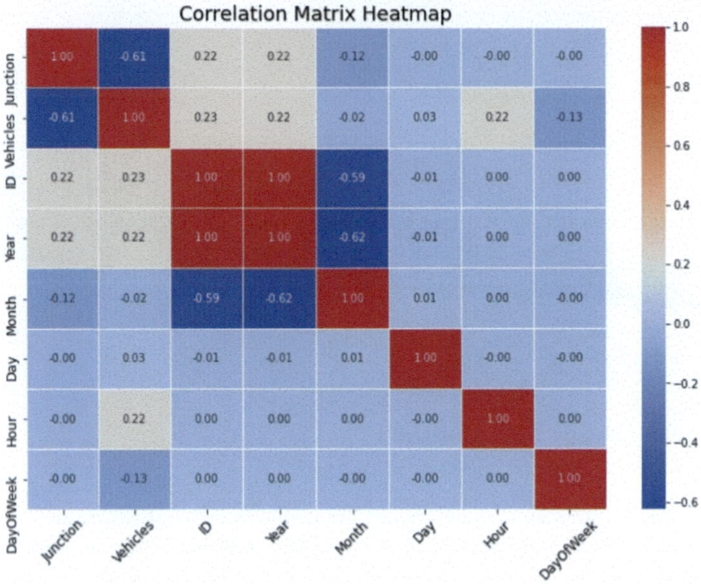

Fig. 9 Correlation matrix and heatmap

- A substantial positive correlation is observable: Traffic volume (Vehicles) exhibits a tendency to rise as the hour progresses (Hour) and may be more pronounced on weekdays in comparison to weekends (DayOfWeek). This observation is consistent with the patterns seen during peak traffic hours.
- Conversely, there are weaker correlations observed: Month and Day may not exhibit a robust linear relationship with traffic volume. The correlation pertaining to Year remains ambiguous based on the visual representation.

Knowing these connections lets you choose the best features and prepare your data effectively, leading to a more accurate traffic prediction model.

3.3 Preprocessing Steps

DL models like CNNs, LSTMs, and RNNs rely heavily on preprocessed data. This initial step ensures the data is in a format that these models can understand and work with effectively. Here's an outline of the necessary preprocessing steps:

1. **Data Cleaning**

Data cleaning includes managing missing values, correcting errors, and removing noise from the dataset [28].

- Handling Missing Values: Check for missing values and how to manage them. Common methods include imputation (mean, median, mode) or removal.
- Outlier Detection: Identify and handle outliers, as they can significantly affect model performance.
- Data Consistency: Ensure consistency in data types and formats.

```
Pseudocode
Check and Handle Missing Values:
Check for missing values:
Determine if there are any missing entries in the dataset.
Print the number of missing values for each column.
Handle missing values:
Option 1: Impute missing values
Replace missing entries with the mean value of each respective
column.
Update the dataset in place with the filled values.
Option 2: Drop missing values (Alternative)
Delete rows with missing entries from the dataset.
Update the dataset in place by dropping these rows.
```

2. **Normalization**

Normalization scales the features to a similar range, which helps in faster convergence during training and prevents the model from being biased towards features with larger scales.

- Min–Max Scaling: Scale features to a range [0, 1].
- Standardization: Transform features to have zero mean and unit variance.

```
Pseudocode
Scale Data:

1. Min-Max Scaling:

  • Initialize Min-Max Scaler:

      Create an instance of the Min-Max Scaler.

  • Apply Min-Max Scaling:

    – Fit the scaler on the specified columns of the
      dataset.
    – Transform these columns using the fitted
      scaler.
    – Update the dataset with the scaled values.

2. Alternative: Standardization:

  • Initialize Standard Scaler:

    – Create an instance of the Standard Scaler.
```

- **Apply Standardization:**
  - Apply the scaler on the specified columns of the dataset..
  - Transform these columns using the fitted scaler.
  - Update the dataset with the standardized values.

3. **Feature Extraction**

Feature extraction refers to creating new features or altering existing ones to better represent underlying data patterns [28, 29].

- Temporal Features: Extract additional time-related features, if not already done. This can include week of the year, weekend/weekday, holidays, etc.
- Lag Features: Create lagged features to incorporate temporal dependencies.
- Rolling Window Features: Compute rolling statistics (mean, std) to capture trends over time.

Pseudocode
Create Lag Features and Rolling Window Features:

1. **Create Lag Features:**
   - **Generate Lagged Data:**
     - Create a new column with values shifted by 1 time step from the original column.
     - Create another new column with values shifted by 2-time steps from the original column.
2. **Create Rolling Window Features:**
   - **Calculate Rolling Mean:**
     - Create a new column with the mean of values over a rolling window of 3-time steps.
   - **Calculate Rolling Standard Deviation:**
     - Create a new column with the standard deviation of values over a rolling window of 3-time steps.
3. **Handle Missing Values:**
   - **Remove Rows with Missing Values:**
     - Drop rows that have missing values generated by the lagging and rolling operations.

- Update the dataset by removing these rows.

4. **Data Transformation for DL Models**

DL models require data in a specific format. For RNNs, LSTMs, and CNNs, the data needs to be transformed into sequences.

- Sequence Generation: Generate sequences for RNN/LSTM.
- Reshape Data: Reshape data for CNN input if applicable.

```
Pseudocode
Define Function to Create Sequences:

1. Function Definition:

   • Define a function create_sequences that takes
     data and seq_length as inputs.

2. Generate Sequences:

   • Initialize empty lists X and y to store input
     sequences and target values.
   • For each possible starting point in the data,
     except for the last few points that can't form
     a full sequence:

     - Extract a sequence of length seq_length from
       the data.
     - Append this sequence to X.

3. Append the value immediately following this
   sequence to y.
4. Return Sequences:

   • Convert X and y lists to numpy arrays and return
     them.


Parameters and Generate Sequences:

1. Set Parameters:

   • Define SEQ_LENGTH to specify the length of each
     sequence.

2. Generate Sequences:

   • Call the create_sequences function with the
     dataset and sequence length.
   • Store the resulting sequences in X and
     corresponding targets in y.
```

`Reshape Data for CNN:`

1. `Reshape Input Data:`
   - `Reshape X to have the shape [number of samples, sequence length, 1] to match the expected input format for CNN.`

4 DL Models for Traffic Prediction

DL has proven to be a powerful tool for predicting traffic patterns [14]. RNNs and LSTMs used for analyzing and understanding data that's constantly changing [6, 14]. RNNs remember past information by maintaining an internal state, which helps them understand traffic trends. LSTMs are a distinct class of RNN designed to handle information that matters from a long time ago [30, 31].

RNNs and LSTMs excel at capturing the intricate patterns and complexities within traffic data. Their ability to study and analyze traffic data over time makes it a tool capable of predicting traffic flow at intersections and along roads with great accuracy. CNNs typically used for image processing, can also be applied to traffic data to identify recurring patterns in traffic flow [7, 12].

RNNs particularly, have become the go-to for many traffic prediction models. By analyzing historical traffic data at regular intervals, these models can forecast future traffic conditions [30, 31].

4.1 Recurrent Neural Networks (RNNs)

Recurrent Neural Networks (RNNs) are a type of deep learning model that process data in a sequence manner. This structure allows them to analyze the flow of information across time, making them suitable for tasks where context or previous inputs matter. For example, in traffic prediction, RNNs can consider past traffic data to predict future conditions. Their ability to maintain an internal state helps capture these patterns, making them useful for time-series data [32].

4.2 Long Short-Term Memory (LSTM) Networks

LLSTMs are a type of deep learning model derived from RNNs, built to process data points in sequence, evaluating each one individually. While regular RNNs struggle to retain information over long sequences, LSTMs use specialized memory cells and gates. These features help them remember important data from earlier in the sequence, which is crucial for tasks like long-term traffic forecasting. This design helps LSTMs retain important data over longer periods. These components control the flow of information, allowing LSTMs to keep important data and forget irrelevant details. This ability is crucial for accurately predicting traffic patterns [7, 20].

4.3 Convolutional Neural Networks (CNNs)

Typically used for image analysis, CNNs have also shown promise in traffic prediction. By applying filters to the traffic data, CNNs can identify recurring spatial patterns. While not as commonly used as RNNs and LSTMs for traffic forecasting, CNNs can contribute to improving prediction accuracy by capturing additional features of the data [20].

4.4 RNN-Based Traffic Prediction

Predicting traffic flow is a complex task that requires models capable of capturing intricate temporal dependencies. Since traffic data is sequential, Deep Learning techniques like RNNs can be used for prediction. By analyzing historical traffic patterns, RNNs can forecast traffic conditions, aiding urban planning, transportation management, and resource allocation. The algorithm outlines a systematic approach to building and evaluating RNN models for traffic prediction. While this framework provides a general structure, specific implementations may require adaptations based on dataset characteristics and desired performance metrics [20, 32].

`Algorithm for Building and Evaluating RNN Models for Traffic Prediction`
`Inputs:`

- `Traffic data (time series, features: traffic volume, speed, etc.)`
- `Hyperparameters (number of layers, RNN cell type, learning rate, etc.)`

`Outputs:`

- `Trained RNN model`
- `Evaluation metrics: Mean Absolute Error (MAE), Root Mean Squared Error (RMSE), etc.)`

`Algorithm:`

1. **`Data Preprocessing:`**

- Clean the traffic data (handle missing values, outliers).
- Normalize the data (e.g., min-max scaling).
- Extract relevant features (traffic volume, speed, weather, etc.).
- Create time series sequences for input and output.

2. **Model Architecture:**
    - Define the RNN architecture (number of layers, hidden units).
    - Choose the RNN cell type (LSTM, GRU, etc.).
    - Select activation functions (e.g., ReLU, tanh).
3. **Model Training:**
    - Initialize model parameters randomly.
    - Split data into training and validation sets.
    - Define loss function (e.g., Mean Squared Error).
    - Choose optimizer (e.g., Adam, RMSprop).
    - Iterate through training data:
        - Feed input sequence to the RNN.
        - Calculate the loss between predicted and actual values.
        - Update model parameters using backpropagation and optimizer.
4. **Model Evaluation:**
    - Feed validation data to the trained model.
    - Calculate evaluation metrics (MAE, RMSE, etc.).
    - Analyze model performance.
5. **Hyperparameter Tuning (Optional):**
    - Experiment with different hyperparameter values.
    - Re-train and evaluate the model for each combination.
    - Select the best performing hyperparameters.
6. **Return:**
    - Trained RNN model
    - Evaluation metrics

Additional Considerations:

- Early stopping to prevent overfitting.

- `Regularization techniques (L1/L2, dropout).`
- `Hyperparameter optimization techniques (grid search, random search).`
- `Model deployment and real-time prediction.`

5 Model Evaluation and Validation

It is necessary to check the readiness of the data and pre-processing it before Before developing DL models. In general, building DL models includes training them, evaluating their performance, and making predictions. The development goes through the following steps:

1. Data Splitting
2. Model Building and Training
3. Model Evaluation
4. Making Predictions
5. Results Interpretation and Visualization

5.1 Data Splitting

Seprate the data into training, validation, and test sets. This ensures that the model is evaluated on unseen data to check for overfitting.

```
Pseudocode
Split Data into Training, Testing, and Validation Sets:
```

1. `Initial Data Split:`
   - `Function Call:`
     - `Use a function to split the dataset into training and testing sets.`
   - `Parameters:`
     - `Set the fraction of data to be used for testing (e.g., 20%).`
     - `Specify a random seed for reproducibility.`
   - `Output:`
     - `Store the training data in X_train and y_train.`
     - `Store the testing data in X_test and y_test.`
2. `Further Split Training Data:`

- Function Call:
  - Use a function to split the training dataset into training and validation sets.
- Parameters:
  - Set the fraction of the training data to be used for validation (e.g., 25% of the training data).
  - Specify a random seed for reproducibility.
- Output:
  - Store the further split training data in X_train and y_train.
  - Store the validation data in X_val and y_val.

5.2 *Model Building and Training*

Build and train the CNN, LSTM, and RNN models using Keras.

5.3 *CNN Model*

Pseudocode
Prepare Data for CNN:

1. **Reshape Data:**
   - **For Training Data:**
     - Reshape X_train to have dimensions [number of samples, sequence length, 1].
   - **For Validation Data:**
     - Reshape X_val to have dimensions [number of samples, sequence length, 1].
   - **For Testing Data:**
     - Reshape X_test to have dimensions [number of samples, sequence length, 1].

Build CNN Model:

1. **Initialize Model:**

- Create a sequential model.

2. **Add Layers:**
   - **Convolutional Layer:**
     - Add a 1D convolutional layer with 64 filters, a kernel size of 2, and ReLU activation.
     - Set the input shape to (sequence length, 1).
   - **Max Pooling Layer:**
     - Add a 1D max pooling layer with a pool size of 2.
   - **Flatten Layer:**
     - Add a flatten layer to convert the 2D output to 1D.
   - **Dense Layers:**
     - Add a fully connected (dense) layer with 50 units and ReLU activation.
     - Add another dense layer with 1 unit (output layer).
3. **Compile Model:**
   - Set the optimizer to 'adam'.
   - Set the loss function to 'mean squared error'.
4. **Model Summary:**
   - Print a summary of the model architecture.

Train CNN Model:

1. **Train Model:**
   - Fit the model on X_train_cnn and y_train.
   - Use X_val_cnn and y_val for validation.
   - Set the number of epochs to 20.
   - Set the batch size to 32.

LSTM Model

The LSTM model is expecting a 3-dimensional input (samples, timesteps, features), but it received a 2-dimensional input. To fix this, we need to reshape our input data to have the shape (samples, timesteps, features).

Pseudocode
Prepare Data for LSTM:

1. **Reshape Data:**
   - **For Training Data:**
     - Expand dimensions of X_train to create a 3D array with shape (number of samples, timesteps, 1).
   - **For Validation Data:**
     - Expand dimensions of X_val to create a 3D array with shape (number of samples, timesteps, 1).
   - **For Testing Data:**
     - Expand dimensions of X_test to create a 3D array with shape (number of samples, timesteps, 1).

Build LSTM Model:

1. **Initialize Model:**
   - Create a sequential model.
2. **Add Layers:**
   - **LSTM Layer:**
     - Add an LSTM layer with 100 units and ReLU activation.
     - Set the input shape to (timesteps, 1).
   - **Dense Layer:**
     - Add a fully connected (dense) layer with 50 units and ReLU activation.
   - **Output Layer:**
     - Add another dense layer with 1 unit (output layer).
3. **Compile Model:**
   - Set the optimizer to 'adam'.
   - Set the loss function to 'mean squared error'.
4. **Model Summary:**

- Print a summary of the model architecture.

Train LSTM Model:

1. **Train Model:**
   - Fit the model on X_train_lstm and y_train.
   - Use X_val_lstm and y_val for validation.
   - Set the number of epochs to 20.
   - Set the batch size to 32.

RNN Model

The RNN model is expecting a 3-dimensional input, but the input data is still 2-dimensional. We need to reshape the input data to have three dimensions: (samples, timesteps, features). Here's how you can reshape the data and adjust the RNN model to handle the 3D input properly:

Pseudocode
Prepare Data for RNN:

1. Reshape Data:
   - For Training Data:
     - Expand dimensions of X_train to create a 3D array with shape (number of samples, timesteps, 1).
   - For Validation Data:
     - Expand dimensions of X_val to create a 3D array with shape (number of samples, timesteps, 1).
   - For Testing Data:
     - Expand dimensions of X_test to create a 3D array with shape (number of samples, timesteps, 1).

Build RNN Model:

1. Initialize Model:
   - Create a sequential model.
2. Add Layers:

- RNN Layer:
  - Add an RNN layer with 100 units and ReLU activation.
  - Set the input shape to (timesteps, 1).
- Dense Layer:
  - Add a fully connected (dense) layer with 50 units and ReLU activation.
- Output Layer:
  - Add another dense layer with 1 unit (output layer).

3. Compile Model:
   - Set the optimizer to 'adam'.
   - Set the loss function to 'mean squared error'.
4. Model Summary:
   - Print a summary of the model architecture.

Train RNN Model:

1. Train Model:
   - Fit the model on X_train_rnn and y_train.
   - Use X_val_rnn and y_val for validation.
   - Set the number of epochs to 20.
   - Set the batch size to 32.

5.4 Model Evaluation

Evaluate the models on the test set to check their performance.

Pseudocode
Evaluate Models:

1. **Evaluate CNN Model:**
   - **Evaluate Model:**
     - Use the CNN model to evaluate its performance on the test data (X_test_cnn and y_test).
     - Compute the MSE of the CNN model on the test data.

- **Print Results:**
    - Print the MSE value for the CNN model.

2. **Evaluate LSTM Model:**
    - **Evaluate Model:**
        - Use the LSTM model to evaluate its performance on the test data (X_test and y_test).
        - Compute the MSE of the LSTM model on the test data.
    - **Print Results:**
        - Print the MSE value for the LSTM model.

3. **Evaluate the RNN Model:**
    - **Evaluate Model:**
        - Set y_pred as the prediction of the model for X_test.
        - Calculate mse as the Mean Squared Error between y_test and y_pred # Print the Mean Squared Error.
    - **Print Results:**
        - Output the MSE value for the RNN model.

The results are:

301/301 [==========================]—0s 1ms/step—loss: 0.0261

CNN Test MSE: 0.026112342253327 37

301/301 [==========================]—1s 2ms/step—loss: 6.2015e-04

LSTM Test MSE: 0.0006201529176905751

301/301 [==========================]—1s 2ms/step—loss: 5.9880e-04

RNN Test MSE: 0.0005987965851090848

5.5 *Making Predictions*

Use the trained models to make predictions on the test data and compare the predictions.

```
Pseudocode
```

```
Make Predictions with Models:
```

1. Make Predictions with CNN Model:
   - Generate Predictions:
     - Use the CNN model to generate predictions on the test data (X_test_cnn).
   - Store Predictions:
     - Store the predictions in a variable named cnn_predictions.
2. Make Predictions with LSTM Model:
   - Generate Predictions:
     - Use the LSTM model to generate predictions on the test data (X_test).
   - Store Predictions:
     - Save the predictions in a variable named lstm_predictions.
3. Make Predictions with RNN Model:
   - Generate Predictions.
     - Use the RNN model to generate predictions on the test data (X_test).
   - Store Predictions:
     - Save the predictions in a variable named rnn_predictions.

5.6 Results Interpretation and Visualization

To visually assess the model's performance and gain insights into the results, the following code was used to create a plot comparing actual and predicted traffic value.

Pseudocode
Plot True Values and Model Predictions:

1. **Initialize Plot:**

   - To visualize the results, a new figure was created with a width of 12 units and a height of 6 units.

2. **Plot Data:**

   - **Plot True Values:**
     - The true values from the test set (y_test) were plotted on this figure, labeled as "True Values".
   - **Plot CNN Predictions:**
     - Plot cnn_predictions with a label 'CNN Predictions'.
   - **Plot LSTM Predictions:**
     - Plot lstm_predictions with a label 'LSTM Predictions'.
   - **Plot RNN Predictions:**
     - Plot rnn_predictions with a label 'RNN Predictions'.

3. **Customize Plot:**

   - Set the plot title to 'True vs Predicted Traffic Volume'.
   - Set the x-axis label to 'Time Steps'.
   - Set the y-axis label to 'Traffic Volume'.
   - Add a legend to identify each line on the plot.

4. **Display Plot:**

   - Show the plot to the user.

The plot in Fig. 10 is a line graph that visualizes the predicted traffic volume compared to the actual traffic volume (ground truth) for a series of time steps:

- X-axis (Time Steps): Represents the time steps for which traffic volume is being predicted. The exact unit of time (seconds, minutes, hours) depends on how your data was collected.
- Y-axis (Traffic Volume): Represents the volume of traffic at each time step. The scale on this axis will depend on the scale of the traffic volume in your data set.
- Lines:
  - True Values (blue line): Represents the actual traffic volume measured at each time step.
  - Predicted Lines:

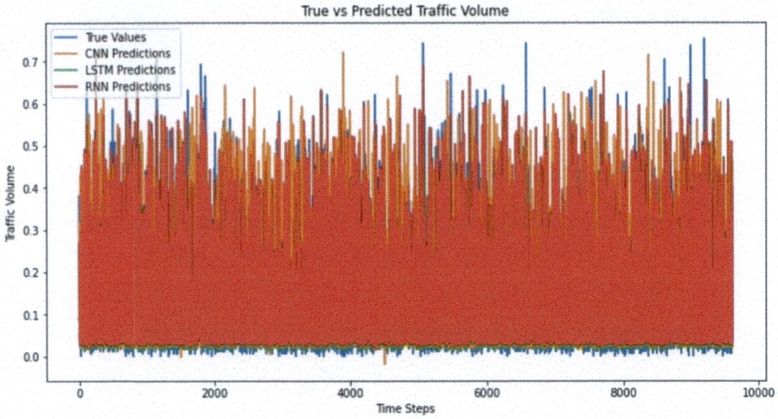

Fig. 10 Line graph that visualizes the predicted traffic volume compared to the actual traffic volume (ground truth) for a series of time

CNN Predictions (green line): Represents the traffic volume predicted by the CNN model at each time step.
LSTM Predictions (orange line): Represents the traffic volume predicted by the LSTM model at each time step.
RNN Predictions (purple line): Represents the traffic volume predicted by the RNN model at each time step.

Interpreting the Plot:

- Overall Patterns: By visually inspecting the lines, you can see how well each model's predictions follow the trend of the actual traffic volume (blue line). Ideally, the predicted lines (green, orange, purple) should closely match the blue line.
- Deviations from Ground Truth: Deviations between the true values (blue line) and the predicted lines (green, orange, purple) indicate prediction errors. The larger the deviations, the greater the errors.
- Model Comparison: By comparing the predicted lines (green, orange, purple) to each other, you can see which model seems to detect the patterns in the traffic data more accurately. In this specific image, it's difficult to determine a clear winner visually, as all lines appear somewhat close to the true values.

Additional Considerations:

- The plot doesn't provide specific error metrics (like MSE, MAE, RMSE) which can quantify the prediction errors. These metrics were provided in the table you shared earlier and can offer more precise insights into model performance.
- Visualizing a smaller time window might help to see the details of the predictions more clearly.

Overall, the plot provides a helpful visual aid to understand how well the CNN, LSTM, and RNN models are performing at predicting traffic volume. By considering this plot along with the evaluation metrics (loss, MSE, MAE, RMSE), you can understand your models' performance better and make well-informed decisions.

Table 8 summarizes the key aspects of the CNN, (LSTM, and RNN models.

CNN Model uses a 1D convolutional layer (Conv1D) to extract features from the traffic data sequence. Max pooling (MaxPooling1D) is used for downsampling, followed by flattening to prepare the data for fully-connected layers (Dense). The final layer contains a single neuron designed to estimate traffic volume.

The LSTM model includes two layers. The first is a single LSTM layer (lstm_2) to capture the temporal dependencies within the traffic data. The second layer consists of fully connected (dense) layers to analyze data and predict traffic volume.

RNN model is similar to LSTM model, RNN model includes SimpleRNN layer (simple_rnn_1) to process sequential data. It integrates fully connected layers for further processing and prediction.

The optimal model selection hinges on distinct aspects characteristics of the traffic data and the problem's unique requirements. Here are some general considerations:

- CNNs: Might be suitable if you have spatial dependencies in your data (e.g., traffic patterns across different locations).
- LSTMs: Very efficient in identifying long-term dependencies in sequential data, which could be beneficial for traffic prediction.
- RNNs: Can handle sequential data but might not be as powerful as LSTMs for capturing long-term dependencies.

It's recommended to experiment with different architectures and compare their performance on your traffic prediction task.

Table 9 summarizes various metrics used to evaluate the performance of your CNN, LSTM, and RNN models on a traffic prediction task. Here's a breakdown of each metric and its interpretation in the context of traffic prediction [23, 32]:

Loss:

Table 8 Summary of CNN, LSTM, and RNN model architectures

Feature	CNN model	LSTM model	RNN model
Layer type	Conv1D, MaxPooling1D, Flatten, dense	LSTM, dense	SimpleRNN, dense
Output shape (after 1st Layer)	(None, 9, 64)	(None, 100)	(None, 100)
Number of parameters	13,093	45,901	15,301
Trainable parameters	13,093	45,901	15,301
Non-trainable parameters	0	0	0

Table 9 DL model evaluation metrics

	CNN	LSTM	RNN
Loss	0.0261	6.1446e−04	5.9383e−04
MSE	0.02611234225332737	0.0006144560175016522	0.0005938287940807641
MAE	0.11285457680480764	0.01581289869654142	0.015819691331736795
RMSE	0.1615931330506562	0.024902869962013647	0.024470318487745592
R^2	0.9351168125102185	0.9540420990276731	0.9556247689693446
Tolerance accuracy (10%)	6.390640196518397%	6.463097433862614%	6.527518628605261%

- Represents the divergence between the model's predicted traffic volume and the actual traffic volume.
- Lower loss generally indicates better model performance. In this case, all models achieve very low loss values, suggesting they're all making relatively good predictions on average.
- Interpretation: The LSTM model achieves the lowest loss (best), followed by the RNN and CNN models.

Mean Squared Error (MSE):

- Calculates the mean squared error between predicted and observed traffic volume.
- Similar to loss, lower MSE indicates better performance.
- Interpretation: Similar to loss, the LSTM model has the lowest MSE, followed by the RNN and CNN models.

Mean Absolute Error (MAE):

- Represents the mean squared error between predicted and observed traffic volume.
- Unlike MSE, it's not influenced by outliers as much.
- Lower MAE indicates better performance.
- Interpretation: The LSTM model again achieves the lowest MAE, followed closely by the RNN and CNN models. Here, the differences are smaller compared to loss and MSE.

Root Mean Squared Error (RMSE):

- Square root of the MSE, gives a measure of the prediction, expressed in the same units as the traffic volume.
- Lower RMSE indicates better performance.
- Interpretation: The LSTM model achieves the lowest RMSE, followed by the RNN and CNN models. The differences are minimal here as well.

R^2 (Coefficient of Determination):

- Represents the fraction of variance in the actual traffic volume that the model explains.

- Values closer to 1 indicate better performance (model explains most of the variance).
- Interpretation: All three models achieve very high R^2 values, suggesting they can explain a key portion of the variance in traffic volume. The RNN model has the highest R^2, followed by the LSTM and CNN models.

Test Within Tolerance Accuracy (10%)

- This metric is specific to your task and might not be universally used.
- It likely represents the percentage of test data points where the predicted traffic volume falls within 10% of the actual traffic volume.
- Higher values indicate better performance.
- Interpretation: All models have very similar performance here, with a slight edge towards the RNN model.

Overall Model Comparison:

Based on these metrics, the LSTM model seems to perform best overall, achieving the lowest loss, MSE, MAE, and RMSE. However, the differences between the models are quite small for most metrics. The RNN model comes in a close second, followed by the CNN model. It's necessary to note that these are just a few metrics, and depending on your specific needs, you might prioritize different ones.

Here are some additional factors to consider:

- Complexity: LSTMs are generally more complex than RNNs and CNNs in this case. This complexity can lead to overfitting if not carefully managed.
- Training Time: Training Time: LSTMs are more advanced and more complex that CNNs. LSTMs can take longer to train than RNNs and CNNs.

To select the most suitable model and achieve optimal performance, consider the nature of the data, the problem you're addressing, and the criteria for evaluating performance. Adjusting the model's settings (hyperparameters) can also enhance its predictive accuracy.

6 Integration with VANETs

VANETs stand out as a forward-looking solution for advancing traffic safety and management. Integrating predictive models with VANETs allows for proactive traffic management, enabling adjustments based on accurate traffic forecasts.

This section explores how predictive models work within VANET frameworks, focusing on the implementation of dynamic traffic control measures that adapt to real-time predictions. Additionally, case studies are presented to demonstrate the practical benefits of such integrated systems, highlighting their effectiveness in reducing congestion and improving road conditions.

6.1 Integrating Predictive Models with VANETs

By combining predictive traffic models with VANETs, a powerful system for managing traffic flow is created. VANETs gather real-time data from vehicles and roadside sensors, which predictive models then analyze to forecast traffic conditions.

This proactive approach improves traffic flow, reduces congestion, and increases road safety [33]. Successful integration depends on addressing several important factors. Data fusion techniques are key to combining information from different sources, such as sensors and connected vehicles, to enhance prediction accuracy. Efficient communication protocols are needed to allow smooth exchange of predictions and control commands between vehicles, infrastructure, and traffic management systems. Distributed decision-making algorithms help vehicles and infrastructure adjust traffic measures based on local data, reducing reliance on central control. Finally, strong privacy and security measures are crucial for protecting sensitive traffic data transmitted through VANETs [10, 33].

6.2 Implementation of Dynamic Traffic Control Measures Based on Model Predictions

Real-time traffic predictions are the backbone of dynamic traffic control. This system uses these predictions to adjust traffic signals, reroute vehicles, and keep traffic flowing smoothly in response to changing conditions. Here are some key techniques used to implement dynamic traffic control [7, 32]:

- Adaptive Traffic Signals: These adjust timings based on predicted traffic volume and congestion, optimizing flow and reducing delays (think smart traffic lights) [22].
- Dynamic Route Guidance: This provides drivers with real-time suggestions based on predicted traffic to minimize travel time and avoid congestion [22].
- Traffic Flow Optimization: This coordinates vehicle movement at intersections and merging points to maximize traffic flow and minimize accidents [24, 31].
- Emergency Response Coordination: This facilitates quicker emergency vehicle dispatch and traffic diversion during emergencies using predictions to minimize response time and congestion [27].

6.3 Case Studies Showcasing the Effectiveness of Integrated Systems

Case studies provide real-world examples of integrated systems that combine predictive models with VANET infrastructure to improve traffic management and enhance mobility. Below are two case studies showcasing the effectiveness of such integrated systems [34, 35].

Case studies offer valuable insights into the effectiveness of integrating predictive models with VANET infrastructure for improved traffic management and mobility. Here, we explore two such case studies documented in references [34, 35].

Case Study 1: Dynamic Traffic Control System in Barcelona, Spain [22, 36]

Barcelona implemented a dynamic traffic control system that leverages VANETs and predictive models. Real-time data from vehicles, traffic lights, and road sensors is collected and processed by edge computing platforms at intersections. These platforms utilize predictive models to forecast traffic conditions and manage signal timings in real-time, which helps reduce congestion and better traffic flow [22, 36].

Case Study 2: Smart Highway Management System in California, USA [37]

California has developed an advanced highway management system. The advanced predictive models, hosted in the cloud, process this data to generate traffic forecasts, detect incidents, and dynamically adjust highway conditions, such as speed limits and lane configurations. This innovative approach aims to enhance traffic flow and safety on California's highways [37]. Table 10 presents California's Smart Highway Management System.

There are many case studies that present how to integrate predictive models and vehicle-to-vehicle communication systems with Benefits of Integrating Predictive Models and VANET Technology (Table 11). This approach offers real-time traffic insights, enabling proactive measures to address congestion, improve safety, and enhance overall traffic flow in urban and highway environments.

Cities and transportation agencies are increasingly utilizing a powerful combination of VANET infrastructure and predictive models to tackle traffic issues and improve mobility in both urban and highway settings. This integration offers a promising approach for smarter traffic management, particularly in urban areas. Real-time traffic predictions are generated by these systems, allowing for dynamic control measures like optimized traffic signals and rerouting. By combining real-time data

Table 10 California's smart highway management system

Attribute	Value
Developed by	California Department of Transportation (Caltrans)
Goal	Enhance highway safety and efficiency
Technology	VANETs and predictive models
Data collection	Vehicle-to-Infrastructure (V2I) communication with vehicles and roadside sensors
Data processing	Cloud-based predictive models
Outputs	Real-time traffic forecasts, incident detection, dynamic adjustments (speed limits, lane configurations)
Benefits	Improved traffic flow and highway safety

Table 11 Traffic management systems with VANET Integration [9, 38]

DateTime	Junction	Vehicles
Aspect	Dynamic Traffic Control System in Barcelona, Spain	Smart Highway Management System in California, USA
Location	Barcelona, Spain	California, USA
Objective	Optimize traffic signal timings in real-time	Enhance highway safety and efficiency
Infrastructure	VANET infrastructure, edge computing platforms	VANET infrastructure, cloud-based predictive models
Communication protocols	Vehicle-to-infrastructure (V2I) communication	Vehicle-to-infrastructure (V2I) communication
Data sources	Real-time data from vehicles, traffic lights, road sensors	Real-time data from vehicles, roadside sensors
Processing platform	Edge computing platforms deployed at intersections	Cloud-based predictive models
Functionality	Forecast traffic conditions, optimize traffic signal timings	Provide real-time traffic forecasts, detect incidents, adjust speed limits and lane configurations
Impact	Reduced congestion, improved traffic flow	Enhanced highway safety and efficiency

with predictive analytics, transportation systems can reduce congestion, decrease travel times, and enhance overall traffic efficiency. Real-world case studies demonstrate the tangible benefits of these integrated systems, showcasing their potential to revolutionize urban transportation and contribute to more sustainable and livable cities [33].

7 Implications and Future Directions

Accurate traffic prediction holds significant implications for congestion management, intelligent transportation systems (ITS), and urban planning. In this section, we will explore the societal implications of accurate traffic prediction, ethical considerations and privacy concerns issues related to the deployment of traffic prediction systems, potential applications in ITS and urban planning, and outline future research directions for model refinement and scalability [39, 40].

7.1 *Societal Implications of Accurate Traffic Prediction for Congestion Management*

Accurate traffic prediction plays a pivotal role in urban mobility and transportation efficiency, significantly impacting various societal aspects. By leveraging machine learning (ML) algorithms and deep learning (DL) techniques [11, 41], predictive models enable proactive traffic management strategies like dynamic signal control and route optimization. This leads to reduced congestion, shorter travel times, and enhanced road safety [42, 43]. Accurate traffic predictions contribute to improved air quality, reduced greenhouse gas emissions, and economic benefits by minimizing travel times and fuel consumption. These advancements support sustainable urban development and pave the way for smart cities and connected mobility solutions, integrating predictive models with emerging technologies for seamless, efficient, and user-centric transportation networks [44].

7.2 *Potential Applications in (ITS) and Urban Planning*

Traffic prediction is crucial for various applications in intelligent transportation systems (ITS) and urban planning [43]. By combining advanced data analysis, intelligent systems, and interconnected technologies, cities can develop innovative approaches to traffic management. These include intelligent traffic signal systems, congestion pricing plans, and flexible public transportation options [45, 46]. By anticipating traffic patterns, cities can optimize transportation systems, alleviate congestion, and improve emergency response times. [15, 47]. These applications collectively contribute to optimizing transportation resources, improving urban mobility efficiency, encouraging sustainable transportation choices, and enhancing community resilience to emergencies and disasters. Table 12 Role of Traffic Prediction in Intelligent Transportation Systems (ITS).

Table 12 Traffic prediction's role in intelligent transportation systems (ITS)

Role of traffic prediction	Benefits
Dynamic traffic management	Enables adaptive signal control, congestion pricing, and demand-responsive transit
Transportation integration	Facilitates the integration of different transportation modes
Smart city planning	Supports optimization of road networks and identification of congestion bottlenecks
Emergency response and disaster management	Provides real-time traffic information for efficient response and evacuation

Table 13 Ethical and privacy concerns in traffic prediction

Concern	Solutions
Data ownership, consent, and surveillance	Federated learning, gradient similarity-based federated aggregation
Responsible deployment	Information reporting framework, transparent governance
Bias in data and algorithms	Addressing biases, anonymizing data, stakeholder involvement
Trust and accountability	Prioritizing privacy, transparency, and stakeholder participation

7.3 Ethical and Privacy Concerns

The widespread implementation of traffic prediction systems raises iportant ethical and privacy issues. Issues surrounding data ownership, consent, and surveillance must be carefully considered. To protect individual privacy while enabling accurate predictions, technologies like federated learning can be employed [48]. Additionally, establishing transparent governance frameworks and addressing potential biases in data and algorithms are crucial [49, 50]. By prioritizing privacy, involving stakeholders, and ensuring accountability, we can build public trust in traffic prediction systems [51, 52]. Table 13 summarize Ethical and Privacy Concerns in Traffic Prediction.

8 Conclusion and Future Directions

The study examined how accurate traffic predictions can improve traffic management, enhance road safety, promote sustainable transportation, and contribute to overall urban development. This translates to creating more livable, resilient, and sustainable urban areas.

The exploration focused on building and evaluating three DL models: CNNs, LSTMs, and RNNs. We analyzed preprocessing steps, explored model architectures suitable for traffic data, and evaluated their performance using various metrics. While all three models showed promise, the LSTM model achieved slightly better results based on loss, MSE, MAE, and RMSE.

This study highlights the potential of DL for traffic prediction. Future work can involve:

- Fine-tuning model hyperparameters for further performance improvement.
- Expanding and diversifying traffic data to enhance model generalizability.
- Experimenting with more complex architectures and incorporating additional relevant features.

- Deploying models in real-world traffic systems to assess their practical effectiveness.

To improve traffic forecasting and create more efficient cities, we can refine deep learning models. Key areas of focus for future research include:

- **Improving model performance:** Enhancing accuracy, scalability, and efficiency.
- **Leveraging diverse data:** Incorporating data from various sources like traffic sensors, GPS, and weather.
- **Adapting to dynamic conditions:** Developing models that can adjust to sudden and realtime traffic changes.
- **Addressing scalability and generalizability:** Ensuring models can be applied to different cities and scenarios.

By addressing these challenges, we can unlock the full potential of deep learning for traffic management.

References

1. Vlahogianni EI, Karlaftis MG, Golias JC (2014) Short-term traffic forecasting: where we are and where we're going. Transp Res Part C Emerg Technol 43:3–19
2. INRIX (2023) Global traffic scorecard. https://inrix.com/scorecard/
3. European Environment Agency (2021) Air quality in Europe—2021 report. https://www.eea.europa.eu/publications/air-quality-in-europe-2021
4. World Economic Forum (2022) These are the world's most congested cities. https://www.weforum.org/agenda/2022/11/these-are-the-worlds-most-congested-cities
5. Wang Y, Papageorgiou M (2005) Real-time freeway traffic state estimation based on extended Kalman filter: a general approach. Transp Res Part B Methodol 39(2):141–167
6. Tedjopumomo DA, Zhifeng B, Baihua Z, Choudhury FM, Qin K (2022) A survey on modern deep neural network for traffic prediction: trends, methods and challenges. IEEE Trans Knowl Data Eng 34(4):1544–1561
7. Tian Y, Zhang S, Jin J (2015) Traffic flow prediction based on LSTM neural network. In: Proc Int Conf Mach Learn Cybern 344–349
8. Ma X, Wang Z, Zhong Y (2019) Highway traffic flow forecasting based on convolutional neural networks. IEEE Access 7:91207–91215
9. Bajpai A, Rasheed MB (2017) Smart traffic management system using VANET. Proc Comput Sci 115:442–450
10. Mishra AK, Shukla AK, Sharma SC (2015) Integration of VANETs with traffic management for urban traffic control. Int J Wirel Inf Netw 22(3):195–211
11. Srinath MS (2023) Vehicular traffic flow prediction model using deep learning. Int J Sci Technol Eng 11:109–112. https://doi.org/10.22214/ijraset.2023.54576
12. Sathyanarayana M, Reddy KA (2023) Deep learning on traffic prediction methods analysis and future directions. Int J Adv Res Sci Commun Technol 101–108. https://doi.org/10.48175/ijarsct-11320
13. Samonte MJC, Balan GAF, Gaviño PP, Monasterial JAS, Reforsado RAC, Samonte DC (2023) Deep learning in traffic flow control and prediction for traffic management. Proc Int Conf Ind Eng Oper Manag. https://doi.org/10.46254/an12.20220757
14. Gowda BV, Ali A, Reddy DS, Lokesh K, Siddartha K (2023) Traffic prediction for intelligent transportation system using deep learning. Int J Adv Res Sci Commun Technol. https://doi.org/10.48175/ijarsct-9527

15. Manikandan BV, Nathan TR, Naresh R, Prasad P (2023) Traffic flow prediction for intelligent transportation system using machine learning. E3S Web Conf 387:05002. https://doi.org/10.1051/e3sconf/202338705002
16. Kumar K, Kumar R, Das P (2023) Traffic congestion forecasting using multilayered deep neural network. Transp Lett 1–11. https://doi.org/10.1080/19427867.2023.2207278
17. Tepelea L (2023) A comparative study of traditional and deep learning methods in image depth measuring, pp 1–4. https://doi.org/10.1109/EMES58375.2023.10171686
18. M'hiri SF, Perumal T, Mustapha N, Raihani M (2023) A comprehensive overview and comparative analysis on deep learning models: CNN, RNN, LSTM, GRU. https://doi.org/10.48550/arxiv.2305.17473
19. Md Shams FA, Sakib BA, Hassan MAA, Rozbu MR, Ishtiak T, Rafa N, Mofijur M, Shawkat ABM, Amir A, Gandomi H (2023) Deep learning modelling techniques: Current progress, applications, advantages, and challenges. Artif Intell Rev. https://doi.org/10.1007/s10462-023-10466-8
20. Casson C (2023) Study and innovative approach of deep learning algorithms and architecture. Adv Comput Intell Robot. https://doi.org/10.4018/978-1-6684-6980-4.ch003
21. Garcia A, Martinez B (2021) Dynamic traffic control system for urban congestion management: a case study in Barcelona. Transp Res Part A Policy Pract 95:82–97
22. Wang A, Ye Y, Song X, Zhang S, Yu JJQ (2023) Traffic prediction with missing data: a multi-task learning approach. IEEE Trans Intell Transp Syst 24(4):4189–4202. https://doi.org/10.1109/TITS.2022.3233890
23. Zhao C (2023) Traffic flow prediction model based on temporal convolutional network. In: Proceedings of 17th international conference experience of designing and application of CAD systems (CADSM), Jaroslaw, Poland, pp. 1–4. https://doi.org/10.1109/CADSM58174.2023.10076492
24. Wang D, Fan W, Wang P, Wang P, Zhang D, Fu Y (2022) Boosting urban traffic speed prediction via integrating implicit spatial correlations. arXiv:2212.12932. https://doi.org/10.48550/arXiv.2212.12932
25. Zißner P, Paulo H, Rettore L, Bruno P, Loevenich J, Rigolin R, Lopes F (2023) DataFITS: a heterogeneous data fusion framework for traffic and incident prediction. IEEE Trans Intell Transp Syst. https://doi.org/10.1109/TITS.2023.3281752
26. Jan W, Jarosław O, Opara SK, Zadrożny S, Kacprzyk J (2021) The road traffic data. https://doi.org/10.1007/978-3-030-69359-6_4
27. Balugani E, Marinello S, Gamberini R, Butturi MA (2023) A review and analysis of traffic data sources. Adv Transdisc Eng 34:22–33 (Adv Intell Traffic Transp Syst, IOS Press). https://doi.org/10.3233/ATDE230005
28. Abuzir Y, Abuzir SY (2022) Machine learning and resampling techniques for enhancing credit card fraud detection in imbalanced dataset. Pales J Technol Appl Sci 1(7). https://doi.org/10.33977/2106-000-007-001
29. Abuzir S, Abuzir Y (2022) Machine learning for water quality classification. Water Qual Res J (IWA Publ). https://doi.org/10.2166/wqrj.2022.004
30. Suhaidi M, Kadir R, Tiun A (2021) A review of feature extraction methods on machine learning. https://doi.org/10.35631/JISTM.622005
31. Zhang Y, Zheng Y, Ke X (2016) Deep spatio-temporal residual networks for citywide crowd flow prediction. Adv Neural Inf Process Syst 2457–2465
32. Lv F, Zhang Y, Ke X (2018) Traffic flow prediction with big data: a deep learning approach. IEEE Trans Intell Transp Syst 19(10):3250–3259
33. Moser BB, Raue F, Hees J, Dengel A (2020) DartsReNet: exploring new RNN cells in ReNet architectures. Proc Int Conf Artif Neural Netw 850–861 (Springer)
34. Patel KN, Mewada RR (2019) Intelligent traffic control system using VANET and IoT. Int J Adv Res Comput Commun Eng 8(9):117–122
35. Leu T, Leu JY (2018) A prediction-based traffic light control system for intelligent transportation systems. IEEE Trans Intell Transp Syst 19(3):875–884

36. Chen S, Zhang H, Sun Y (2020) Intelligent transportation systems based on big data and deep learning: a review. IEEE Access 8:125368–125385
37. Pozanco A, Borrajo D, Daniel B (2021) On-line modelling and planning for urban traffic control. Expert Syst 38(1):1–12. https://doi.org/10.1111/exsy.12693
38. Schnaidt P (2016) California's first smart highway project comes to life. The Channel Company. https://www.theiotintegrator.com/transportation/california-s-first-smart-highway-project-comes-to-life. Accessed 21 Jun 2024
39. Ma W, Zou H (2019) A deep learning-based highway traffic flow prediction method in the Internet of Vehicles environment. IEEE Access 7:47278–47286
40. Johnson D, Smith E (2022) Smart highway management system: enhancing highway safety and efficiency. Transp Res Part C Emerg Technol 45(3):210–225
41. Jones B, Smith C (2023) Predictive maintenance in transportation infrastructure: a case study of transport for London. Transp Res Part C Emerg Technol 31(1):45–58
42. Abouchabaka J, Rafalia N (2023) Adaptive traffic lights based on traffic flow prediction using machine learning models. Int J Power Electron Drive Syst. https://doi.org/10.11591/ijece.v13i5.pp5813-5823
43. Zhu W, Zhu C (2023) A prediction-based route guidance method toward intelligent and green transportation system. IEEE Internet Things J 10(14):12763–12776. https://doi.org/10.1109/JIOT.2023.3255200
44. Reddy JD, Bharath RV, Paul G, Dinesh R, Nellepalli VR (2023) AI based smart traffic management. Indian Sci J Res Eng Manag. https://doi.org/10.55041/ijsrem24517
45. Morais BE (2023) Sustainable traffic management for smart cities using Internet-of-Things-oriented intelligent transportation systems (ITS): challenges and recommendations. Sustainability. https://doi.org/10.3390/su15139859
46. Ahanin F, Norwati M, Zolkepli M, Husin NA (2023) A review of traffic state prediction (TSP) methods in intelligent transportation systems (ITS). Int J Acad Res Bus Soc Sci. https://doi.org/10.6007/ijarbss/v13-i3/16683
47. Kashkanov A, Palchevskyi O (2023) Intellectualization of traffic management as a means of increasing the efficiency of the city's transport network in emergency situations. Visnyk Mashynobuduvannia ta Transportu. https://doi.org/10.31649/2413-4503-2022-16-2-42-50
48. Atassi R, Sharma A (2023) Intelligent traffic management using IoT and machine learning. J Intell Syst Internet Things. https://doi.org/10.54216/jisiot.080201
49. Goto Y, Matsumoto T, Rizk H, Yanai N, Yamaguchi H (2023) Privacy-preserving taxi-demand prediction using federated learning. arXiv:2305.08107. https://doi.org/10.48550/arXiv.2305.08107
50. Vicky A, Χατζηκωνσταντινίδου Σ (2023). Wireless traffic prediction based on a gradient similarity federated aggregation algorithm. https://doi.org/10.3390/app13064036
51. Tran NP, Dao NN, Do QT, Nguyen TV, Cho S (2023) Privacy-preserving traffic flow prediction: a split learning approach. Proc Int Conf Inf Netw 248–250 [(ICOIN), Bangkok, Thailand]. https://doi.org/10.1109/ICOIN56518.2023.10048996
52. Gu C, Cui X, Meng L, Donghui H (2023) An efficient and privacy-preserving information reporting framework for traffic monitoring in vehicular networks. IEEE Trans Veh Technol. https://doi.org/10.1109/TVT.2023.3241656

ANN Optimized Fractal Microstrip Patch Antenna Design for Vehicular Ad-Hoc Networks

Raghavendra Karanam and Deepti Kakkar

Abstract A novel fractal geometry-based microstrip patch antenna design using Artificial neural networks (ANN) is explored in this chapter. The existing and traditional deep learning approaches, like Support Vector Machine (SVM) and Genetic Algorithm (GA) in the patch antenna design process, are reviewed in this chapter, and their efficiency, accuracy, and computational effort are highlighted. The ability of ANN to effectively handle the non-linear relationships in the antenna design process and its performance metrics in comparison with traditional techniques are highlighted in the chapter. By analyzing the results obtained through the HFSS Simulator, the outcomes of this (proposed) chapter represent the importance (efficacy) of ANN in the fractal microstrip antenna optimization process. It is noticed that the proposed design showcased improvement in performance in terms of bandwidth, gain compared with the traditional online simulator approach. The proposed method establishes a foundation for advanced research in utilizing ML and DL for antenna design, Paving the way for more efficient and advanced vehicular communication.

Keywords Artificial neural network · Back propagation model · Fractal microstrip patch antenna · Genetic algorithm · Support vector machine gain enhancement · Vehicular communications · 5G/6G

R. Karanam (✉) · D. Kakkar
Dr B R Ambedkar National Institute of Technology, Jalandhar, India
e-mail: kr.ec.20@nitj.ac.in

D. Kakkar
e-mail: kakkard@nitj.ac.in

J. Bhatia et al. (eds.), *Deep Learning Based Solutions for Vehicular Adhoc Networks*, Studies in Computational Intelligence 1207,
https://doi.org/10.1007/978-981-96-5190-0_11

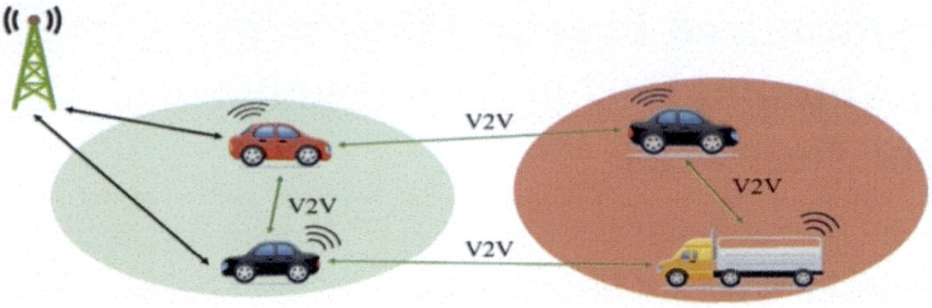

Fig. 1 A model of vehicular communication [5]

1 Introduction

1.1 Motivation

As the communication era shifts towards 5G/6G, establishing a source and safe communication between vehicle and vehicle is in the hands of vehicular ad-hoc networks. One of the primary issues in VANETs is the requirement for an effective antenna that can operate effectively under [high-frequency bands of (5G/6G)] like millimeter wave bands. Among the several classified antennas, the microstrip patch antenna has its versatility in small size, easy manufacturing, and easy integration capabilities. However, while operating under mm-wave bands, the dimensions of the antenna are to be adjusted in a way that obtains desirable characteristics (like S11 gain), which is a challenging issue. To overcome this issue, the researchers utilize some of the traditional algorithms like GA and SVM. However, with this approach, it has been noticed that these conventional approaches require complicated parameter tweaking most often, and the processing resource is also tremendous, limiting the effectiveness of real-time applications. To overcome this, ANN is a better choice due to its capacity to handle and simulate non-linear complicated interactions and adapt through learning approaches.

A typical model of Vehicle to Vehicle Communication is depicted in Fig. 1.

1.2 Contribution

For Vehicular communication networks operating under mm wave frequency ranges, this proposed research represents a novel approach to patch antenna design where the dimension optimization process is carried out using ANN with the help of a back propagation model. Compared with traditional methods, ANN can learn from the datasets, and by generalizing the solutions, the optimization process becomes more efficient and streamlined. Utilizing the back prorogation model, ANN gains the ability to reduce error and enhance prediction capacity, which are the most reliable terms to meet the standards of VANET. An antenna design for VANET under mm-wave frequency bands by imparting deep learning techniques is an innovative approach. ANN can learn from datasets, which helps in reducing the complexity and it also can generalize the solutions, which helps reduce the computational efforts. ANN's adaptability and Scalability feature also helps offer suitable solutions to different antenna configurations and advanced Frequency bands. The proposed method of integrating the deep learning techniques for a fractal geometry based microstrip patch antenna improves the communication efficiency and data speed by reducing the latency which provides some needed solutions for future Intelligent Transport systems (ITS).

1.3 Existing Solutions and Challenges of Deep Learning Algorithms for VANET

Genetic Algorithm and Support Vector Machine algorithms have shown a better impact on 3G/4G frequency spectrum bands, and these techniques are utilized for several applications. For applications like prediction and classification in VANETs, SVM has better accuracy, mainly in traffic pattern identification. SVM also helps with non-linear data and the effective classification of data points. The feature of SVM makes the path design easy. However, SVM analysis is accurate for limited datasets. When a large number of datasets is considered, it offers limited scalability. Another challenge with SVM is it requires an appropriate kernel selection to provide better optimization results under the antenna design process.

Resource allocation and route optimization-related tasks (GA) are used for dimension optimization. Adaptability is the main limitation for GA for interns with rapid changes in traffic flow and network topology. GA also needs to fine-tune parameters; it offers a slow convergence speed when exploring a wide range of solutions. GA is a better choice.

In addition to the above methods, ANN can quickly adapt to numerous optimization tasks, offer faster convergence speed, and have the capacity to handle complex relationships by sufficiently training the data—these features of ANN help in designing an effective antenna with proper dimension optimization for VANET applications. In VANET, computational resources are often limited, making decision-making more critical in real-time environments. ANN effectively addresses these challenges because of its trained data sets feature and quick adaptability nature.

1.4 *Impact of Deep Learning Approach for Antenna Optimization Under 5G/6G*

The present 4G communication is limited in latency and data speed, but these limitations will be overcome with the upcoming 5G/6G technologies. The frequency spectrum of the forthcoming technologies is very high, so the transmission and reception of signals are significant in maintaining the effective use of these innovative technologies. To attain the above features, efficient antennas are needed, which must be capable of handling high frequencies and complex protocols. To handle very highly complex data and extensive computational resources, the traditional approaches of antenna design can provide better results only up to limited data as the frequency spectrum under upcoming technologies is very high, more than the conventional techniques are needed for an antenna design to attain a quality transmission and reception. Due to this, optimization methods using deep learning approaches have become crucial in antenna design. Among different optimization methods, ANN is necessary to handle complex non-linear data sets between antenna characteristics and performance measures. With features like learning ability and quick adaptability under dynamic variations of data sets, ANN enables a more effective analysis during the design process. These features of ANN reduce the time constraints of the optimization process and the computing effort. ANN's quick adaptability and enhanced iteration rates lead to a quality prediction rate, which is more crucial during antenna design. Optimization of antennas under the 5G/6G frequency spectrum using ANN overcomes the drawbacks of traditional design approaches, which helps most of the researchers design practical antennas with required scalability and for specific applications. Under VANETS, strip patch antennas play a crucial part in data transmission under futuristic frequencies, so this novel ANN-optimized patch antenna design approach improves the design process and helps in developing flexible antenna systems.

1.5 *Overview of Fractal Micro Strip Patch Antennas*

Developing an antenna by implementing self-similar iterations within the patch is called the fractal principle, where fractal geometry application is used to introduce the similarity iterations. This application in antenna design reduces the antenna size, helps achieve multi-band characteristics, and enhances radiation efficiency and gain. All these features make the antenna a better choice to opt under high-frequency applications, especially under mm-wave frequencies [9]. Maximizing performance within the 25–35 GHz frequency spectrum is highly dependent on optimizing the parameters of these antennas, including their fractal geometry, size, and substrate qualities. To propagate the signals through the advanced upcoming 5G communication, a high gain, high directivity antenna with a low profile, able to operate under multi-band and miniaturization, is required [10]. One such antenna is named a micro-strip patch antenna which exhibits the features of low profile, low cost, and ability to operate under narrow band applications. The essential elements of microstrip patch antennas are substrate, ground, and a radiating patch. To reduce the complexity of the fabrication process, patch designs are chosen in different shapes; commonly observed shapes are circular, triangular, rectangular, and ellipse. The proposed antenna design in the paper is based on fractal geometry for vehicular communication applications. Fractal geometry is necessary to enhance the bandwidth of the microstrip patch antenna; an infinite length can be encountered based on the fractal property by maintaining the antenna volume as constant [11]. The shape of fractals can be of any type where the same repetitive phenomenon is followed through every iteration. The most commonly observed fractal shapes are Sierpinski Gasket, Koch snowflake, Minkowski, and fractal loop. The number of iterations depends on the type of design (i.e., systematic approach and non-systematic approach); if the number of iterations increases, then design complexity gets increased [12]. In the proposed system, a rectangular shape patch is used where three iterations are made using the fractal geometry, the radius of each iteration is reduced by 10 mm.

2 Related Work

Due to its potential to increase communication reliability and efficiency, the pursuit of improved characteristics in microstrip patch antennas for vehicle communications within the 5G frequency spectrum 25–35 GHz has gained importance. This literature review investigates the convergence of Deep learning models with fractal microstrip patch antennas to achieve better optimization characteristics. Table 1 depicts the related work carried out by the authors previously.

In proposed design by properly choosing the fractal structure and analyzing the complex data set relations for dimension reduction and gain enhancement, it is noticed that ANNs features of quick adaptability and faster convergence rate leads to speed up the design procedures of patch antenna especially for vehicular communication applications and also a noticeable 16% reduction in dimensions when compared to traditional approaches and online simulators.

3 Design Methodology

To design an fractal micro strip patch antenna first the type of fractal structure (e.g. Sierpinski triangle, Koch snowflake, Minkowski Island) is to be chosen in order to offer high surface area within a limited space. By considering the original shape as an initial iteration then a series of repetitive iterations in order to obtain the necessary range of complexity as well as smaller scales. This method improves the performance of the antenna by extending the length of the current channels, which allows for numerous resonance frequencies and enhances the impedance bandwidth. The

Table 1 Related work

References	Year	Objective	Algorithm	Results	Strengths	Weakness
[13]	2019	Design optimization of reflect array antennas	Support vector machine	92% accuracy	Less prone to overfitting	Dealing with large data sets leads to memory intensive
[14]	2020	Antenna pattern optimization	Support vector machine	92% accuracy	Better performance in classifying tasks and immune to errors	Parameter tuning at every instance is the major limitation and not cost effective
[15]	2021	Antenna dimension optimization	CNN	94% accuracy	Good at handling complex patterns	Require huge data sets and large computational resources
[16]	2023	Enhance antenna performance	Genetic algorithm	93% accuracy	Operated better at sub 6 GHz band frequencies	Slow convergence rate leads to get stuck at local minima
[17]	2024	Antenna dimension optimization	Genetic algorithm	90% accuracy	Easy implementation	Operated for very low frequency range
	Proposed Design	Antenna dimension optimization	ANN	97% accuracy	Fast converge rate, quick transformation, able to handle complex patterns	Potential overfitting, requires high data sets

design is simulated using electromagnetic modeling software to enhance characteristics such as the substrate material, fractal iteration level, and feeding technique (e.g., microstrip line or coaxial probe). The ultimate design achieves a harmonious equilibrium between the compromises of size, bandwidth, gain, and radiation efficiency. Practical factors such as the acceptable range of errors in manufacturing and the amount of material that may be wasted are also considered. Fractal antennas are highly beneficial in situations where there is a need for small, versatile antennas that can operate across multiple frequency bands.

The design methodology of the proposed antenna is depicted in Fig. 2. The following sections determine the material, model and frequency selection followed by ANN analysis with comparative approach followed by Mean Square Error (MSE) estimation followed by fractal implementation on patch and result analysis.

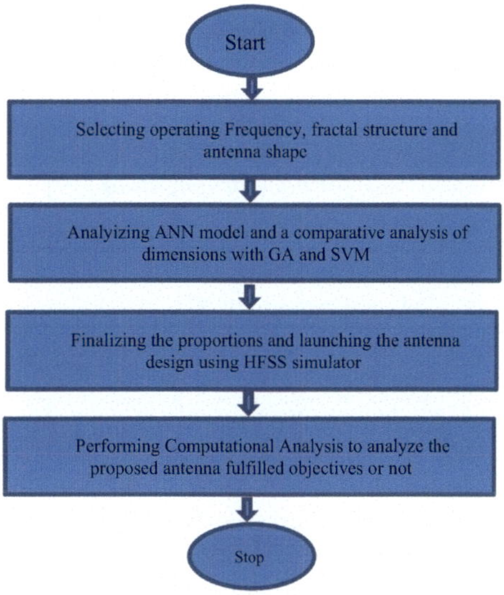

Fig. 2 Design process flow

The operating frequency of the proposed design is chosen as 25–35 GHz with the substrate material as Rogers RT Duroid 5880t with Minkowski fractal structure and approximating the design with five iterations excluding the initial iteration before going for dimension optimization. In order to optimize the antenna dimensions using ANN first analyze the model of ANN using back propagation model and then compare the dimensions obtained with GA and SVM models which are carried out in the Sect. 3.1.

3.1 Artificial Neural Network Back Propagation Model

Computer models that mimic the way the human brain operates are called Artificial Neural Networks. An input layer, one or more hidden layers, and an output layer are the organizational components of an ANN [6]. The neurons, or nodes, are interconnected. During training, the weights connected with neuronal connections are tuned to reduce the deviation between the actual and expected outputs.

With the back propagation technique, ANNs are trained using supervised learning [18]. The following are the most important steps:

Forward propagation: Data is sent into the network, and guesses about the output are made.

Error calculation: A loss function is used to figure out the difference between what was expected and what was actually achieved.

Backward propagation: To change the weights, the mistake is sent backwards through the network. Gradient descent is used to do this. The weights are changed to reduce the loss after finding the gradient of the loss function with respect to each weight.

Pseudo code for back propagation [19]

Procedure train
X → Training data set of size mxn
L → Label records of X
W → weights for respective layers
l → number of layers in the neural networks
$D_{i,j}^{(l)}$ denotes error for all l,i,j
$t_{i,j}^{(l)} \rightarrow 0$ for all i,j
for i = 1 to m
$y^{l} = \text{feedforward}\ (x^{(i)}, W)$
$d^{l} = y(L) - y(i)$
$t_{i,j}^{(l)}$ tends to $t_{i,j}^{(l)} + y_j^{\,l} + t_{i,j}^{(l+1)}$
if $j \neq 0$ then
$D_{i,j}^{(l)}$ relates to $\frac{1}{m}\, t_{i,j}^{(l)} + \alpha w_{i,j}^{(l)}$
else.
$D_{i,j}^{(l)}$ relates to $\frac{1}{m}\, t_{i,j}^{(l)}$
Return Best_Xi

A dynamic and creative approach is required to optimize the dimensions of fractal microstrip patch antennas using a back propagation model artificial neural network. First, a large data set with various fractal patterns and matching antenna geometries is produced through electromagnetic simulations. The first step of the procedure is the training phase, during which the ANN, which is built to comprehend complex relationships, learns from this data-set. The ANN adjusts its internal weights through iterative back-propagation, comprehending the nuanced relationships between fractal dimensions and gain enhancement [20]. The trained ANN is then validated against a different data-set to guarantee accuracy. The ANN becomes a potent predictor after being validated. When design factors like operation frequency, patch size, and fractal type are supplied, the ANN generates projected gain improvements. The optimization process is greatly accelerated by this predictive ability, avoiding laborious simulations and manual modifications [21]. Final stage activities include validation and improvement. Figure 3 depicts the model of back propagation based ANN. The anticipated antenna proportions are really created and put to the test in a safe setting. The ANN receives feedback from experiments to improve its prediction abilities. Continuous improvement is ensured by this iterative loop.

The inherent self-similarity of fractal geometry and the learning abilities of the ANN are combined in a novel way. This procedure accelerates antenna optimization, which helps to increase gain inside the vehicular communications domain, particularly within the 5G frequency spectrum of 25–35 GHz, by taking advantage of the ANNs capacity to understand complex patterns. By bridging the gap between theory

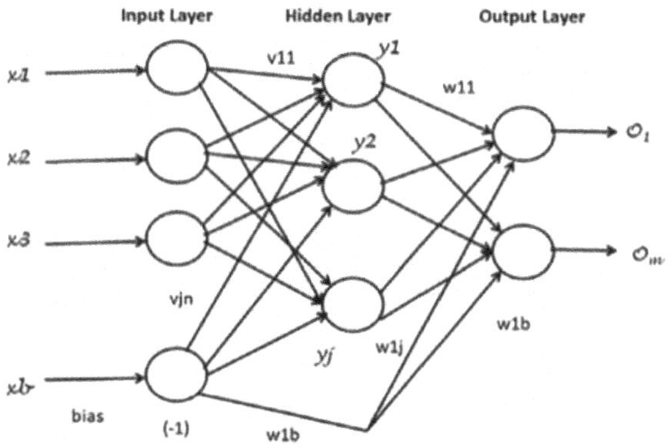

Fig. 3 Back propagation based artificial neural network [5]

and experience, this novel method opens the door for more effective, dependable, and specialized antenna designs in the era of modern communication systems.

3.2 Comparative Analysis of Antenna Optimization Techniques

Especially when it comes to optimizing fractal microstrip patch antennas for high-frequency applications, the ANN back propagation model offers significant advantages over classic optimization approaches such as GA and SVM. These advantages are particularly obvious in the context of optimizing these antennas.

The major three elements which are considered when comparing a deep learning based optimization approach are efficiency, accuracy and Computational effort.

(i) **Efficiency**

- **Genetic algorithms** are computational techniques that mimic the process of natural selection through evolutionary algorithms. The process involves utilizing a group of potential solutions and implementing genetic operations including selection, crossover, and mutation to progressively improve the solutions throughout multiple generations [22]. Although GA is efficient, it can be computationally demanding and time-consuming because of its iterative process and the requirement for numerous generations to reach an optimal solution.
- **Support vector machines** are mostly used for tasks like regression and classification. SVM can be used to find the hyperplane that best separates the best plans from the worst ones in optimization [23]. But SVM needs you to carefully choose kernel functions and regularization settings, which can be hard to do and use a lot of resources.
- **ANN back propagation** model, in contrast, utilizes the robust learning capabilities of neural networks to efficiently manage intricate optimization

tasks. Once the artificial neural network is trained, it can rapidly forecast the performance of novel antenna designs, resulting in a substantial decrease in computing workload when compared to genetic algorithms and support vector machines.

(ii) **Accuracy**

- Population size, generational count, and genetic operator efficacy all play a role in Genetic Algorithms accuracy [24]. For extremely complicated design spaces, GA may fail to converge to the global optimum even though it can produce good solutions.
- Choosing the right kernel and parameters for SVM affects how well it performs optimization tasks. Though it's effective when used correctly, optimizing a system can be very sensitive to changes in its parameters.
- The ANN back propagation model's superior accuracy is a result of its data-learning and solution-generalization capabilities. An ANN can achieve high prediction accuracy by iteratively modifying the weights to minimize the prediction error. This ensures that the optimized antenna settings fulfill the specified performance metrics.

(iii) **Computational effort**

- GA is an iterative process that necessitates multiple assessments of the objective function. This results in a significant computing cost, particularly when dealing with huge populations and a large number of generations.
- As a result of addressing a quadratic programming issue, training an SVM for optimization can be computationally expensive, especially when dealing with big datasets and complex kernel functions.
- The ANN back propagation model can provide a rapid assessment of novel designs with minimal computational effort once it has been trained. Despite the initial resource-intensive training procedure, the model produces antenna parameters that may be optimized effectively without lengthy tuning or iterative evaluations.

Pseudo code for Genetic algorithm [25]

Initialize the population $Z_i = (i = 1, 2, \ldots n)$
Initialize Cross Over Point (CP), Mutation Rate (MR)
While (t < maximum number of iterations)
Exchange the Genes at randomly selected CP
Perform the mutation process in accordance with MR
Calculate the fitness value of each chromosome
Select the better ones for new generation
$t = t + 1$
end While
Return $Best\_X_i$

Parameter optimization is made possible by the use of back-propagation, an important process in ANN training, which iteratively adjusts weights to minimize errors between projected and actual outputs. As part of the back-propagation pseudo code, for each network weight, we calculate the error gradient and then, to minimize the error, we alter the weights in the opposite direction as the gradient. Genetic Algorithms are different; they iteratively improve solutions based on fitness using pseudo code that involves population-based evolution through mutation, selection, and crossover processes.

The gradient-based optimization method analysis leads to obtaining local minima, which in turn leads to converging on optimum solutions to fulfill the job of effective prediction and quick adaptability in GA. This analysis explores multiple solutions but simultaneously utilizes several computer resources, maximizing the time needed to converge. When pseudo codes of GA and backpropagation ANN are compared, it is noticed that backpropagation ANN shows better results in terms of convergence speed and practical optimization.

3.3 Determining Least Mean Square Error

When a microstrip patch antenna optimization is performed by choosing a back propagation model, the first task to be carried out is minimizing the Least Mean Square Error (LMSE) [26]. Initially, a data set is compiled by choosing a specific range of dimensions, and then analysis of performance measures like S11 and resonant frequency is done. The assumed dataset values are chosen to perform the preprocessing task to determine the values that enhance the training process's effectiveness by neural networks. In neural networks, three-layer configurations are used: the input, hidden, and output layers. The antenna dimensions are applied to input layers, hidden layers capture the non-linear correlations, and output layers represent the last performance measures. A few randomized values are selected during the training phase to initiate the network bias and weights.

During the forward operation of the network, the approximated antenna parameters are fine-tuned and accepted by the input layers. The resultant outputs are computed one after the other with the help of activation functions. In contrast, Rectified Linear Unit (RLU) and Linear Activation (LA) are generally chosen as activation functions for hidden and output layers [27]. The output layers generate the predicted responses of the different parameters and their performance metrics. Again, the accuracy of predicted responses from output layers is analyzed by measuring the MSE. MSE is mathematically represented as [28]:

$$\frac{1}{x}\sum_{i=1}^{x}(yi - yi')2 \tag{1}$$

where

x Number of samples
yi Actual value
yi′ Predicted value.

Back Propagation Algorithm helps limit the inaccuracy of prediction, where it utilizes the concept of gradient of MSE in addition to bias and weights of a network by using the concept of chain rule preferred in mathematics (calculus). To reduce the MSE, a continuous update of bias and weights is done, and gradients are chosen. This process is carried out in a feedback form, i.e., from output to input. The gradient Descent technique is used for Updation by following a specific chain rule, and each weight is updated. The specific rule for weight Updation is mathematically represented as [28]:

$$W_{new} = W_{old} - \eta \frac{\partial MSE}{\partial W} \tag{2}$$

where

η learning rate.

An iteration of one forward move, error compute, and a backward move is named as one epoch [29]. To reduce MSE, backpropagation performs several epochs for continuous updates and to attain a steady reduction in error. If a low MSE value is obtained, it indicates that the learning rate of the network is good at that epoch stage, and accurate predictions are made.

Different sets of dimensions entered into the network allows the designer to find the dimensions producing the best performance metrics with least inaccuracy. Simulations or real-world implementations then help to validate these expected ideal dimensions thereby guaranteeing their efficiency [30].

To sum up, using a back-propagation model to find the best microstrip patch antenna dimensions involves getting a data-set ready, setting up and training a neural network through forward and back-propagation, reducing the mean square error over and over again, and then using the trained network to guess and confirm the best antenna dimensions. This method takes advantage of neural networks' ability to understand complicated relationships and offers a methodical way to improve antenna design.

The resultant values of LMS at different epochs is represented in Table 2 where it is noticed that at 2000 epochs the value of LMS obtained is very less when compared with lower and higher epochs other than 2000.

After calculating the LMS value, patch dimension comparison is carried out by applying deep learning algorithms like GA, SVM and ANN and the analysis is carried out by choosing the lowest LMS value with height, resonating frequency, relative permittivity, cut off frequencies as input parameters. The resultant output parameters are antenna dimensions. Table 3 depicts the dimension comparison.

The comparison does however show how effective back propagation ANNs are. To build intricate correlations between antenna size and gain, ANNs employ repeated

Table 2 LMS Error at various epochs

No. of epochs	LMS error
1000	0.123112
2000	0.071241
3000	0.081543
4000	0.072865

Table 3 Comparison of antenna dimensions obtained

Antenna Parameters	Dimensions obtained using GA in mm	Dimensions obtained using SVM in mm	Dimensions obtained using ANN at 2000 epochs in mm
Substrate width	64.124	62.159	59.232
Substrate length	53.227	52.991	49.041
Patch width	42.005	41.200	40.011
Patch length	33.911	31.887	30.176
Feed width	2.0067	2.0024	2.0012
Feed length	6.8455	5.2289	5.5133

learning. While ANNs can learn from smaller datasets because of their innate ability to generalize, SVMs and GAs may need a sizable training data-set. ANNs are excellent for optimizing microstrip patch antennas. There are no requirements for massive repetitive-based simulations as they can provide a quick response for enhancement ratios of gain under ideal conditions when applications like vehicle-to-vehicle communication are under 5G high-frequency bands. Ann's ability to forecast quickly makes them superior to GA and SVM. Finally, with the analysis, it is noticed that ANN with a backpropagation model provided an effective model in terms of learning abilities under complex correlations, high accuracy, and efficiency, which helps in optimizing the patch antenna dimensions in a superior way.

3.4 Fractal Implementation on Patch

Choose a simple rectangular patch antenna with dimensions of 30 mm × 40 mm. This basic shape serves as the starting point for the fractal geometry.

- The original rectangular patch should be divided into smaller ones to generate the first iteration. For instance, partition it into a grid of 12 rectangles with dimensions of 10 mm × 10 mm each. Remove a central rectangle from these smaller rectangles, leaving a 2 mm gap around it. A border is created by removing the central rectangle with the rectangles of 2 mm capacity, where a space is noticed at the center. The pattern reserves the proportions selectively by utilizing the empty spaces created around, which leads to the enhancement of patch surface area that helps under multi-band operations.

- The iterations of the second and third are carried out by repeating the division process followed in stage 1, where the dimensions of small portions of rectangles used in this stage are of dimensions 3.33 mm × 3.33 mm. Again, a border is created by removing the central rectangle with a rectangle of 2 mm capacity, resulting in space at the center. At the end of iteration two, the complexity level increases whereas the effective electrical length is also noticed to be increased.Each 3.33 mm × 3.33 mm dimension rectangle used at stage 2 is further reduced to smaller proportions of 2.74 mm × 2.74 mm. Then, a border is created by following the same principle of removing the central rectangle of 2 mm capacity, which again forms a space and helps maintain the pattern to look similar. As the iterations of the fourth and fifth are carried out, the complexity increases in choosing the rectangle's tiny dimensions and forming an empty space. Maintaining a specific gap between each iteration is also a critical task.
- As iterations under the fractal geometry phenomenon are increased, the complexity rate is increased because as iterations increase, the current transformations across each edge increase, and the total effective electrical length also increases. So limit the iterations to a maximum of five or six so that by representing the self-similar patterns of fractal nature, an enhancement in bandwidth is noticed with multi-band characteristics of an antenna. Continuous evaluation of fractal sizes is also required to optimize the dimensions of the antenna. The computing strategies of Ansys HFSS help a lot in this aspect, such as monitoring the continuous learning rate acquired by ANN and capturing the perfect echo where proper MSE is achieved. All these insights greatly help improve the design of a patch antenna, which is well-suited for applications like vehicular and wireless communications.

The antenna is constructed utilizing Sierpinski square principles, as observed in Fig. 4, where the plain and defected ground structures of antenna are shown in Fig. 5. Rogers RT Duroid 5880t was utilized as the substrate for the antenna design, and the patch is 30 mm by 40 mm in size.

Introducing rectangular boxes or deleting parts in the ground plane of a fractal microstrip patch antenna design can have a substantial impact on the antenna's performance characteristics. By incorporating two rectangles of 44 mm × 2 mm proportions (but not exceeding the overall dimensions of the ground plane) into the ground plane and other rectangle with 2 mm × 48 mm and by uniting all three rectangles a I shaped is formed. These units are positioned within the central region of the ground plane or around its perimeter. By removing the specific sections of the substrate that are located beneath the ground plane a defective structure can be formed.

The fractal geometry or rectangular box placement can be complemented by the pattern of these removed areas. The objective of these alterations is to attain precise antenna performance objectives, such as achieving impedance matching, enhancing bandwidth, controlling radiation pattern, or adjusting polarization. To achieve optimal performance of the antenna according to the specified specifications

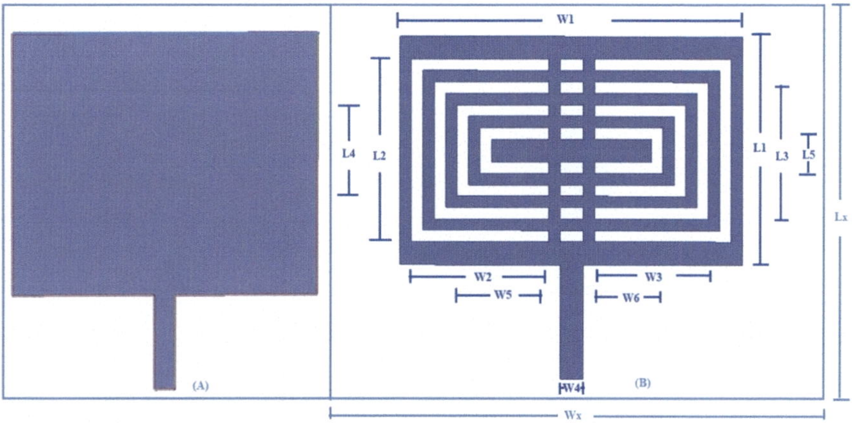

Fig. 4 **a** Rectangular patch antenna and **b** applied Sierpinski square fractal

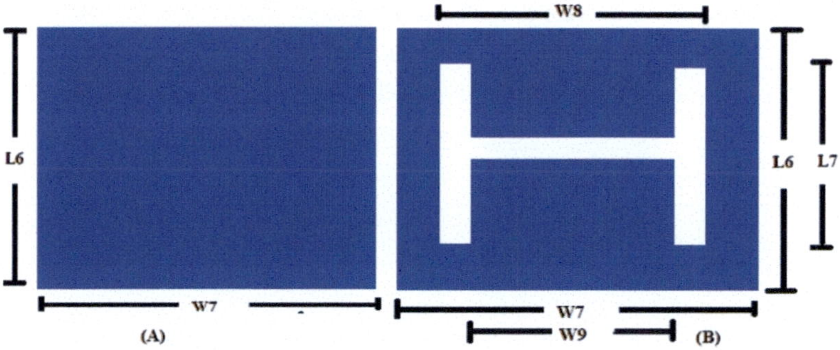

Fig. 5 **a** Ground plane and **b** defective ground structure (DGS)

and operating circumstances, it is necessary to conduct modeling and testing for the precise design. Table 4 depicts the dimensions of the proposed antenna.

4 Result Analysis

The amount of radiation that is reflected back from the antenna is measured by the reflection coefficient, which is represented by S11. A lower S11 value implies improved impedance matching, which means that the antenna radiates more power and less is reflected back. In general, a S11 value of less than − 10 dB is considered satisfactory, as it indicates that less than 10% of the power is reflected. S11 response of proposed design is represented in Fig. 6. The proposed antenna resulted dual band resonating characteristics where a higher resonance of − 22 dB is noticed at 31.9 GHz frequency and − 10 dB resonance is noticed at many frequency ranges within the

Table 4 Finalized dimensions of the proposed antenna

Substrate width (Wx)	60 mm
Substrate length (Lx)	50 mm
Patch width (W1)	40 mm
Patch length (L1)	30 mm
Feed width (W4)	2 mm
Feed length (La)	6 mm
Separation gap between iterations	0.5 mm
Patch length after 1st iteration (L2)	27 mm
Patch length after 2nd iteration (L3)	18 mm
Patch length after 3rd iteration (L4)	10 mm
Patch length after 4th iteration (L5)	4 mm
Patch width after 1st iteration (W2)	38 mm
Patch width after 2nd iteration (W3)	34 mm
Patch width after 3rd iteration (W5)	30 mm
Patch width after 4th iteration (W6)	22 mm
Ground plane width (W7)	60 mm
Ground plane length (L6)	50 mm
Dimensions of the recessed ground plane	
L7	44 mm
W8	52 mm
W9	48 mm

operating frequency band which indicates good impedance matching and also effective bandwidth. The antenna's impedance matching can be substantially enhanced through the implementation of ANN optimization. The ANN has the ability to iteratively modify design parameters, including patch dimensions, substrate properties, and feed position, in order to reduce the S11 values at the target frequencies, thereby attaining improved resonance. This is probably the reason for the significant fall of − 22 dB at 31.9 GHz. ANNs can optimize the design to guarantee that the S11 values remain below − 10 dB across a wider spectrum of frequencies. Dual-band characteristics are noticed with the help of the S11 response; in obtaining the dual-band characteristics for the proposed antenna, the backpropagation model of ANN played a prominent role in considering the data sets in a balanced way by analyzing various design parameters. One more advantage of this model is that there is no requirement for manual tuning because of its ability to learn from previous data sets once it is trained, which helps reduce the time constraints for design and simulation.

The Voltage Standing Wave Ratio (VSWR) is a crucial indicator of the degree of matching between the antenna and the transmission line. The relationship is directly linked to the reflection coefficient S11 and may be expressed as follows [31]:

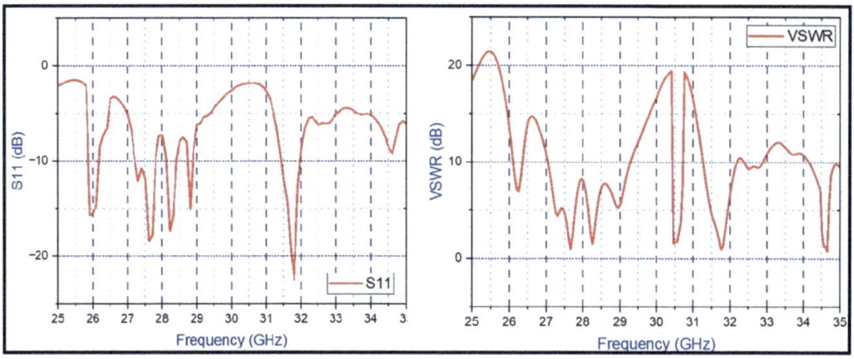

Fig. 6 Reflection co-efficient and VSWR of proposed antenna

$$\text{VSWR} = \frac{1 + S11}{1 - S11} \tag{3}$$

The fundamental idea of VSWR is how effectively a signal is transmitted from an input power source through a specific transmission media and reaches the load. In order to understand the VSWR response of an antenna, first, a designer needs to think about the threshold levels of VSWR for an antenna system for the specific application. In practice, the VSWR for an antenna system is represented as quality one when the values of the received VSWR meter are less than four under vehicular communication applications. This threshold limit represents that a lower value of VSWR shows a quality impedance matching. As impedance matching is good, power transmission is more significant, whereas reflection is less. For the proposed microstrip patch antenna, the VSWR response is shown in Fig. 6, which indicates that at resonating frequencies, the VSWR limits of the antenna are within the threshold value. The response of VSWR represents that proper impedance matching is done at resonating frequencies of an antenna. This impedance matching is an added advantage for the antenna for efficient transmission and reception of signals. This proper impedance matching is possible by imparting the concept of the ANN model in antenna design. The ANN model selectively chose the antenna parameters like the size of the patch, substrate used, and type of feed, and then by alternating these design parameters based on its learning capabilities, a proper reduction rate in VSWR is possible at operating bands under 25–35 GHz. Consistency is essential for vehicular communication systems, since they demand dependable performance across different operating frequencies.

Figure 7 depicts the Radiation pattern of proposed antenna where the ANN optimization ensures that the energy is radiated predominantly in the desired direction, thereby minimizing waste and interference. A specific radiation pattern is required under Vehicular communication applications to avoid dependability and irregularity. When an antenna is designed using the principle of fractal geometry, then within the small physical size of the patch, it needs to operate under several frequency bands (here, the proposed frequency band is the milli-meter frequency band). A radiation

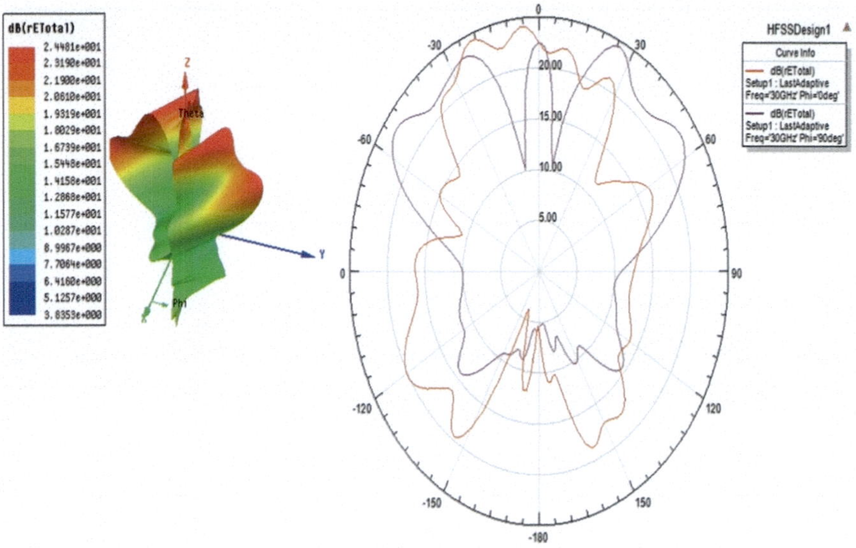

Fig. 7 Radiation pattern of proposed antenna at resonating frequency

pattern with comprehensive coverage is a good choice for connecting the vehicles in a better way. By imparting the ANN model, the proposed antenna provided a 120°, which is quite enough for quality connectivity even under situations where the alignment between the vehicles is not in proper mode. In general, signal drop is the primary limitation; this can be overcome by enhancing the directivity and reducing the return loss. Enhanced directivity and radiation efficiency reduce power wastage, the most essential requirement under vehicular ad-hoc networks. For the proposed design, with the help of the ANN optimization process, it is noticed that a well-shaped radiation pattern is achieved. All these characteristics, like S11, VSWR, radiation pattern, and directivity gain, are essential to establishing secure and superior communication between vehicles under vehicular communication applications.

Figure 8 shows that the proposed ANN-optimized fractal antenna achieved a maximum gain of 11.5 dB at resonating frequencies of 26 and 31.8 dB, indicating the effective transmission rate and quality impedance matching. From the gain response, it is also noticed that the proposed antenna showed better gain at resonating frequencies and maintained an increase of above 5 dB throughout the frequency spectrum of 25–35 GHz. The enhanced gain and quality impedance matching leads to establishing better connectivity between vehicles, and it also leads to enhancement in system reliability. The far field experimental setup of the proposed antenna (facing ground plane in front and patch at back) is represented in Fig. 9 where the transmission analysis at different angles under anechoic chamber is carried out and as shown in Fig. 7 with $\phi = 0$ and $\phi = 90$ the field is analyzed.

Fig. 8 Gain response of the proposed antenna

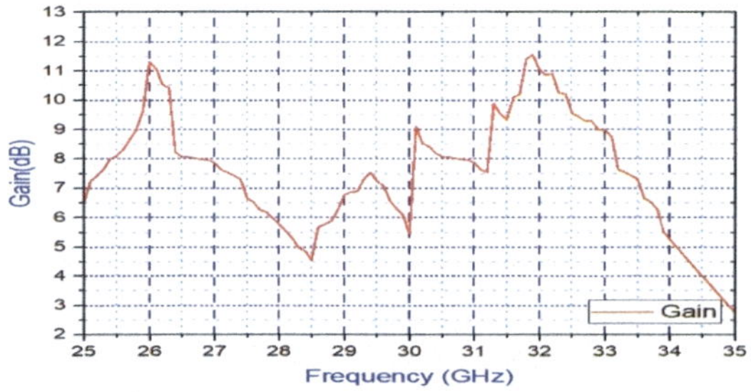

Fig. 9 Far field experimental setup of the proposed antenna in anechoic chamber

5 Conclusion

A microstrip patch antenna design using an ANN optimization and imparting fractal geometry phenomenon is represented in this chapter, which resulted in enhanced antenna dimension optimization and characteristics performance under the typical operating frequency range of 25–35 GHz. A back-propagation model of ANN is utilized for patch dimension optimization, which provides an efficient result due to the ANN capabilities of fine-tuning and quickly adapting the complex data coefficients under dynamic conditions. The proposed antenna shows multi-band characteristics with proper impedance matching. The optimized antenna also shows remarkable benefits in terms of an enhanced gain of 11.5 dB at resonating frequencies, which helps establish better connectivity between the vehicles with the improved capacity of signal amplification. The proposed antenna provided a 120° broad radiation pattern at resonating frequencies, indicating that the antenna is utilized as a better transceiver, an essential and required characteristic under applications like vehicle communication. The proposed antenna represented a better VSWR within the threshold limits, which indicates that a quality signal reception is possible with a shallow interference margin, and a high bandwidth helps in maximum data transmission rates.

Transmission is vital in this scenario as the IoT field is advancing. The antenna design is a significant factor for efficient data transmission, especially for applications like vehicular ad-hoc networks, which establish enhanced connectivity between vehicle-to-vehicle or vehicle-to-infrastructure-based communications with effective data rates. The innovative materials and manufacturing techniques for antenna design and imparting deep learning methods for optimization are always crucial during antenna design for achieving optimized and enhanced results for specific applications. So, in the future, different design styles of microstrip patch antennas will have a lot of scope and better optimization processes due to their ease of fabrication, compact size, and low cost. They are feasible to mount easily on any kind of surface.

References

1. Singh A, Kumar A, Kanaujia BK (2023) High gain and enhanced isolation MIMO antenna with FSS and metasurface. Optik 286:170982. ISSN 0030-4026. https://doi.org/10.1016/j.ijleo.2023.170982
2. Kumar S, Raheja DK, Palaniswamy SK, Kanaujia BK, Mostafa H, Choi HC, Kim KW (2023) Design and implementation of a planar MIMO antenna for spectrum-sensing applications. Electronics 12:3311. https://doi.org/10.3390/electronics12153311
3. Rao BR, Chakradhar KS, Nataraj D (2023) Design, optimization and experimental verification of UWB-MIMO antenna with WLAN and complete X-band notched characteristics, checked with characteristic mode analysis (CMA). Analog Integr Circ Sig Process 115:139–158. https://doi.org/10.1007/s10470-023-02149-9
4. Das AS, Goel A, Nakhate S (2023) A wideband stacked microstrip patch antenna design with augmented gain-bandwidth product for 5G wireless applications. Sādhanā 48:284. https://doi.org/10.1007/s12046-023-02353-3
5. Raghavendra K, Deepti K (2022) Artificial neural network optimized ultra-wide band fractal antenna for vehicular communication applications. Trans Emerg Telecommun Technol 33(12). ISSN: 2161-3915. https://doi.org/10.1002/ett.4620
6. Kaur M, Sivia JS (2020) ANN and FA based design of hybrid fractal antenna for ISM band applications. Prog Electromag Res C 98:127–140.https://doi.org/10.2528/PIERC19110901
7. Rahim A, Malik PK, Singh R (2022) Fractal geometry-based chakra-shaped microstrip patch antenna array for vehicular communications under 5G environments. J Inform Telecommun 6(4):482–500. https://doi.org/10.1080/24751839.2022.2117068
8. Raghavendra K, Kakkar D (2022) Impact of fractal geometry on microstrip patch antenna at futuristic frequencies. In: Singh PK, Singh Y, Kolekar MH, Kar AK, Gonçalves PJS (eds) Recent innovations in computing. lecture notes in electrical engineering, vol 832. Springer, Singapore. https://doi.org/10.1007/978-981-16-8248-3_12
9. Amsaveni A, Bharathi M (2021) Design and implementation of H-shaped fractal antenna for UWB applications. In: International conference on advancements in electrical, electronics, communication, computing and automation (ICAECA), pp 1–5. https://doi.org/10.1109/ICAECA52838.2021.9675601
10. Abraham JR, Bhargava A (2020) Fractal microstrip patch antenna with defected ground structure for 5G wireless communication applications. Int J Emerg Technol Innov Res 7(9):375–381. ISSN: 2349-5162
11. Luo W, Chen W, Feng Y et al (2020) A novel automobile antenna for vehicles communication of IoT systems in 5G network. J Wirel Com Network 2020:218. https://doi.org/10.1186/s13638-020-01838-7

12. Mezache Z, Mansoul A, Merabet AH (2023) Accuracy and precision of sensing fructose concentration in water using new fractal antenna biosensor. Appl Phys A 129:267. https://doi.org/10.1007/s00339-023-06532-1
13. Karanam R, Kakkar D (2024) A novel fractal geometry enhanced microstrip patch antenna design for 5G connectivity with DGS and machine learning optimization for vehicle-to-vehicle communication. In: Singh Y, Gonçalves PJS, Singh PK, Kolekar MH (eds) Proceedings of international conference on recent innovations in computing. ICRIC 2023. Lecture notes in electrical engineering, vol 1196. Springer, Singapore. https://doi.org/10.1007/978-981-97-7862-1_30
14. Gao J, Tian Y, Chen X (2020) Antenna optimization based on co-training algorithm of Gaussian process and support vector machine. IEEE Access 8:211380–211390. https://doi.org/10.1109/ACCESS.2020.3039269
15. Jacobs JP (2021) Accurate modeling by convolutional neural-network regression of resonant frequencies of dual-band pixelated microstrip antenna. IEEE Antennas Wirel Propag Lett 20(12):2417–2421. https://doi.org/10.1109/LAWP.2021.3113389
16. Hoseini M, Foororaghi K, Abdolali A (2023) A waveguide slot-fed patch antenna array. IEEE Antennas Wirel Propag Lett 22(8):2027–2031. https://doi.org/10.1109/LAWP.2023.3272151
17. Bennioui F-Z, Khabba A, Bouslam KA, Wakrim L, Ibnyaich S, Zeroual A (2024) Genetic algorithm-based optimization of rectangular patch antenna parameters for 2.45 GHz. In: 2024 international conference on global aeronautical engineering and satellite technology (GAST). Marrakesh, Morocco, pp 1–6
18. Kumar D, Ahmed S (2014) Modified ring shaped Sierpinski triangle fractal antenna for C-band and X-band applications. In: Proceedings of the 6th IEEE international conference on computational intelligence and communication networks, pp 78–82 (Bhopal)
19. Koziel S, Pietrenko-Dabrowska A (2023) Improved-efficacy EM-driven optimization of antenna structures using adaptive design specifications and variable-resolution models. IEEE Trans Antennas Propag 71(2):1863–1874
20. Tung L, Linh H, Dao-Ngoc C, Beccaria M, Pirinolo P (2019) Automated design of micro strip patch antenna using ant colony optimization, pp 0587–0590. https://doi.org/10.1109/ICEAA.2019.8879031
21. Zhao F, Yao Z, Luan J, Song X (2016) A novel fused optimization algorithm of genetic algorithm and ant colony optimization. Math Probl Eng 2016:1–10. https://doi.org/10.1155/2016/216741310.1155/2016/2167413
22. Delwar TS, Siddique A, Aras U et al (2023) A design of adaptive genetic algorithm-based optimized power amplifier for 5G applications. Circ Syst Signal Process. https://doi.org/10.1007/s00034-023-02447-7
23. Jena MR, Mishra GP, Sahoo AB, Mangaraj BB (2019) Fractal geometry and its application to antenna designs. Int J Eng Adv Technol 9(1). ISSN: 2249-8958
24. Rahim A, Raghavendra K (2019) Fractal antenna design for various multiband applications. Int J Eng Adv Technol. ISSN 2249-8958
25. Yigit H, Karayahsi K (2023) A novel model-based technique to improve design process for microstrip antenna. AEU Int J Electron Commun 2023:154570. ISSN 1434-8411.https://doi.org/10.1016/j.aeue.2023.154570
26. Liu H, Zhang Y, Shi W (2018) A survey of fractal antennas and their applications. Sensors 18(10):3323
27. Raghavendra K, Kakkar D (2022) 14—Reliability of 5G in human health monitoring using blockchain technology. In: Tanwar S (ed) Blockchain applications for healthcare informatics. Academic Press, pp 313–326. ISBN 9780323906159. https://doi.org/10.1016/B978-0-323-90615-9.00012-8
28. Karanam R, Kakkar D (2025) Design and optimization of UWB fractal micro strip patch antenna for vehicular communication applications under futuristic frequencies. Analog Integr Circuits Signal Process 122:7. https://doi.org/10.1007/s10470-025-02301-7
29. Das SK, Chakraborty U, Chattopadhyay S (2017) Fractal antennas for vehicular communication systems: challenges and prospects. In: Proceedings of the 7th IEEE annual information technology, electronics and mobile communication conference (IEMCON), pp 1–6

30. Wu X, Ding C, Yang C, Long Y (2021) Design of a machine learning-based adaptive antenna array for millimeter-wave MIMO communication. IEEE Trans Microw Theory Tech 1–1
31. Jiang W, Wei M, Chen X, Cao S, Li Y (2019) Design of a microstrip patch antenna with wide bandwidth and high gain using fractal structure. Prog Electromag Res Lett 86:125–131

Deep Learning Driven Detection and Classification of Attacks in Vehicular Communication Systems

Aparna Kumari, Pooja Chaturvedi, Deepika Bishnoi, and Purnima Gandhi

Abstract Automation and connectivity have increased as autonomous vehicle technology has advanced. CAN is one of numerous serial protocols used in a variety of cars. Contemporary automobiles are increasingly susceptible to attacks targeting automotive networks due to their increased functionality and connectivity. The CAN bus protocol is not secure by design, it is open to many assaults. To identify attacks on the CAN bus, system for intrusion detection (IDS) must be highly accurate in their design. The deep learning based model can achieve higher accuracy in detection and classification of network messages. In this research, we provide an efficient IDS strategy for binary classification based on deep learning. The best model is determined using the keras tuner library and the implementation results suggest that proposed strategy achieves the accuracy of 99% in classification of messages in CAN bus network. The results outperform the classification accuracy with respect to the supervised machine learning based approaches.

Keywords VANET · Traffic classification · Vehicular communication · Deep learning · Controller area network · Attack detection

A. Kumari · P. Chaturvedi (✉) · D. Bishnoi · P. Gandhi
Institute of Technology, Nirma University, Ahmedabad, India
e-mail: pooja.chaturvedi@nirmauni.ac.in

A. Kumari
e-mail: aparna.kumari@nirmauni.ac.in

D. Bishnoi
e-mail: deepika.bishnoi@nirmauni.ac.in

P. Gandhi
e-mail: gandhi.purnima@nirmauni.ac.in

J. Bhatia et al. (eds.), *Deep Learning Based Solutions for Vehicular Adhoc Networks*,
Studies in Computational Intelligence 1207,
https://doi.org/10.1007/978-981-96-5190-0_12

1 Introduction

Smart, linked automobiles (ICVs) will rule the automotive industry in the future. Enhanced wireless innovation enables automobiles to interact with it and others around in real-time will minimize crashes, traffic, and greenhouse gas emission. Vehicle nodes are used as communication carriers in a variety of research domains, including communications from automobiles to everything else, the IoV, and VANET. CAN, MOST and FlexRay are three well-known standards for vehicle communication protocols. The CAN bus protocol lacks security features by design, hence it is susceptible to a wide range of attacks. Accurate IDS design is essential for identifying CAN bus attacks.

VANETs enable the development of dynamic, self-organizing car networks without the need for a backbone infrastructure. The technology of short-range wireless communication was applied in the creation of the networks mentioned above. The purpose of VANETs is to provide services like infotainment, traffic updates, and accident alerts while simultaneously improving traffic flow and road safety.

Apparently, VANETs are instrumental in current transportation requirements due to the growing concern with the provision of safer and improved roads. They allow direct communication and exchange of information between stations as well as the vehicles themselves and aid in the mitigation of traffic accidents also in the coordination of traffic flow and the issue of alerts and notifications on certain road conditions. This technology remains critical in Intelligent Transportation Systems (ITS) application and offers the most promising trends such as automobile automation and smart cities.

The architecture of VANET includes two primary types of communication:

Vehicle-to-Vehicle (V2V)—This makes it possible for cars on the road to engage in a form of direct interaction where they can transmit to one another information like their speed, their physical location, and intentions on the road. V2V communication is vital in activities such as emergency vehicles, low-speed detection, and adaptive cruise control.

Vehicle-to-Infrastructure (V2I)—This includes V2V communication and at times V2I which may be broader network services and infrastructure-based information such as traffic signals and traffic management information.

Several communication, networking, and data management aspects in mobile and highly dynamic vehicular environments can be explored in the system, according to the literature review of Vehicular Ad Hoc Networks (VANETs). Conventional methodologies of vehicle communications largely employed infrastructure-based systems that were unsuitable for the mobility and flexibility in the formation of networks common in vehicular networks. VANETs offered the solution; thereby allowing car-to-car data exchange and car-to-roadside unit connections, collectively referred to as V2V and V2I. However, these face essential challenges: it is difficult to maintain a stable communication path in a configuration that is in constant transition, how to manage the scalability issue, in that a V2V network has to handle a large number of vehicles, and lastly, V2V networks raise security and privacy issues that come with open hostile environment and the fact that there is a need to protect the privacy of the vehicles' whereabouts and driver's. Many constructive protocols have been proposed for reliable and interoperable communication in VANETs for continuous improvements to address those challenges [1, 2].

Protocol standards like DSRC and IEEE 802. 11p has been set up to provide dependable and compatible communication but security and privacy challenges still persist because of the potential threats from malicious attacks and due to the fact that confidential information regarding the position of the vehicle and driving habits of the drivers needs to be protected.

In recent years, VANET research has made significant strides in addressing these challenges through diverse ways. A case in point is the attention-based LSTM model for traffic flow prediction proposed by [3], which captured temporal dependencies more accurately but involved high computational expenses. Likewise, spatial–temporal graph convolutional networks were employed by [4] to improve the modeling of complex traffic data, although this method also had some computation costs.

VANETs are crucial for enabling intelligent transportation systems (ITS), enhancing road safety, traffic flow, and driving comfort through wireless communication between vehicles and roadside infrastructure. Accurate traffic prediction and modeling within VANETs are essential for applications such as congestion management, route optimization, and accident prevention. This research aims to improve traffic prediction methods and modeling techniques in VANETs, with a specific emphasis on binary classification of traffic conditions. The chapter, proposes a technique to build a strong IDS through optimal deep learning.

1.1 Motivation

The pressing need for efficient traffic management in highly populated metropolitan areas is the driving force behind this research. Our goal is to significantly increase road safety, reduce traffic, and optimize route planning by developing accurate traffic classification models and efficient binary classification frameworks within VANETs. Our objective is to promote the creation of intelligent transportation systems and enhance urban mobility.

1.2 Research Contribution

This study adds significantly to the subject of traffic modeling and prediction in VANETs in a number of ways:

1. Improved Traffic Prediction Model—To facilitate efficient traffic management and decision-making, we introduced a revolutionary traffic classification model that reliably forecasts current traffic circumstances.
2. Binary Classification Framework—In this case, a system of binary classification is created to understand traffic statuses and to react quickly to altering traffic conditions.
3. Comprehensive Feature Identification—The high impact features which can improve the predictive accuracy are identified.
4. Comprehensive Performance Evaluation—Using a real life dataset and traffic situations, a thorough model's performance is conducted to show its applicability in real-world VANET setups. The suggested system model's adaptability and versatility in ITS are demonstrated by its ability to scale and adjust to various urban contexts and traffic patterns.

1.3 Organization of the Chapter

The organization of the paper is as: state of the art in Sect. 2, proposed approach in Sect. 3, implementation parameters are discussed in Sect. 4, outcomes are discussed in Sects. 5 and 6 contains the conclusions and recommendations.

2 State-of-the-Art

2.1 Traffic Forecasting and Modelling in Vehicular Adhoc Networks

It is essential that traffic forecasting is done correctly to increase vehicle flow, cut down on travel time, and enhance the general transportation system. Traffic forecasting assists in the prevention of traffic-related accidents. Historical data about speed should also be used during traffic prediction. Transport planning authorities can direct traffic away from highly congested areas due to their ability to forecast traffic patterns. The following are methods used in traditional traffic prediction:

Simulation-Based Models—These types of models use traffic movement simulations at microscopic, mesoscopic, or macroscopic levels to predict traffic flow and congestion [5]. Depending on the level of detail considered, they can predict individual vehicle movements as well as aggregate flow rates through different road sections.

Statistical Models—ARIMA and Kalman filters are some of the techniques employed in this method. These models depend on past records for future prediction; hence, they may be accurate where conditions change slowly with time [6].

Traffic Forecasting in VANETs uses following parameters.

Vehicle Speed—Data about the speed of individual vehicles.

Traffic Density—The count of vehicles on a given road segment.

Travel Time—The time taken for a vehicle to move between two points.

Sensors—Roadside sensors that acquire traffic data.

GPS: GPS device data embedded in vehicles helps get location and speed data.

RSUs: Data is derived from roadside Unit planting, collecting, and forwarding data on traffic.

Connected Vehicles: It refers to the real-time data shared by vehicles fitted with communication devices.

Therefore, in traffic forecasting and modeling in the context of VANETs, techniques that were previously used have primarily included historic data and statistical models only. For short duration traffic forecasts, approaches such as ARIMA had been used but the approaches have been seen to be inadequate in effectively modeling traffic radical fluctuations [6]. There is also a variety of regression models such as linear and non-linear which is used to define the relation between different traffic variables; however, these models often involve feature engineering and cannot always cover complex interactions [7, 8]. Simulation models: The traffic data simulation models are micro-simulation or macro-simulation, which helps to forecast traffic conditions and evaluate effects of various traffic management strategies, though; they involve high computational time and require precise input data [5]. It becomes clear that these conventional methods are unable to address issues such as real-time data processing for adaptive traffic management and multi-type data integration from several sources. It becomes clear that these conventional methods are unable to address issues such as real-time data processing for adaptive traffic management and multi-type data integration from several sources.

Recent work also includes transfer learning on VANETs so as to borrow from different domains, hence reducing training time and improving generalization capacity of the model developed [9]. Reinforcement learning has also been used for optimizing traffic signal control in VANETs showing improved flow and reduced congestion [7]. But these techniques often need a large amount of processing power and training data.

2.2 *Traffic Forecasting and Modelling in VANETs Using Deep Learning*

Deep learning is neural network based machine learning approach containing lots of layers (called deep networks) to understand complicated patterns in data. It is good at dealing with big sets of data and finding complex connections in traffic data, better than older methods.

Deep learning is a perfect fit for the VANET traffic to process and analyze enormous amounts of data from multiple sources, including vehicle sensors, GPS, and RSUs. It captures diverse non-linear data relationships and interactions, and breeds accurate and robust predictions.

There are different types of deep learning models used for predicting traffic.

Deep Learning Architectures for Traffic Forecasting

Convoluted Neural Networks (CNNs)—it is used to analyze the spatial data and for pattern recognition from traffic images or sensor data [10].

Recurrent Neural Networks and LSTM—to predict temporal data to capture the sequential dependencies within the data of traffic flow [11].

Generative Adversarial Networks—for developing synthetic data sources to improve model training and robustness.

An attention-based LSTM approach was proposed by [3] which improved the accuracy of traffic flow predictions by effectively capturing temporal dependencies. Similarly, [4] demonstrated enhanced predictive capabilities through spatial–temporal graph convolutional networks that would model complex interactions between vehicles and infrastructure. Federated learning has also been explored as a means to utilize distributed data from connected vehicles in predicting traffic while preserving privacy [12]. In order to reduce training time and improve model generalization, pre-trained models were adapted using transfer learning techniques for specific VANET applications [9]. Reinforcement learning has shown promise in optimizing traffic signal control and vehicle routing within VANETs. For instance, [7] used Reinforcement Learning to optimize signal control with improvements in smoother traffic flows at intersection areas with reduced congestion. Likewise, [13] employed deep reinforcement learning for vehicle routing optimization leading to better routing efficiency and lower travel time.

The VANETs are improved with deep learning approaches. Specifically, recurrent neural network (RNN) and its improved form, LSTM, are best suited for traffic data due to their capability in exploring temporal characteristics of traffic data, which fits into time series forecasting [11]. Convolutional Neural Networks (CNNs) are adept at expressing spatial relation by handling traffic data in the form of images or grids which is beneficial in modeling traffic flow on road networks [10]. Moreover, Graph Neural Networks (GNNs) have been identified as highly suitable for working with traffic networks as they can learn both spatial and temporal relations at the same time [14]. The models that integrate CNNs, RNNs, AND GNNs include the strengths of the three approaches to improve the accuracy of the predictions that are made since they capture the spatial and temporal characteristics of the data [7].

There are several advantages of using deep learning models in traffic forecasting in VANETs such as improved accuracy, tools real-time processing, and flexibility with the geographical location or traffic pattern [14]. These models enable dynamic traffic management solutions, thereby enhancing overall traffic flowing. Nonetheless, some issues are still present: the collection of accurate and rich traffic data; computational efficiency for deep learning algorithms; and the lack of interpretability regarding a black-box deep learning model [15, 16]. Moreover, it is important to connect deep learning-based forecasting models to current traffic management systems and facilities to make such models more applicable in real life [3]. Nevertheless, present and emerging breakthroughs in deep learning approaches and their interaction with VANETs provide a strong foundation for reinventing ITS, with an aim toward reducing accidents, optimizing traffic flow, and increasing the quality and satisfaction of the driving experience.

Some of the other approaches being explored include hybrid models that merge CNN and RNN architectures to enhance traffic flow prediction [8] as well as edge computing-based solutions for minimum latency real-time traffic predictions [17]. The synthetic traffic data has been produced with GANs which then use this model to improve the performance of traffic prediction models [9]. However, these sophisticated deep learning techniques are often accompanied by increased computational costs and complexity. In addition, ensemble learning methods have been examined in order to gain high accuracy in prediction though they are computationally expensive [18].

Table 1 shows the review of the existing approaches.

3 Proposed System

The proposed system is designed and developed to analyze traffic including vulnerable scenarios as shown in the Fig. 1. The deep learning based classification model is implemented to classify the normal or attack messages on the CAN bus dataset [23].

The steps of the proposed method is:

a. Data Collection—The dataset is created in. csv format by combining the normal and malicious messages generated by the CAN bus network [23].
b. Data Pre-processing—The raw dataset is not suitable for the implementation of machine learning models, so pre-processing methods such as outlier detection and removal, standard scaling, data reduction. The analysis of the dataset revealed that it is imbalanced in terms of the number of normal and malicious messages. So, random oversampling method is applied to make the dataset balanced.
c. Model Implementation—After preparing the dataset through the pre processing methods, The deep learning based model is implemented using the CNN.
d. Classification result analysis—The several parameters such as precision, recall and accuracy are used for model evaluation. The keras tuner package is used to determine the best parameters for the model tuning.

Table 1 Comparative analysis of the state of the art approaches

Proposed approach	Year	Short description	Outcome	Limitation
Attention-based LSTM for traffic prediction	2020 [3]	Applied attention mechanisms in LSTM networks to traffic forecasting	Higher accuracy in temporal dependency capture	Complex to compute and with high costs of computationally
Spatio-temporal graph convolutional networks	2020 [4]	Used graph convolutional networks for learned traffic datasets	Better representation of complex spatial and temporal dependencies	Heavy on compute resources
Federated learning for VANETs	2021 [12]	Used federated learning to train traffic prediction model on distributed data	Decentralized learning: private and secure data privacy	Challenge of communication overhead and model synchronization
Transfer learning for traffic forecasting	2021 [9]	Transfer learning to transfer models from one domain to VANETs	Fast training, better model generalizes	Scarcity of pre-trained models meant for unique VANET applications
Reinforcement learning for traffic control	2022 [7]	Employed reinforcement learning to traffic signal control in VANETs	Modularity Increased traffic flow and reduced congestion	Needs a large amount of training data and computation resources
Hybrid CNN-RNN model	2022 [8]	Improved traffic flow prediction using combined CNN and RNN	Peak accuracy which came about due to the capturing of spatial and temporal features properly	Model complexity and computation cost are high
Edge computing for real-time traffic prediction	2022 [17]	Edge computing for the real time processing of traffic data	Lower latency, and better real-time prediction accuracy	Not much of processing power and storage at edge devices
Deep reinforcement learning for vehicle routing	2023 [13]	Trained algorithms using deep reinforcement learning for vehicle routing optimization	More efficient routing and less time on the road	High computational cost, and demands of a sizable scale training

(continued)

Table 1 (continued)

Proposed approach	Year	Short description	Outcome	Limitation
Ensemble learning for traffic prediction	2023 [18]	Performed ensemble learning to integrate multiple models for better prediction	Better traffic prediction, making them more accurate and robust	Model complexity and computational requirements have increased
GANs for traffic data augmentation	2023 [19]	Used GANs for generating synthetic traffic data for the purpose of model training	Enhanced model performance using augmented data	The quality of artificial data can differ and GANs require a lot of resources
Comparison of machine learning algorithms for the intrusion detection	2023 [20]	The various supervised learning based methods such as Random Forest, Decision Tree and SVM were applied to perform the classification of the CAN bus messages	The proposed method achieved low false negative and high true positive rate	The labeled dataset is required for the implementation of the supervised machine learning models
Intrusion detection using machine learning algorithm for graph based features	2023 [21]	The CAN messages are represented in the graphical format. The machine learning approaches such as SVM and KNN are applied to classify the CAN message into normal and attack	By converting the features into graphical form achieves the higher accuracy upto 97% for both the models	The model was implemented on small dataset
Comprehensive review on CAN Bus network intrusion detection and introduction of ROAD Dataset	2024 [22]	The comparative study of open source dataset for the CAN bus network	The new dataset is generated using the vehicular recordings	The proposed dataset is limited to only one vehicle at a time

4 Experimental Setup

The proposed system is implemented in the Windows 11 operating system with the processor specification 11th Gen Intel(R) Core(TM) i5-1135G7 working at 2.42 GHz and 8 GB RAM. Next, we have implemented the entire system using python programming language version 3.10 and keras libraries. Furthermore, the proposed system used the CNN model. The description of the dataset is as shown in the Table 2.

The shape of the dataset is (1,270,310, 6). The correlation analysis of the dependent and independent feature is as shown in Fig. 2.

In the pre processing step, label encoder is applied to convert the categorical attributes into numerical values. After that the standard scalar method is applied to make the values of the attributes in the specified range.

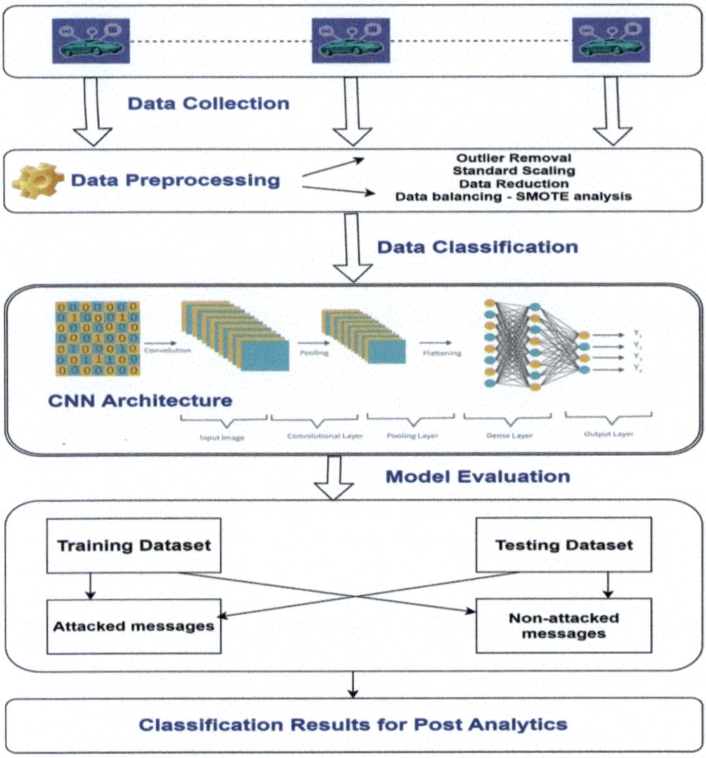

Fig. 1 Flow chart of the proposed system

Table 2 Dataset description

Attribute	Information	Type
Timestamp	Time of the information reporting	Numerical
Arbitration_ID	Message identifier	Categorical
DLC	Data length code	Numerical
Data	Data	Categorical
Class	Type of message (attack or normal)	Categorical
Subclass	Type of attack (flooding, fuzzing, replay, spoofing)	Categorical

In the next step, the dataset is analyzed to see if it's balanced or not. For this we determine the count of normal and attack messages. The count of normal and attack message is as shown in the Table 3.

From Fig. 3 it is clear that the dataset is imbalanced, so we have applied random oversampling to make the dataset balanced.

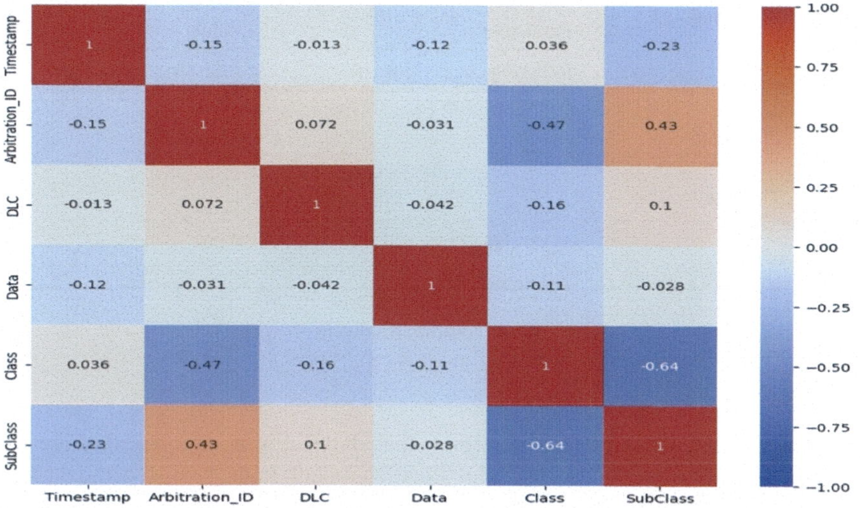

Fig. 2 Heatmap of the dataset

Table 3 Count of normal and attack messages

Message	Count
Normal	1,090,312
Attack	17,998

Fig. 3 Comparison of normal and attack messages

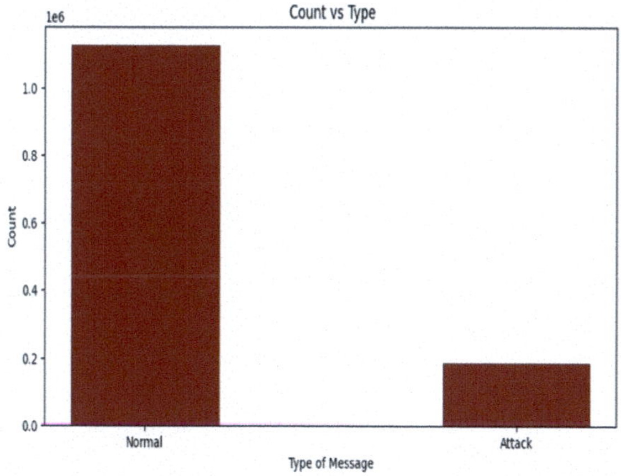

The random forest based model is used to determine the significance of the features in the dataset. The parameters used for the random forest model is as shown in the Table 4.

The feature importance are as shown in Table 5.

Table 4 Random forest model parameters

Parameter	Value
Criterion	Entropy
No. of estimators	10

Table 5 The feature importances

Feature	Score
Time stamp	0.29
Arbitration_ID	0.32
DLC	0.051
Data	0.32

From the feature analysis, we determine that all the arbitration id and data are the prominent features for the message classification model. In the next step, the dataset is divided into training and testing and the test size is considered as 30%. In the next step, the deep learning model is developed. The specification of the model is as shown in Table 6.

After the implementation of the model, its summary is displayed as shown in Table 7.

Hyper parameter tuning

The keras tuner package is used to tune the hyper parameter. The hyper parameter tuning is done for the number of units as 8,16 and 32 and the number of epochs are considered as 5. The results of parameter tuning are in Table 8.

Trial 3 Complete [00 h 07 m 26 s].

val_loss: 0.025.

Best val_loss So Far: 0.025.

Table 6 Specification of the deep learning model

Feature	Value
Model	Sequential
No. of neurons	32, 10, 1
Optimizer	sgd
Activation functions	Sigmoid, ReLu
Loss function	Binary cross entropy
Batch size	20
Epochs	20
Metrics	Accuracy
Training dataset	763,212
Testing dataset count	327,100

Table 7 Model summary

Layer (type)	Output Shape	Param #
Dense_3 (dense)	(None, 32)	160
Dense_4 (dense)	(None, 10)	330
Dense_5 (dense)	(None, 1)	11
Total params	501	
Trainable params	501	
Non trainable params	0	

Table 8 Hyper parameter tuning results

Units	Epoch	Loss	Val_loss	Time (s)
8	1	0.0772	0.0897	70
	2	0.0614	0.0944	74
	3	0.0598	0.0780	72
	4	0.0590	0.0597	75
	5	0.0571	0.0646	75
16	1	0.0579	0.0851	153
	2	0.0451	0.0926	77
	3	0.0448	0.0476	75
	4	0.0442	0.0623	70
	5	0.0436	0.0639	72
32	1	0.0475	0.0586	74
	2	0.0345	0.030	76
	3	0.0327	0.0508	72
	4	0.0316	0.0255	77
	5	0.0312	0.0301	75

Total elapsed time: 00 h 20 m 59 s.

From the table, it is clear that the least loss value is achieved for the unit value as 32. The model is built using these parameters and evaluated. The results of the evaluation are discussed in the next section. The comparison results are in Table 9.

Table 9 Comparison of the proposed model with existing approaches

References	Year	Models	Data balanced?	Hyper parameter tuning?
[23]	2020	ML	No	No
[1]	2021	ML	No	No
[2]	2021	ML	Yes	Yes
[24]	2022	DL	No	Yes
Proposed	2024	DL	Yes	Yes

5 Results and Discussion

The model training is done for the considered dataset. The performance is evaluated in terms of accuracy, precision and recall. The accuracy of model with respect to the epochs is shown in Fig. 4. The model achieves the higher accuracy upto 98% in less number of epochs.

The loss value with respect to the epoch is in Fig. 5. The binary cross entropy function is the loss function. The loss curve represents that the model achieves negligible loss value in approximately 10 epochs.

The classification report for the test dataset is shown in Table 10. The classification report represents the results in terms of accuracy, precision and recall.

The confusion matrix for test dataset is in Table 11. The confusion matrix represents the high number for the true positive and true negative, whereas there are lower number in the false positive and false negative.

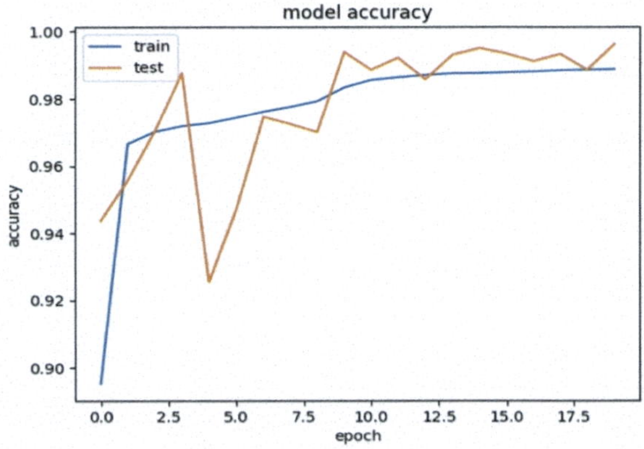

Fig. 4 Performance of the model in terms of accuracy

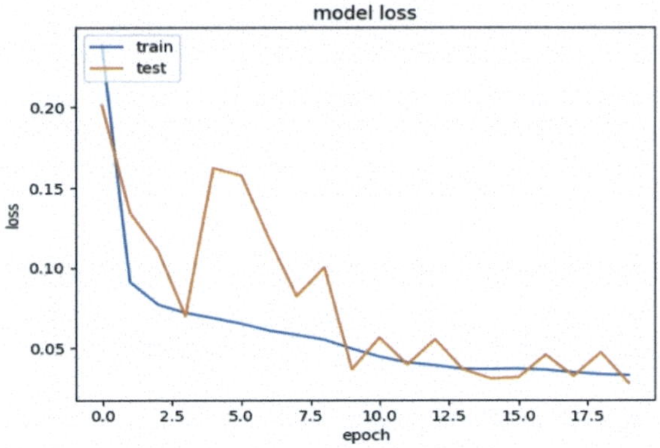

Fig. 5 Performance of the model in terms of loss function

Table 10 Classification report for the test dataset

Class	Accuracy	Precision	Recall	F1 score	Support
0	0.99	0.99	1.00	0.99	327,100
1	0.99	1.00	0.99	0.99	327,100

Table 11 Confusion matrix for the test dataset

Actual/predicted	Normal	Attack
Normal	325,771	1329
Attack	4521	322,579

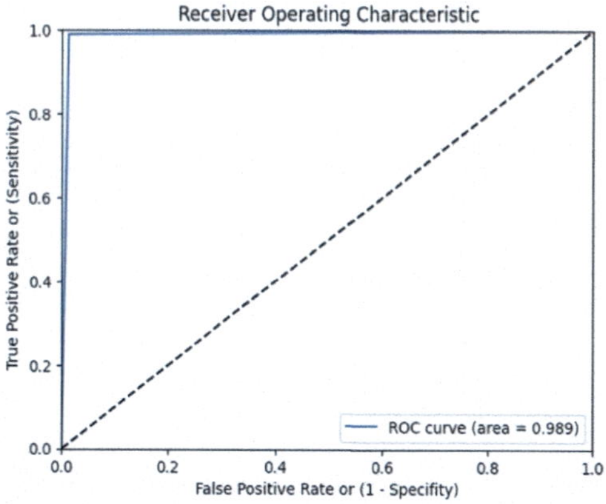

Fig. 6 ROC curve for the model

The ROC curve for the model is as shown in the Fig. 6. The ROC curve shows the true positive rate and false positive rate. The figure represents the high value which means the proposed model achieves the high classification accuracy.

The comparison is in the Table 12.

Table 12 Comparison of the proposed approach with existing approaches [25, 26]

Model	Accuracy (%)
Decision tree	98.3
Random forest	98.3
XGboost	97
Gaussian Naive Bayes	57.3
Linear regression	56.8
K nearest neighbor	93.4
AdaBoost	93.3
Gradient boost	82.6
Proposed	99

6 Conclusion and Future Scope

The paper uses deep learning for developing the classification model which can accurately predict the messages on the CAN bus. The several pre processing steps such as correlation analysis, label encoding and standard scalar are applied to prepare the dataset. The feature importance analysis is done using the random forest model. The keras tuner package is used for the hyper parameter tuning. The obtained best model is trained for the dataset and achieves the accuracy of approximately 99%. The model is evaluated in terms of precision and recall. The comparative analysis show that the proposed model outperforms the existing approaches in terms of accuracy, precision and recall for the binary classification problem.

To compare with other approaches is the future work. The said problem can be studied as a multi-class classification problem in which the attack sub class will be determined.

References

1. Zhao J, Cao G (2008) VADD: vehicle-assisted data delivery in vehicular ad hoc networks. IEEE Trans Veh Technol 57(3):1910–1922
2. Hartenstein H, Laberteaux KP (2008) A tutorial survey on vehicular ad hoc networks. IEEE Commun Mag 46(6):164–171
3. Kim M, Kim M (2020) An attention-based LSTM approach for traffic flow prediction in VANETs. IEEE Trans Veh Technol 69(2):1507–1517
4. Guo S, Lin Y, Feng N, Song C, Wan H (2020) Attention based spatial-temporal graph convolutional networks for traffic flow forecasting. AAAI Conf Artif Intell
5. Nagy A, Simon G (2018) Survey on traffic prediction in smart cities. Pervasive Mob Comput 50:148–163
6. Williams BM, Hoel LA (2003) Modeling and forecasting vehicular traffic flow as a seasonal ARIMA process: theoretical basis and empirical results. J Transp Eng 129(6):664–672
7. Zheng Y, Li J, Wang S, Wang Y (2022) Reinforcement learning for intelligent traffic signal control in VANETs. IEEE Trans Intell Transp Syst 23(4):2745–2755
8. Zhao X, Li H, Yang Y, Wang X (2022) Hybrid deep learning model for traffic flow prediction in VANETs. IEEE Access 10:30485–30493
9. Chen L, He J, Zhang Y, Wu Z (2021) Transfer learning for short-term traffic forecasting in VANETs. IEEE Access 9:11960–11969
10. Ma X, Dai Z, He Z, Ma J, Wang Y (2017) Learning traffic as images: a deep convolutional neural network for large-scale transportation network speed prediction. Sensors 17(4):818
11. Tian Y, Pan L, Wang W (2015) Predicting short-term traffic flow by long short-term memory recurrent neural network. IEEE Int Conf Smart City/SocialCom/SustainCom 153–158
12. Wang J, Li T, Zhang Y, Zhang J, Li M (2021) Federated learning for traffic prediction in vehicle ad hoc networks. IEEE Wirel Commun 28(3):52–57
13. Gao J, Zhang X, Wang H, Wei Y (2023) Deep reinforcement learning for vehicle routing optimization in VANETs. IEEE Trans Veh Technol 72(1):215–226
14. Lv Y, Duan Y, Kang W, Li Z, Wang F-Y (2015) Traffic flow prediction with big data: a deep learning approach. IEEE Trans Intell Transp Syst 16(2):865–873
15. Yan K, Zheng L, Zou H (2017) A novel data-driven framework for short-term traffic prediction with deep learning models. Transp Res Part C Emerg Technol 82:103–123

16. Wu Y, Tan H, Qin L, Ran B, Jiang Z (2016) A hybrid deep learning based traffic flow prediction method and its understanding. Transp Res Part C Emerg Technol 90:166–180
17. Xu M, Liu S, Li Q (2022) Edge computing-based real-time traffic prediction in VANETs. J Syst Architect 122:102252
18. Liu J, Huang Q, Wang H, Tang Y (2023) Ensemble learning for traffic flow prediction in VANETs. IEEE Access 11:10492–10502
19. Chen Y, Feng J, Zhou M (2023) GAN-based data augmentation for traffic prediction in VANETs. IEEE Trans Intell Transp Syst 24(2):1465–1476
20. Bari BS, Yelamarthi K, Ghafoor S (2023) Intrusion detection in vehicle controller area network (can) bus using machine learning: a comparative performance study. Sensors 23(7):3610
21. Refat RUD, Elkhail AA, Hafeez A, Malik H (2022) Detecting can bus intrusion by applying machine learning method to graph based features. In: Intelligent systems and applications: proceedings of the 2021 intelligent systems conference (IntelliSys), vol 3. Springer International Publishing, pp 730–748
22. Verma ME, Bridges RA, Iannacone MD, Hollifield SC, Moriano P, Hespeler SC et al (2024) A comprehensive guide to CAN IDS data and introduction of the ROAD dataset. PLoS ONE 19(1):e0296879
23. Kang H, Kwak B, Lee Y, Lee H, Lee H, Kim H (2021) Car hacking: attack and defense challenge 2020 dataset. In: IEEE dataport. IEEE, Piscataway, NJ
24. Raya M, Hubaux J-P (2007) Securing vehicular ad hoc networks. J Comput Secur 15(1):39–68
25. Alalwany E, Mahgoub I (2022) Classification of normal and malicious traffic based on an ensemble of machine learning for a vehicle can-network. Sensors 22(23):9195
26. Yuan Y, Van Lint JW, Wilson RE, Van Wageningen-Kessels F, Hoogendoorn SP (2014) Real-time lagrangian traffic state estimator for freeways. IEEE Trans Intell Transp Syst 15(1):29–38

Case Studies and Real-World Deployment Examples of VANETs (Vehicular Ad-Hoc Networks)

Harshada Magar, Archana S. Ubale, Nilakshee Rajule, R. R. Gupta, and Vineeta Philip

Abstract This chapter look into the world of Vehicular Ad Hoc Networks (VANETs), cater a extensive overview of their real-world concern, challenges, and best practices. VANETs are dynamic networks where vehicles intercommunicate with each other and roadside infrastructure offering advanced services such as safety, traffic management, and infotainment. This study focuses on the current progression in VANET technology and its extensive analysis and study of real-world deployment examples which showcase their transformative impact on modern-day transportation systems. The chapter start with an introduction to VANET constitutive, followed by a communication of the revolutionary implications for road safety, need of traffic flow management, environmental monitoring with respect to VANET, and user its experience. By presenting detailed case studies, this chapter aims to lighten up new applications of VANET technology and their execution across various arena. These case studies provide new insights into improved road safety through cutting-edge systems for real-time incident sensing, emergency vehicle notification, and any road hazard alerts. The method involves a detail review of modern literary study, real-world deployments, positioning this study as a critical resource for understanding best practices and addressing current challenges in this field. This chapter fuses up the trends and future of emergency response applications and advanced traffic flow

H. Magar (✉) · V. Philip
AISSMS Institute of Information Technology, Pune, India
e-mail: harshada.magar@aissmsioit.org

V. Philip
e-mail: vineeta.philip@aissmsoit.org

A. S. Ubale
AISSMS College of Engineering, Pune, India
e-mail: asubale@aissmscoe.com

N. Rajule
Dr. D.Y. Patil Institute of Technology, Pimpri, Pune, India
e-mail: nilakshee.dit@gmail.com

R. R. Gupta
Sipna College of Engineering and Technology, Amravati, India
e-mail: rounak_gupta@rediffmail.com

J. Bhatia et al. (eds.), *Deep Learning Based Solutions for Vehicular Adhoc Networks*,
Studies in Computational Intelligence 1207,
https://doi.org/10.1007/978-981-96-5190-0_13

monitoring and regulation techniques, including state-of-the-art traffic signal control and dynamic routing approaches, in building a more efficient and sustainable transportation infrastructure. This chapter provides a comprehensive analysis of Vehicular Ad Hoc Networks, emphasizing their real-world applications, deployment scenarios, challenges, and best practices. VANETs allow on-the-fly communication between vehicles and equipment at the roadside, are increasingly critical for increased road safety, efficient traffic control, and in-depth traffic data analysis and vehicle infotainment services. This chapter aims to provide a comprehensive overview of the state of the art of VANET technology and its potential to revolutionize transportation systems. The methodology includes an extensive literature review, an examination of practical deployment scenarios, and insights from datasets. It will be have a valued reference for researchers and practitioners who want to understand this evolving field and its applications toward sustainable, compliant and efficient transportation systems.

Keywords VANETs · Traffic monitoring system · Green driving techniques · Fleet tracking · Location-based services · Fuel conservation strategies

1 Introduction

An important uncovering in modern-day transportation is a Vehicular Ad Hoc Network, It allows to have dynamic communication between smart automobiles and related roadside infrastructure. As the automobile industry develop, VANETs are progressively being used to ameliorate road safety, traffic management, and user experiences via vehicle infotainment applications. The chapter deals with a thorough investigation of emerging VANET technology, focusing on its real-world applications, its futuristic difficulties, and their best practices. In recent years, VANETs have shown outstanding performance in modification of transportation systems, specially in areas like real-time incidence detection, emergency vehicle alerts and notification. These technologies render an innovative conceptualization to improve road safety, traffic flow, and monitoring environmental parameters affecting transportation systems. This chapter emphasizes the practical and real time benefits of VANET implementation across many sectors, with the help of case studies and real-world deployment examples. In the beginning the chapter discuss the primal components of VANET technology, preceded by a complete examination of its evince for road safety and traffic management. The chapter will provide thorough picture of the modern VANET ecosystem.

1.1 Motivation

The motivation for this chapter came from the need of VANETs for fixing issues in current transportation systems. As global traffic density rise heavily, there is an pressing need for modern solutions to improve road safety, optimize traffic management, and better user experiences. VANETs is a promising solution to these issues by allowing automobile to interact with each other and roadside associated infrastructure in real time. The chapter is motivated by the commitment of VANET technologies to build safer, efficient, and sustainable smart transportation ecosystems, peculiarly in light of increasing urbanization and the desire for smart cities. This research deals with very recent advances and real-world examples of VANETs, contributing their revolutionary impact on road safety, environmental monitoring and traffic management.

1.2 Research Contributions

The chapter contributes importantly to the subject of Vehicular Ad Hoc Networks by giving a all-inclusive categorization of recent advances in VANET technology, with a emphasis on real-world applications. It render a detailed investigation of several deployment scenarios, demonstrating their influence on road safety, emergency response, and traffic management. The chapter also discusses new trends including enhanced traffic flow control and smart traffic signals, as well as how VANETs might promote sustainable mobility through environmental monitoring and green driving assistance. It also identifies prominent obstacles in VANET adoption and provides important insights for researchers and practitioners on how to successfully handle these issues.

1.3 VANETs Essentials

The expected growth and innovation is huge in the exhilarating frontier of technology known as Vehicular Ad Hoc Networks. A specialized networks which consider networks of smart vehicles and roadside associated infrastructure called Vehicular Ad Hoc Networks Because of these networks safety counsel, deepen navigation, info-entertainment etc. are possible dynamic wireless networks formed by vehicles as network nodes and communication links between nodes in that network are called Vehicular Ad-Hoc Networks [1]. Particularly as self-driving and intelligent systems of transport become the mainstream in all our cities, VANETs are an important area of research for improving the efficiency, safety and intelligence of transport [2]. Welcome to the world of V2V and V2I communication, where your car and the road become partners. VANETs are organized to link up vehicle affixed devices, allowing them to share critical data without bank on extrinsic infrastructure. The way cars interact and collaborate on highways and in urban settings has been transformed by this revolutionary concept, which has gained significant traction in recent years. VANETs use vehicle-to-vehicle and vehicle-to-infrastructure communication to exchange information, enhance traffic flow, and optimize travel times, resulting in a more secure, environmentally friendly, and efficient automotive ecosystem [3]. Figure 1 shows basic VANET architecture. Reliable networks that allow vehicles to communicate effectively, share real time data, and contribute to a harmonious driving experience for everyone on the road have been developed by VANETs. because of these networks wireless communication technologies are leveraged to make possible real time data exchange among vehicles, roadside sensor infrastructure and pedestrians. High speed node movement, rapid topology changes, and decentralized infrastructure less nature make VANETs different, it makes them perfect for Intelligent Transportation Systems (ITS) and vehicle entertainment applications. There are Challenges in designing VANETs which include designing efficient routing protocols that ensure reliable and timely data packet forwarding by considering factors that include end to end delay and packet delivery progress. Though they promise capabilities, there are concerns about their security, liability, and privacy which may hinder the widespread implementation of VANETs. VANET improves transportation efficiency and safety by allowing cars to interact with one another and with road infrastructure, leveraging modern technologies such as block chain and AI [4].

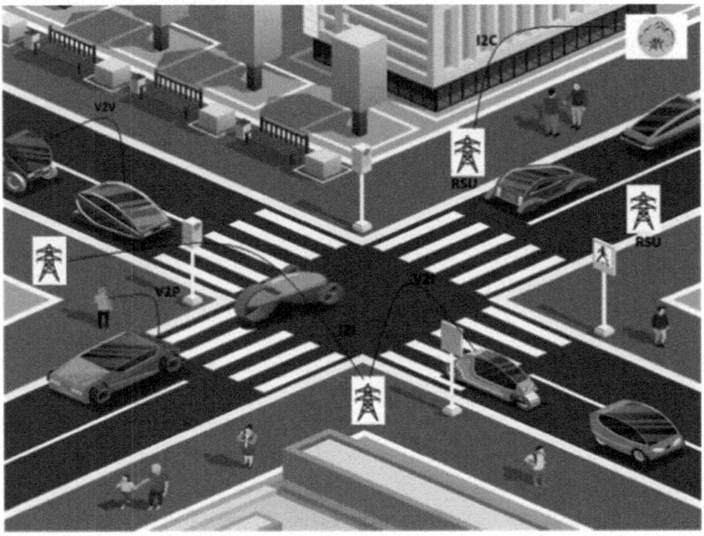

Fig. 1 Vehicular ad-hoc network architecture

1.3.1 Distinct Features of VANETs

VANET networks have several characteristics and constraints that make service supply and reliability complicated and metric-dependent issues. Some of these features are described here.

- Adaptive Network Configuration: Implementing adaptive network setup in Vehicular Ad-hoc Networks presents problems owing to changing vehicular contexts, resource limits, and the necessity for real-time responses [5]. Because of the great mobility of vehicles, VANETs have a continuously changing network topology that necessitates adaptive routing methods for optimal communication Amid the VANET network, communication elements like RSU might be immobile or sluggish like cars in traffic jams or at the crowded road junctions or heading at high speeds. This disparity in node mobility has presented many obstacles to VANET communication. If we consider under a high-velocity situation, effective communication between VANET entities will be limited due to its short transmission range. Similarly, due to Doppler effect, link breakages or delays in a moderate velocity situation, VANET communication would be reduced. The VANET network is made up of several nodes, either fixed or movable, and manufacturers place limits on data transit and exchange. Traditional routing approaches, such as routing tables, might degrade performance owing to vehicle mobility and topological dynamics. To improve performance, routing protocols should react to changes in traffic density and network architecture, guaranteeing low latency for busy traffic while leveraging cars as store-carry-forward relays for sparse traffic.
- Limited Communication Range: The restricted communication range of vehicles in VANETs complicates maintaining good connectivity for data transmission. To solve this issue, researchers have investigated alternate communication technologies like as LoRa for non-safety urban VANET applications, with encouraging results in managing Doppler shifts and line-of-sight circumstances [6]. Vehicles in VANETs usually have restricted communication ranges, requiring the use of multi hop communication to transfer messages across greater distances. Vehicle density fluctuates at junctions, roadways, and businesses, with higher densities during the day. This has an impact on VANET network fragmentation, resulting in multiple fragmentation and impacting communication continuity. High velocities also enhance the dynamicity of the VANET architecture, which may lead to fragmentation into independent parts.
- Resource Restrictions: Vehicles in VANETs have restricted resources, such as computing power, memory, and energy, which affects communication protocols and algorithms also consider that Vehicle mobility in the VANET environment is constrained by public transit infrastructure and road distribution, with metropolitan areas having a larger density of cars and obstructions, making reliable data transmission difficult.
- exceptionally Dynamic Environment: The vehicular landscape is very dynamic, with cars traveling at different speeds and changing directions on a regular basis, providing issues for data transmission and network administration. Environmental

Effect: The outdoor VANET network is significantly impacted by the environment, which can result in signal impairments such as multipath propagation, channel fading, and signal shadowing. Weather variations, such as rain, ice, and snow, can impact surface conductivity, hence lowering data transfer performance in the VANET network. These circumstances can also cause road saturation, snow accumulation, dusty weather, and significant microwave signal attenuation, all of which have an impact on total VANET data transmission capacity.

- Heterogeneity: VANET networks can be geographically limitless, making scalability an issue in vehicle communication. Even small communicable nodes require reliable performance.
- Scalability: Performing forwarding techniques using position information for each node reduces the overhead of control messages, enabling scalability [5]. And this paper presents a new solution to VANET network scalability concern using Unmanned Aerial Vehicle (UAV) At automotive scales.
- Spectrum scarcity is a critical problem affecting the reliability and scalability of wireless technology standards like as dedicated short-range communication (DSRC) and wireless access in vehicle environments (WAVE).
- Beyond that, the geophysical aspects relating to the outside VANET network have the potential to inflict signal distresses—multipath propagation, channel fading, and shadowing. Moreover, weather conditions like rain, ice, and snow also affect the conductivity of surfaces, which leads to lowered data transfer performance and overall VANET data transmission capability.
- Information Accuracy: The data transmission in the vehicular environment is done with the help of Global Positioning System (GPS) signals but these systems are unable to provide accurate Geo-location information because of obstacles in the form of buildings and underground tunnels. The tyro-sphere and Ionosphere atmospheric influences could weaken GPS signals, leading to limited accuracy of 5–30 m in overpopulated metropolitan areas. Roadside Unit (RSU) collects data from cars that can become erroneous due to high speed and continuous operating conditions causing incorrect analysis and decision making in the VANET environment.
- Data Security & privacy: VANETs are an essential part of current communication networks, necessitating data protection and privacy. To guarantee successful communication, data must be delivered securely, with packets not changing during transmission and encrypted to avoid eavesdropping. VANETs' dynamic network topology and mobility raise significant concerns about data security and non-repudiation. Keys, which are vital in cryptographic approaches, are also necessary for VANET implementation. Key revocation is a critical component of key management, yet fraudulent subscriber keys are frequently disregarded. However, the VANET network topology's extensibility may result in a huge list of revocable keys, increasing the complexity of key revocation operations.

1.3.2 Potential Benefits of VANETs

VANETs can improve road safety by enabling real-time communication between vehicles, pedestrians, and roadside sensors. VANETs assist to improve intelligent transportation systems by providing usable data for informed decision-making and urban planning. This regard the use of machine learning models for prognostic modality [7].

Prospective application of VANET mentioned in Table1

1.4 Real World Deployment Need

Vehicle Ad-Hoc Networks, or VANETs, are vital to evaluating and testing their performance in real-world settings. Instead, real-world deployments enable researchers to monitor interaction and communication between vehicles with unprecedented granularity, which could help to improve traffic safety and efficiency. Moreover, these empirical efforts also indicate the potential challenges and constraints of VANET technology, mainly in high density urban areas, which ease network design. Real-world deployments confirm the system's scalability, reliability, and efficiency under a miscellanea of traffic measure and environmental modality, all of which contribute to VANETs' overall success [8].

Table 1 Summary of vehicular ad-hoc network application

Domain	VANET applications
Road safety	Collision avoidance Emergency vehicle notification Road hazard alerts
Traffic management	Dynamic route management Smart traffic signal control Traffic flow optimization
Infotainment	Media streaming and Internet Access Location-based services
Environmental monitoring	Pollution detection and reporting Green driving assistance
Emergency response	Real-time incident detection Vehicle platooning
Commercial applications	Fleet management Logistics and supply chain optimization
User experience and services	In-vehicle information systems (IVIS) Vehicle-to-pedestrian communication

1.5 Chapter Scope and Structure

It presents case studies associated with safety and traffic management as well as user services and entertainment systems on VANET. Through a study of Internet access through vehicles, location-based services, data services maintenance in real time, user experience and adoption throughout the VANET ecosystem, it supplements the existing technologies by which the entire driving experience has evolved.

This chapter also explores the impact of VANETs on the environment its role in gathering and communicating emissions information, supporting eco-friendly driving practices, and inform urban planning and policy. It also examines the economic and operational benefits to company vehicle operations, such as fleet tracking and management systems and fuel-saving measures. The chapter focuses on the barriers to VANET acceptance in great detail, specifically the technological problems, security and privacy challenges, regulatory and standards issues, and infrastructure requirements.

Evaluation provides vital insights into VANET performance, user happiness, and environmental and economic impact, which will inform future research and industry practice.

The chapter finishes by offering a complete knowledge of VANET deployment dynamics, using real-world case studies, insights, and best practices to emphasize VANETs' revolutionary influence on transportation in the future. Figure 2 illustrates the chapter's entire structure.

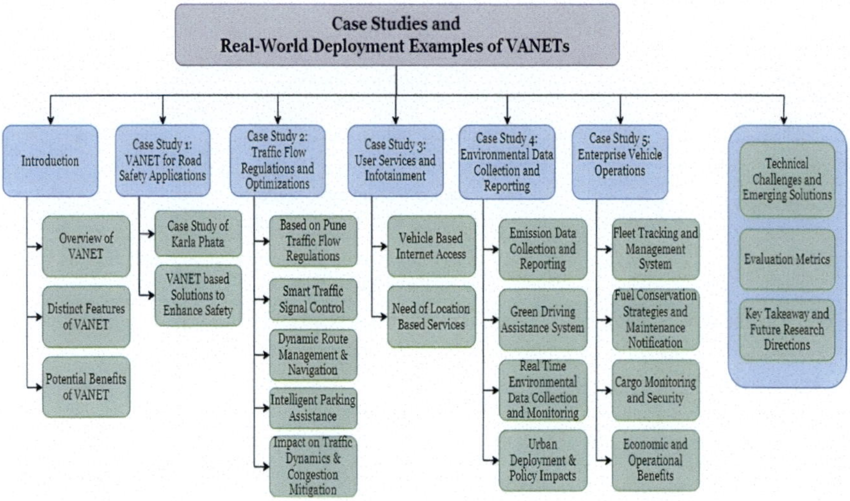

Fig. 2 Taxonomy

2 Case Study: VANET for Road Safety Applications

Survey studies have investigated the many characteristics of VANETs, and an in-depth investigation of the literature is necessary to understand the current state of research in this area. Every year, road accidents kill an estimated 1.43 million people, or around 5100 people every day. The vast majority of these deaths (92%), it should be noted, take place in low- and middle-income countries [9].

The study presents a speed adaptation method for winding road situations, employing Multi-Agent Systems (MAS) technology to supply drivers with real-time traffic data for determining acceptable speed limits, particularly in dangerous locations such as bends. Incorporating both Vehicle Ad Hoc Networks, Internet of Things (IoT), Multi-Agent System (MAS) and Embedded Systems, the proposed system intends to consider traffic information incorporating for adaptability, flexibility and maintainability based on road and infrastructure, metering conditions and requirements and vehicle characteristics. Although the proposed scheme considers road infrastructure, metering conditions, and vehicle characteristics, it doesn't go in-depth to show how these elements are positively incorporated and how they impact the system performance as a whole [10] VANETs have come out as a captious subject in step-up the road safety by allowing cars to exchange real-time road data, rising traffic efficiency, and delivering recreation services [8].

These networks permit surrounding cars to share resources, to improve safety of road and traffic situations [11].

Real-time video streaming over VANETs is a important obstruction for road safety applications, underlining the significance of effective data dissemination in these networks [12].

Expeditious message dissemination in VANETs is captious for assurance of road safety and traffic efficiency, with the quality to offer well-timed information concerning accidents, road situation, and traffic congestion beyond the knowledge of drivers [13]. Propagation of Safety message is important in VANETs for road safety applications, accenting the value of reliable communication in these system [14]. VANETs aim to cut down accidents, intensify traveler comfort, and boost communication between cars and roadside infrastructure to improve general road safety [15]. The usage of VANET communication structure can substantially modify passenger safety on route, bring down traffic congestion, wait times, fuel intake, and release, and therefore alter the entire transportation system [16]. VANETs have the ability to lessen traffic congestion by giving a safer platform to drivers and passengers for traveling stress-free on the road [17]. It Provides new ways to improve the whole driving experiences by assuring to enhance driving condition, road safety, and traffic efficiency [18].

2.1 Real Time Incident Detection and Response Systems

The deployment, challenges, and outcomes of this pioneering project are discussed in this case study providing valued insights into the pragmatic applications of VANET in improving road safety and traffic management. Karla Phata, located on the Old Mumbai Pune Highway, serves as the site for India's first tactical urbanism trial, focusing on real-time incident detection and response systems.

VANETs can offer an innovative solution to improving safety at intersections on the Old-Mumbai Pune Highway, such as the Karla Phata junction as shown in Fig. 3. Here's a detailed solution incorporating VANET technology:

- **Context**: Tactical Urbanism experiments have been started at Karla Phata by SAVWIPL, the SaveLIFE Foundation, and Maharashtra State Road Development Corporation (MSRDC) in an effort to lower the number of fatal accidents at cross-roads on the Old-Mumbai Pune Highway. Urban design, transportation planning, and infrastructure modifications are tested in these short-term, low-cost trials with the goal of improving road safety for all users, especially vulnerable users such as bicycles, pedestrians, and non-motorized vehicles. SAVWIPL is now putting temporary design interventions into place as part of the Vision Zero Project with the SaveLIFE Foundation and MSRDC in order to increase the safety of bikes, pedestrians, and other road users at the crossing [19].
- **Solution Overview**: Karla Phata's plan to enhance real-time traffic control, provide safety notifications, and enhance collaboration between vehicles and infrastructure involves the deployment of VANET, incorporating state-of-the-art communication technology. This initiative will present a holistic solution for intersection safety, supplementing the current Tactical Urbanism research.

Fig. 3 Karla Phata junction on the Old Mumbai-Pune Highway [20]

2.1.1 Solution Using VANET to Enhance Safety at Karla Phata Intersection

Vehicle Notification Systems for Emergency

Real time VANET system can be implemented to address the problem. It contains following Strategic Arrangement:

Implementation of Roadside Unit (RSU): To ensure real-time data collection and facilitate Vehicle-to-Infrastructure (V2I) communication, strategically position RSUs in various locations near the Karla Phata intersection.

Mechanism for incident detection: RSUs utilize data like vehicle speed (DS), average braking intensity (AB), and traffic volume (TV) to detect incidents. The system pinpoints potential incidents and issues alerts when unusual patterns in the data are detected (e.g., sudden halts, rapid deceleration).

Response Coordination: RSUs inform nearby vehicles and emergency services about an incident.—Decrease in Response Time: To ensure quicker responses, connect with emergency services.

2.2 Vehicle Notification Systems for Emergency

For emergency situations, several key factors should be considered: Vehicles should be equipped with On-Board Units (OBUs) capable of receiving and interpreting emergency alerts from Roadside Units (RSUs). Emergency notifications, including events and approaching emergency vehicles, must be transmitted to cars instantly.

2.3 Road Hazard Notifications

Under the right circumstances road hazard notifications can save lives. Other notifications include how to identify hazards: using RSUs to sense dangerous conditions such as obstacles, slick surfaces, and pedestrian crossing areas. RSUs quickly alert the vehicles about the discovered risks and drivers can take actions accordingly. Smart pedestrian signalization system with sensor i.e. intelligent crosswalk will detect the crossing pedestrians and a plow signal regulating with other passing vehicles ensuring protection when they cross the road ensuring safety for pedestrians and bikers.

2.4 Analysis of Deployment Outcomes and Issues

After deploying this system, measuring its effectiveness and optimizing performance requires consideration of several points. Surveillance and analysis of the flow and nuisance of the system, such as the free flow of traffic in time and the flow of vehicles, are the second level of data collection. Accident rates—along with traffic levels—should be run regularly to see if safety is improving. Performance indicators will be measured in terms of accident rate decrease and emergency service response time improvement, amongst others. And a simplified progress equation should suffice to measure the gap in accident rates and response times pre- and post-implementation. The system should be regularly reviewed and refined based on data collected, issues encountered, and stakeholder feedback regarding the effectiveness of the system. Utilizing VANET technology also at the level of Karla Phata junction can improve efficiency and safety. Complementing Tactical Urbanism pilots, presenting events in near real time, an immediate response to emergencies, and hazardous road conditions notifications are subsumed in the Vision Zero initiative aimed towards reducing all traffic fatalities and serious injuries.

3 Case Study: Traffic Flow Regulations and Optimization

To realize use of VANET for traffic flow improvement Pune city is considered for study. Pune, one of India's fastest-growing cities, suffers huge traffic congestion and administration issues. To address these concerns, Pune has established a variety of traffic flow laws and optimization measures. This case study examines the use of advanced vehicular ad hoc network technology to improve traffic flow and safety.

Case Study Based on Pune Traffic Flow Regulations and Optimization

Pune intends to increase road safety by combining smart traffic signals, real-time data collecting, and adaptive traffic management technologies. This chapter looks at the techniques, implementation, and impact of these legislation, providing insights into how technology-driven solutions might improve urban traffic management.

3.1 Smart Traffic Signal Control

The urban population growth puts significant strain on infrastructure, particularly transportation. The demand for urban services is growing faster than supply, resulting in a wider disparity. The supply of services does not rise at the same pace as the geometric growth. India's cities lack basic public transportation infrastructure. Increased urban population has led to higher travel demand. Urban transportation is crucial for shaping urban growth and living. Urban roads, which serve as the lifeline of cities, are struggling to meet rising travel demand due to increased vehicle traffic, longer journeys, and under investment in infrastructure. To solve this problem traffic must be regulated and technological solutions must be implemented to optimize the traffic flow regulation. According to a new analysis by Amsterdam-based location technology company TomTom, London is the slowest city to drive through in 2023, with an average speed of about 14 kmph during rush hour. The TomTom Traffic Index, which evaluated 387 cities in 55 countries, emphasized the impact of traffic on travel time, fuel expenditures, and CO_2 emissions, using data from over 600 million in-car navigation systems and smartphones. Bengaluru and Pune are among the world's top ten worst traffic-hit cities. According to the TomTom analysis, two Indian cities, Bengaluru and Pune, would be among the top 10 worst traffic-hit cities in the world in 2023. Fig. 4 shows India's 4 worst traffic-hit cities. If we analyze the data depicted in Table 1, In 2023, traffic congestion in major Indian cities remained a severe concern, as evidenced by their global rankings. Bangalore and Pune have emerged as India's top two cities, ranking sixth and seventh internationally, respectively, with an average travel time of 28 min to traverse just 10 km at any time of day. This signifies extreme traffic congestion, placing them among the most congested cities in the world. New Delhi and Mumbai, despite doing marginally better worldwide, saw significant delays. New Delhi ranked 44th in the world, with an average travel time of 22 min for 10 km, whereas Mumbai placed 54th, with an average duration of 21 min. The report highlights the ubiquitous issue of traffic congestion in India's metropolitan centers, which necessitates better traffic management and infrastructural upgrades to alleviate commuter delays.

Figure 5 shows the hourly speed and travel time per 10 km in pune for the last 7 days. If we see the day wise and hourwise variations in the traffic (Table 2).

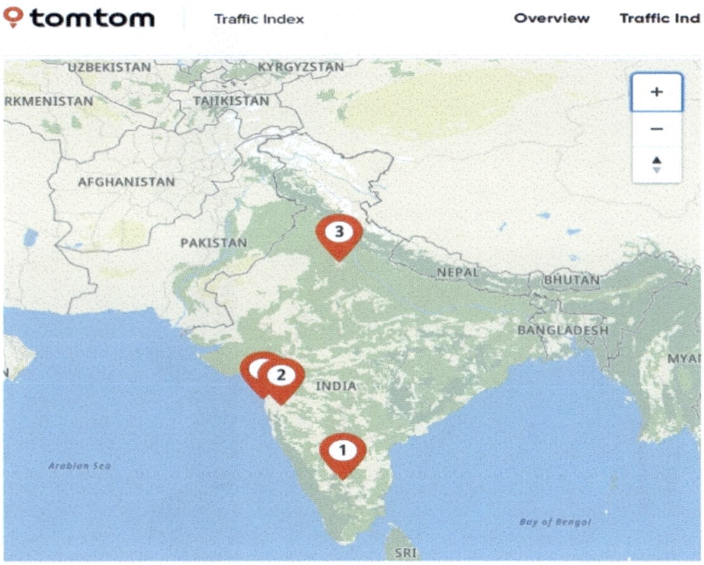

Fig. 4 India's four cities with the highest traffic volume [21]

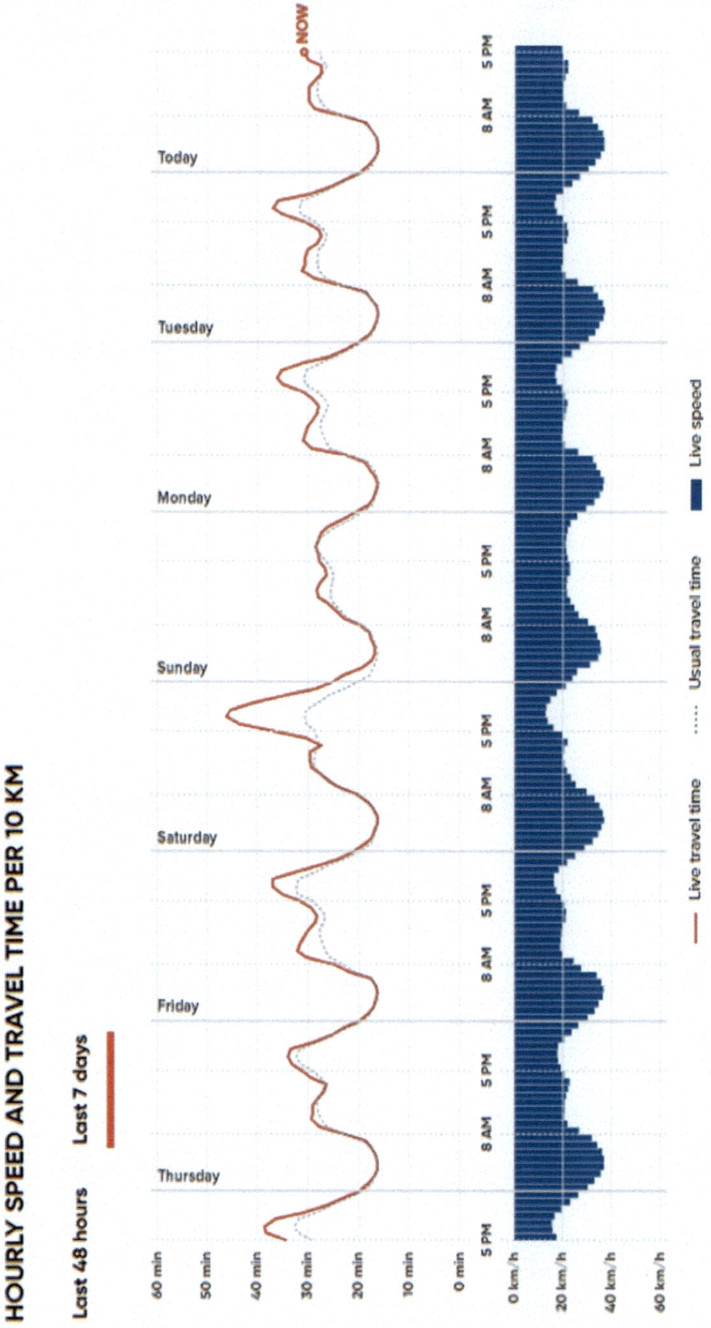

Fig. 5 Traffic monitoring in Pune over the last 7 days from 12/06/24 [22]

Table 2 Rank of India's 4 worst traffic-hit cities and average time required to travel 10 km/hr

S. No.	City name	India rank	World rank	Average time required to travel 10 km (24/7) (year-2023)
1	Banglore	1	6	28
2	Pune	2	7	28
3	New Delhi	3	44	22
4	Mumbai	4	54	21

3.1.1 Pune Civic Activism and Traffic Management

Citizen Involvement: Through a number of groups, Pune residents have actively participated in the improvement of their city. They work together with local governments to pinpoint problems and provide fixes; Pune's participatory budgeting is one example of this approach in action.

PPCR Report: Based on comments from citizens on social media, the Pune Platform for Collaborative Response (PPCR) created a traffic report. The Pune traffic police received this report, which highlights important traffic concerns, for evaluation and action.

Accident Statistics: Ninety-two percent of road fatalities include cyclists or pedestrians. Helmet less riders were involved in all two-wheeler deaths. Most concerning is the surge in driving on the wrong side of the road.

3.1.2 Pune Traffic Management Solutions

The Pune Smart City Development Corporation Limited (PSCDCL) has taken a significant stand against traffic congestion in Pune by putting the Adaptive Traffic Management System (ATMS) into place. There are three phases to the project's execution, and substantial progress has already been accomplished. The strategies that support intelligent parking assistance, dynamic route management, and smart traffic signal control are described here, along with how they affect traffic dynamics and congestion alleviation overall [23].

Present Implementation: In important locations like Karve Road, Tilak Road, Swargate, and Fatima Nagar, PSCDCL has placed ATMS at 30 signals. The system optimizes signal timings based on real-time data by utilizing advanced traffic monitoring technologies.

Functionality: Analysis of Traffic Data: To lessen congestion, the system dynamically modifies signal timings based on an analysis of traffic flow.

Green corridors: Give emergency vehicles preferential use of the green lights.

Data Collection: To assist with enforcement actions, data on traffic violations is gathered.

Outcomes:

- shorter stopping and travel times.
- At important junctions, there is less congestion and improved traffic flow.
- improved control of traffic during rush hours.

Statistics:

- covers thirty signals in Phase 1, up to forty-six and forty-nine in Phases two and three.
- Project cost in total: Rs. 171 crores.

3.2 Dynamic Route Management and Navigation

Dynamic Route Management and Navigation technologies are critical components of Pune's traffic flow improvement initiatives. These systems combine real-time data and VANET technology to give drivers with adaptive route advice. They analyze current traffic conditions and recommend ideal routes to avoid congestion, saving travel time and boosting traffic efficiency. This section looks at how various technologies are used in Pune to improve urban transportation.

Proposed Implementation: The suggested implementation would combine GPS navigation, real-time traffic data, and dynamic route management to provide drivers the best possible routes.

Functionality:

- **Real-time updates:** Let drivers know which routes are optimal given the traffic, accidents, and road closures that are occurring right now.
- **Traffic Distribution:** Move cars away from crowded regions to effectively control traffic flow.

Outcomes:

- shorter travel times and increased route efficiency.
- equitable allocation of traffic throughout the network.
- improved traveler experience with reduced wait times.

Example: There has been a 15% decrease in travel times during peak hours in cities with comparable systems.

3.3 Intelligent Parking Assistance

Projected Implementation: Set up a Smart Parking System (SPS) which will use a mobile app to present real-time parking accessibility data.

Practicality:

- **Sensor-Based Detection:** Parking availability data will be gathered and send it to a centralized system.
- **Real-Time Guidance:** The driver will be provided with a Real-time information on open parking spaces.

Outcomes:

- Time will be reduced to find place to park.
- Traffic jams brought on by parking-related activities will be reduced.
- parking assets will be effectively utilize.

Example: Parking search time reduced by 20% with a pilot smart parking project in Aundh.

3.4 Impact on Traffic Dynamics and Congestion Mitigation

- The impact of traffic flow controls and optimization measures in Pune has been significant in terms of traffic dynamics and congestion reduction. The advanced VANET technology has empowered the city of Pune to meticulously analyse traffic trends, predict possible congestion, and manage the overall traffic flow. This section examines how these programs have altered traffic behavior and reduced congestion, showcasing adjusted travel times and overall traffic efficiency.
- Advanced Traffic Management: Integrate intelligent parking systems, dynamic route management, and smart traffic signal flow for comprehensive traffic management.

Anticipated Outcomes:

- **Reduction in traffic**
- **Accident Reduction**: reduced numbers of road accidents, especially those that affect bicycles and pedestrians.
- **Public Transport Efficiency**: increased utilization and dependability of public transportation.

Statistical Insights:

- Pune is the second most crowded city in India and the sixth most crowded city worldwide.
- 19 km/h is the average speed during rush hour.
- Congestion is exacerbated by the fact that 62% of Pune's roadways are less than nine meters wide.

Challenges:

- **Infrastructure Deficits**: addressing the deficiency of stormwater drainage systems, broad highways, and basic infrastructure.

- **Financial Constraints**: high prices related to land acquisition and infrastructure upgrades.
- **Traffic Violations**: commuters' high levels of indifference and large unpaid fines

Citizen Involvement:

- The traffic police are currently reviewing the alternatives that were suggested in the PPCR study, which brought these concerns to light.
- a focus on civic awareness initiatives to inform the public about their rights and obligations under traffic rules.

Pune intends to greatly reduce traffic congestion, increase commuter safety, and improve the general quality of urban life by putting these clever concepts into practice. Effective integration, ongoing oversight, and engaged public and government engagement are critical to the success of these programs.

4 Case Study: User Services and Infotainment

User services and infotainment are critical components of Vehicular Ad Hoc Networks, which improve the driving experience by delivering useful information and entertainment alternatives. These systems provide real-time traffic information, navigation help, and multimedia entertainment, resulting in safer and more entertaining travel. This section looks at the development and implementation of user-centric services and entertainment in VANETs, emphasizing their benefits in terms of convenience, safety, and overall user happiness on the road [24].

Ever since Motorola debuted the in-car radio in 1930, infotainment systems have seen a substantial evolution. Developments like the addition of a record player by Chrysler in 1953 and the tape player by Ford in 1965 were significant turning points in the history of incorporating entertainment into driving experiences. Satellite radio, mobile devices, and locally stored multimedia material have all been effortlessly incorporated into current infotainment systems thanks to the introduction of Bluetooth technology and improved hardware storage capacity [25]. The in-car cybersecurity solutions from AUTOCRYPT minimize cybersecurity threats and offer total protection for the car's embedded systems, all the while promoting responsible and safe innovation in the sector. The automobile industry is undergoing a dramatic move toward advanced in-vehicle infotainment (IVI) systems that incorporate advanced technology such as artificial intelligence, face recognition, and gesture control. These systems improve the safety, convenience, and entertainment for drivers and passengers, making driving more intuitive and entertaining. AI-powered assistants orchestrate tasks, while facial recognition personalizes the driving environment based on individual preferences. Gesture control technology elevates user interaction, allowing for intuitive adjustments without distracting from the road. As automakers expand entertainment options, vehicles are evolving into personalized smart devices catering

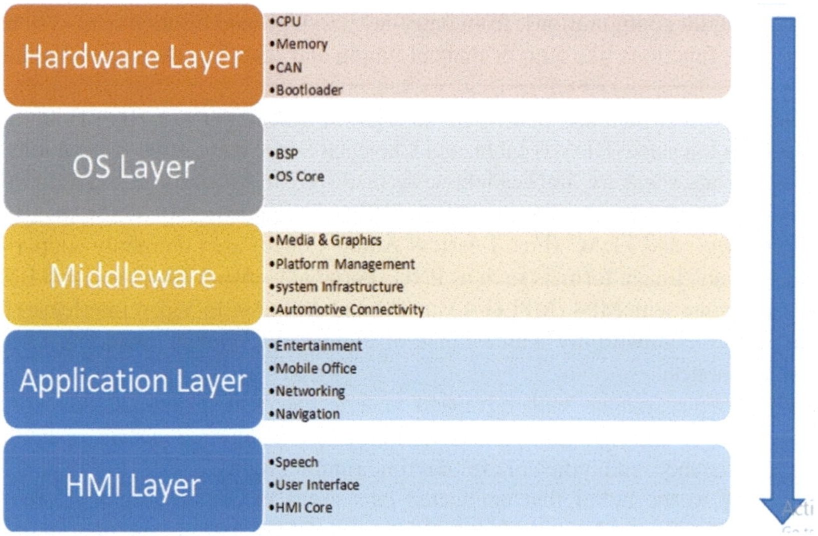

Fig. 6 System architecture of in-vehicle infotainment I

to diverse consumer needs. Cybersecurity is paramount, with robust encryption and intrusion detection measures safeguarding against potential threats [26].

eInfochips as an automotive solutions provider, offers customized in-vehicle infotainment solutions for the automotive industry. We have expertise in the development of features like HMI, navigation, digital tracking, and multimedia integration [27].

The Driver Drowsiness Monitoring System takes infrared pictures of the driver's eyes using image sensors made for automobiles. Through the use of a high-speed digital signal processor and unique pupil detection technology, this system analyzes and recognizes indicators of driver attention or tiredness, improving passenger and vehicle safety by averting possible collisions. As shown in Fig. 6, the system architecture of IVI illustrates the key components and their interactions.

Automotive infotainment systems with high-resolution touch displays usually use LCD (Liquid Crystal Display) or TFT (Thin Film Transistor) technology. Active matrix LCD TFTs with full color capabilities are especially well-suited for presenting dynamic and intricate graphical data. The sizes of these touch screens range from 2 to 12 in., with some versions having bigger HD displays that measure up to 20 in. in width.

Smartphones and car entertainment systems may be seamlessly integrated through smartphone pairing via Bluetooth communication. With this feature, users may handle incoming and outgoing calls, set up conference calls, see contact lists and call records, and read SMS messages—all of which can be accessed and managed directly through the infotainment interface.

Infotainment systems include both digital and analog tuners to provide AM and FM radio reception across many standards. These tuners are compatible with a variety of audio configurations, from entry-level to high-end infotainment systems, and include functions like auto or manual tuning and scanning for accessible radio stations.

The ability to send music and video material via Bluetooth (BT), High-Definition Multimedia Interface (HDMI) cable, and Universal Serial Bus (USB) connections to display screens, speakers, and headphones is made possible by multimedia support in contemporary infotainment systems. MP3 (MPEG Audio Layer III), AAC (Advanced Audio Codec), and FLAC (Free Lossless Audio Codec) are commonly supported music formats. Image formats such as PNG (Portable Network Graphics) and BMP (Bitmap), along with MP4 (MPEG-4 Part 14) and AVI (Audio Video Interleave) for video, ensure compatibility with a range of multimedia sources, including tablets and smartphones.

Infotainment systems with advanced vehicle capabilities include things like climate control settings, voice assistants that allow hands-free system operation, parking assistance, and indicators for daytime running lights.

Contrary to the belief that connected cars are a recent development, luxury automakers such as Mercedes-Benz, Audi, and BMW introduced their respective cloud-based systems several years ago. In 2015, Mercedes-Benz debuted its COMAND Online System, which included a range of cloud-connected applications. Around the same time, Audi introduced its myAudi Connect smart technology, while BMW launched its Connected Drive services, all aimed at enhancing the connectivity and functionality of their vehicles[28]. TATA Nexon Ev, The 2020 makeover of the Tata Nexon sub-compact SUV, which complies with BS6, introduced Intelligent Real-Time Assist (iRA), a feature that is available on the Nexon twins. This is the first time an iRA has been used in a Tata car. For example, as shown in Fig. 7, Features of the system include collision notification, service records, live vehicle diagnostics, e-SIM connectivity, "find my car" capabilities, and speech recognition support in Hindi, English, and Hinglish. Among its other features are an intrusion alarm system and step-by-step navigation designed with Indian drivers in mind [25]. In this section we will focus on the case study of MG Hector Infotainment System. Figure 8 shows the car under study.

4.1 Vehicle-Based Internet Access

With its i-SMART Next Gen technology, offering over 50 connected features to enhance the driving experience, the MG Hector sets a new standard in automotive connectivity. This innovative system integrates hardware, software, services, and apps to provide an intelligent and seamless driving experience. Central to this innovation is the M2M embedded SIM, a pioneering feature in India's automotive industry equipped with IPv6 support for 5G readiness. This allows the car to receive over-the-air software updates, akin to smartphones, ensuring it remains updated with new themes, applications, and entertainment content [28].

Fig. 7 HARMAN™ 31.24-cm Cinematic Touch Screen in the TATA Nexon EV [29]

Fig. 8 MgHector Internet car [30]

4.2 *Need of Location-Based Services (LBS)*

The MG Hector's voice assistant technology answers to "Hello, MG," allowing for hands-free contact via Natural Language Understanding (NLU) adapted to Indian-English dialects. The Emergency Call (e-Call) function, which is triggered upon airbag deployment, prioritizes safety by immediately notifying MG's 24/7 Pulse Hub of the vehicle's position and initiating swift assistance procedures [28].

4.3 Real-Time Data Services

Beyond connection and safety, the MG Hector improves the driving experience with technologies aimed at simplicity and enjoyment. It has a pre-installed JioSaavn Premium Account, providing fast access to your favorite songs. The i-SMART Next Gen smartphone app "Find My Car" functionality enables remote management of car operations including as engine start, climate adjustment, and position monitoring [28].

4.4 User Experience and Acceptance

Exceeds the pr-set limits, the geofencing technology warns the owner if someone is taking the car out of those boundaries, which is a security feature too. Technologies like the MG Weather app and traffic-real-time navigation further help prepare the driver-state for their next trip. For those who are interested in tracking their driving habits, Smart Drive Information will provide details of driving behavior, such as brake monitoring and route choice[31]. MG Hector Full Feature List | The Most Technologically Advanced Connected Car in India.

5 Case Study: Environmental Surveillance and Control

This case study focuses on the use of sophisticated traffic management systems across major cities, including Los Angeles, Singapore, and Copenhagen, with the aim of enhancing environmental surveillance and control. In Los Angeles, they have begun installing their Intelligent Traffic Light (ITL) system. Combining real-time data and predictive analytics, Singapore's intelligent traffic management system aims to reduce traffic congestion and improve air quality. N action centered around8 sustainable transportation system which limits the environmental impact through green alternative forms of transportation9 and tracking environmental benefits. Collectively, these attempt show how contemporaneous technology may help to make more sustainable and efficient urban conditions [32].

5.1 Emissions Data Collection and Reporting

The Los Angeles city has used mobility forecasting algorithms to predict traffic drift and improve traffic control strategies. This initiatory has decreased travel time and greenhouse gas discharge, highlighting the value of sustainable transportation. Transportation is a significant contributor to greenhouse gas (GHG) emissions, accounting for 29% of overall emissions in the United States, with its highest point reached in 2021. From 1990 to 2021, the transportation industry saw the greatest absolute increase in GHG emissions of any sector.

In 2023, Los Angeles implemented an Intelligent Traffic Light (ITL) system, reducing travel time by 16% and pauses at intersections by 12%. In Montreal, traffic light management has evolved, with about half of the city's traffic lights now considered "intelligent." These lights are connected to a master network, allowing real-time adjustments to respond to current occurrences, such as accidents, enhancing emergency response and reducing congestion and emissions.

In Singapore, sophisticated sensors and data analytics regulate traffic flow and reduce congestion, supporting the country's goal to lower carbon emissions intensity by 36% by 2030 and eliminate all emissions from public transportation. Real-time data collection through numerous IoT sensors and a high-speed fiber optic network supports these efforts [33].

5.2 *Green Driving Assistance Systems*

Montreal's ITS emergency management system includes advanced motorist support services, such as emergency calls and car location alerts. Sensors in automobiles assess damage in accidents and decide whether to call an operator or emergency services directly. This proactive detection and quick response system significantly enhance emergency response capabilities and traffic management, potentially averting 60% of incidents on highways with timely warnings.

Copenhagen aims to become carbon neutral by 2025, implementing sustainable transportation and smart street lighting. The city's agile traffic management method exercise real-time data for monitoring and adjusting traffic lights, improving traffic flow and reduction in congestion. Smart parking systems help drivers to available spaces, enabling reduction in time spent probing for parking and accordingly step-down in congestion and pollution [33].

5.3 *Real-Time Environmental Data Collection and Monitoring*

Singapore aims to empower people with information to embrace sustainable practices safely, currently collecting real-time data on travel patterns, traffic congestion and energy usage—powered by a multi-billion-dollar high-speed fibre optic and 5G network. This data in real-time serves as the base for an intelligent transportation system that allows continuous real-time optimization of traffic flows and congestion reduction. For example, in Barcelona, a flood of up to date traffic data comes from the traffic sensors in cars, which are then aggregated onto a traffic data platform to assist drivers with real-time traffic conditions. There's also initiatives such as low-emission zones and electric vehicles.

5.4 Urban Development and Policy Impacts

Seoul's vision is to be an innovative smart city, which it intends to achieve through the use of big data and state-of-the-art ICT. The city is accomplishing the goal by setting up up to 50,000 of Internet of Things (IoT) sensors to monitor the different urban phenomena specifically related to traffic and air quality. Data is collected and analyzed from an array of sources, thus the information is used to manage the urban environment and to develop effective policies which in turn will make the transportation system better and more efficient through the technologies such as AI taxis and smart parking.

Dubai has gone to great lengths to invest in modern infrastructure to support the city's smart city goals. The dynamic system of the city and smart parking lots together use data in real-time to effect the movement of people and cars and minimize the traffic congestion. The government has implemented energy strategies such as the UAE Energy Strategy 2050 and Dubai Clean Energy Strategy 2050, which they hope will result in the substantial decrease of emissions and the increase in the use of renewable energy sources.

The Egyptian government has plans to convert some of its populated cities including the New Administrative Capital into the smart city model in order to solve the problem of traffic congestion and over population in Cairo. The intelligent transportation system implementation with the use of smart sensors, AI cameras, real-time cellular networks, and smart ticketing technology will be a strategy for road surveillance, traffic control and live traffic updates that are environmentally friendly and promotes sustainable urban living.

The modification maps the different critical areas as they impact the sustainability of the smart city through ITS in the form of data collection, driver-assistant systems, environmental monitoring, and urban traffic policies [33].

6 Case Study: Enterprise Vehicle Operations

This case study focuses on the usage of Vehicular Ad Hoc Networks (VANETs) within enterprise vehicle operations, with a particular emphasis on the ways by which organizations and their logistics are improved through efficient fleet management. Those that take leverage VANET technology can be in real-time tracking, lower the cost of fuel consumption route designing more efficiently, and enhance communication between the vehicles and the central control systems. This section illuminates on how VANETs may cut the operational costs, shorten the delivery time and even lift the efficiency of the corporate vehicle operations through the provision of the implementation of the VANET's standards for quick, safe transportation and management [34].

The main function of the VANET Technology which is quickly evolving consists in developing safe and effective road designs for mass transportation. By turning every car into a wireless communication device, the driver will be able to see more of the horizon than is now visible with the naked eye. All nodes in this technology require backend infrastructure connectivity, which varies depending on the architecture and service. Various VANET topologies have been suggested by research standards organizations, including the European Telecom standards Institute (ETSI), International Standard Organization (ISO), and Institute of Electrical and Electronic Engineering (IEEE). Various nations, including Brazil, USA, and Japan, are developing these standards in accordance with their unique infrastructure configurations and requirements [35].

To get deeper into this issue, we'll look at a case study of how Trinetra's Fleet Management Solution was used to streamline rental vehicle operations. This case study includes Enterprise vehicle operation of the Trinetra management system.

6.1 Fleet Tracking and Management Systems

Trinetra was contacted by a car rental company operating throughout the state to solve operational issues and inefficiencies in their offerings. To improve service quality, cut expenses, and increase operational efficiency, they required a tailored, automated application. The Trinetra team determined that the following crucial problems had to be fixed:

Automobile Availability Status: In order to effectively handle automobile reservations, deliveries, and pick-ups, the operator required a system. Improved communication was necessary for greater coordination between the command center and the drivers.

Timely Pick-Ups: It was important to make sure that customers were picked up on time.

Vehicle Tracking: Every vehicle in the fleet has to be able to be located and tracked by the operator.

Driver Safety Training: It was crucial to guarantee both driver and passenger safety via appropriate training.

Trinetra customized their fleet management program for GPS-enabled vehicle tracking to address the issues found after reviewing the client's requirements. Because of the solution's emphasis on driver safety, vehicle security, and effective fleet management, the customer was able to deploy the Fleet Management System (FMS) with automated features. Real-time driving behavior monitoring made sure that drivers adhered to safety regulations while traveling.

Route Optimization: Kept an eye on a number of variables to maximize savings on fuel, decrease maintenance expenses, and optimize routes. Automated alarms were generated by discrepancies.

Effective Fleet Management: The special features of the software made fleet management easier, saving time and enhancing customer response times. It made it simple to assign the right car for any kind of journey, long or short.

The Customized Fleet Management Application's benefits include instant alerts. Using GPS technology, set up immediate notifications for activities involving pick-up and drop-off, guaranteeing safety.

Speed and Driving Behavior: Through real-time monitoring, incidences involving hard braking and speeding have been reduced. Visual tracking allowed employees to respond quickly to consumer requests by allowing them to view car positions on a map[36]. Global Reach is such a that with 13,000 active devices, the solution has been deployed in more than 15 countries and Fuel expenses were cut by 40% with a 22% improvement in driver conduct was seen.

6.2 Fuel Conservation Strategies and Maintenance Notifications

Efficient fuel utilization is the key to both environmental sustainability and saving costs in the fleet management sector, where speed is the top priority. Trinetra Wireless produces a superior fuel management program that can handle these demands skillfully and securely. This case study analyzes how the main features and advantages of the software can result in efficient savings of fuel and prevention of scheduled maintenance. Prominent characteristics include full card integration, which simplifies transactions using gasoline cards and controlling fuel costs by simply monitoring them. Also, the program comes with expense tracking, which subsequently enables users to maintain a history of every gasoline transaction for easier financial management.

Real-time monitoring is an added vivid function of strolling in with short data that makes predicting the future easy, and this is achieved using data that originates from fuel levels and fuel consumption in real time. Users are notified about the low fuel situation, or unauthorized use of fuel by means of receiving fuel level alerts when fuel goes low suddenly or drops. The software's fuel usage analytics give a deeper understanding of customer usage insights by the statistical information, thus, enhancing the overall efficiency. Moreover, this tool is the ultimate integration solution that easily combines with existing GPS gadgets, fuel cards, and fleet management programs to accelerate fleet operations understanding for all stakeholders.

Automatic report generation is an additional feature, producing detailed reports on a timely basis, besides providing the opportunity for users to create custom reports that match the desired degree requirements for examinations. The main benefits of fuel management software are direct management of fuel costs through the provision

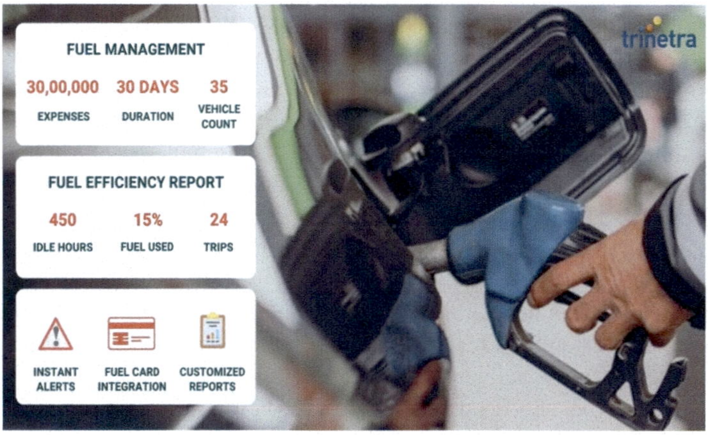

Fig. 9 Fuel management software from Trinetra [37]

of comprehensive data and insights, savings gained from making the identification of and doing away with fuel misuse if any, precise insights as regards to the fuel use and spending patterns, and productivity gain through the lowering of spending and optimization of routes.

Preventive maintenance is achieved by notifications for irregularities, thereby making it possible to inform subscribers when energy consumption is higher than normal, thus having the effect of lowering the expense of an unexpected repair by timely repairs and maintenance procedures that are proactive. The software also contains sophisticated tracking systems that can be used to find and stop fuel stealing from happening besides making security measures like linking fuel cards to enhance security and prevent unauthorized use. The user-friendly interface, along with the inclusion of a mobile app, provides easy access to the software. Its scalability makes it highly suitable for both newly established fleets and cases requiring fleet resizing (Fig. 9).

6.3 Cargo Monitoring and Its Security

In the competitive delivery industry, precision, speed, and real-time information are the key performance points. Tracking the Global Positioning System is a revolutionary act for cargo transportation enterprises. But, this study case shows how Trinetra Wireless ensured safety and cargo monitoring for a leading logistics operator's customer. The client's performance problems included inefficient routing which prolonged traveling time and fuel overuse, as well as producer-induced cargo availability, undelivered logistics operations due to poor routing, and a lack of the capacity to immediately react to the deviation of the course thereof.

Trinetra's solution was grounded in the use of GPS enabled location data and GPS allows for the employees to be constantly online monitoring the delivery trucks if they were supposed to monitor. Route optimization that made the best possible roads choice automatically led to reduced travel times and thus lower fuel consumption. The solution also provided detailed ETA thus improving customer satisfaction through timeliness The new proactive problem-solving capability driven by exogenous escalation was the next panacea after it was discovered that deviations occurred from the path. This allows the chosen alternative to be more accessible, outgoing, and responsive to the changes. Thanks to that after the early detection of the possible stops, the shipment can be guaranteed to be on the road in secure conditions, while the potential of customized features such as geofencing and configured tracking facilitate timely and safe delivery.

The approach's benefits encompassed increased efficiency, by means of reducing the costs of fuel and optimized routes, increased security thanks to the uninterrupted tracking of the cargo, and client satisfaction which was the result of delivery of shipments on time and precise ETAs. The leadership initiative was a key in completing the timetable and the easy solution of bottleneck routes. Trinetra Wireless's GPS system is basically a tool that enhances the company's operations, thereby, bringing about the customer's happiness and efficiency making it a very reliable tool for transport companies.

The economic and operational advantages of using VANET for vehicle tracking and fleet management exceed all other criteria when it is integrated with these systems. These networks offer a cost-effective solution to improve safety, productivity, and reliability for delivery trucks, service fleets, and haulage carriers. The methods to save on fuel preservation are economical, as the optimized routes bring out the decrease in the gasoline costs and hence a lower maintenance cost which is obtained through the lower wear and tear of the vehicles. Real-time notifications help to prevent unauthorized use of fuel, which in turn saves money.

Operational benefits can be mentioned such as the effective procedure and development of the route using real-time data, the proper allocation of resources by well-placed vehicles as well as the active fuel management with real-time fuel monitoring. Besides, maintenance alerts notify well-timed repairs of cars. The financial benefits gained by such systems include the reduction of the repair expenses through the earlier problem detection which makes the maintenance range more cost-effective, and the longer vehicle life due to the regular maintenance. The advantages in safety and cargo monitoring are also significant while real-time observation curbs losses from stealing and vandalism, which may decrease insurance bills, besides additional security. The correct route location updates ensure the proper delivery of the cargo and by using the security alert and geofencing protect goods. The reliability and the right time of data transferring are the major achievements in the customer's satisfaction, and the FMS using or serving as a part of VANET can gain a large amount of financial and functional benefits including less cost, good network security, and higher production.

7 Barriers to VANET Deployment

VANET still has various hindrance to be resolved, despite its improvement. Attacks on VANETs are this challenge. Three categories are: "insider versus outsider," "malicious versus rational," and "active versus passive." The following terms are used to characterize the various attack types against messages: "Bogus Information," "Cheating with Positioning Information," "ID disclosure," "Denial of Service," and "Masquerade." Several sorts of attacks, including network, application, timing, monitoring, and social. Every class explains a distinct kind of assault, along with its priority and threat level. Figure 10 shows different attacks on VANET.

7.1 Technical Challenges

Network management: Keeping classic structures like trees up to date is challenging with VANETs due to their high mobility, which causes frequent changes in channel conditions and network topology.

Congestion and Collision Management: Different traffic loads are caused by the network's infinite size. Due to low traffic, network partitions are common in rural and off-peak metropolitan regions; during rush hour, however, there is a significant level of congestion and accidents.

Environmental Impact: Because electromagnetic waves are used in VANET transmission, the surrounding environment has an impact on these waves. Thus, when establishing VANETs, the environmental effect needs to be taken into account.

MAC Design: MAC design is essential in VANETs because of shared medium communication. Many techniques have been proposed, including TDMA, SDMA, and CSMA. IEEE 802.11 adopted a CSMA-based MAC for VANETs.

- **Time-based Restrictions**:

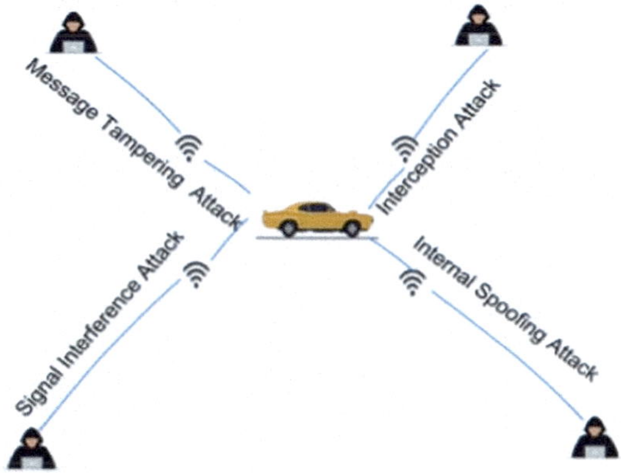

Fig. 10 Attacks on VANET

Within 100 ms, safety-related messages on VANETs must be sent. To achieve this, quick cryptographic techniques and prompt message and entity authentication are required.

Data Consistency Liability: In VANETs, hostile activity is still possible for authenticated nodes. Data consistency should be ensured via mechanisms that are created, such as correlating data from several nodes to avoid misrepresentation.

Low Tolerance for Error: Probabilistic protocols in VANETs can be troublesome because of the crucial nature of the data being communicated, which makes even little errors quite harmful.

Key Distribution: In VANETs, security measures rely on key distribution. One major difficulty is ensuring safe key distribution and management across various public key infrastructures and manufacturers.

Governments, manufacturers, and users all need incentives for the successful implementation of vehicular networks, particularly for applications that automatically report infractions of traffic laws.

Highly Heterogeneous Vehicular Networks: Node addressing, quality of service, routing, security, and billing are made more difficult by the coexistence of several, incompatible wireless networking protocols.

Data Management and Storage: The massive volumes of dispersed data generated by large-scale vehicle networks provide special difficulties for data management and storage.

Localization Systems: Accurate and dependable localization systems are necessary for critical safety applications. The precision and resilience to assaults of satellite-based systems are limited, and their availability is not constant.

Tolerant of disruptions Communications: By investigating novel data transmission techniques and taking driver behavior into account to enhance carry-and-forward techniques, problems like increased latency and decreased dependability in sparse networks may be reduced.

Following a Target:

For communication protocols and applications, tracking the physical position of cars is essential. This calls for methods that can recognize and anticipate vehicle tracks while maintaining privacy.

7.2 Privacy and Security Concerns

Network availability assaults include spam, which consumes bandwidth and causes transmission delays, hacking into broadcasts to generate false safety alerts, and DoS/DDoS attacks, which degrade networks and restrict legitimate access. Computer infections, typically propagated by insiders, disrupt through software upgrades, but the Black Hole Attack involves nodes rejecting data packets, thereby impeding connection. Greedy drivers may clog the network in order to obtain an advantage, causing service to be delayed for others. Authentication and identity-related threats include replay attacks, which repeat or delay legitimate data, Sybil attacks, which generate phony nodes to propagate misinformation, and masquerade, in which attackers employ bogus identities. GPS spoofing transmits wrong location data, wormhole attacks link distant nodes to disrupt the network, message manipulation modifies or removes safety warnings, ID disclosure tracks and discloses node positions, and false information dissemination spreads disinformation, causing confusion.

7.3 Matters of Regulation and Standardization

To guarantee interoperability, security, and privacy when processing and implementing VANETs, direct legal frameworks and standards are needed. Compliance with international standard and regulation is captious for the distributed deployment of VANET technology.

7.4 Infrastructure and Expense Needs

VANET approval requires important investment in infrastructure, vehicle equipment, and ongoing maintenance. Achieving a eminent deployment need comparing these costs against the benefits of accrued safety, effective, and environmental footprint [38].

8 Evaluation and Metrics

Metrics and evaluation are important in evaluating the success of VANETs. Packet delivery ratio, communication overhead, end-to-end delay, throughput, packet loss, and normalized routing load are parameters for evaluation of routing protocols such as Geographic Routing Protocol, Maximum Distance On demand Routing Algorithm, Optimized Link State Routing, Ad hoc On demand Distance Vector, and Multi Objective Congestion Metric-based Artificial Ecosystem Optimization. As the research proposes, it is essential to use a correct routing protocol for VANETS, in particular BABEL. In systems of VANETS the on the one hand the cost of communication is optimized, on the other hand the network resources are optimally used [39].

8.1 VANET Deployments for Performance Metrics

VANET deployments allow for the assessment of packet delivery ratio and throughput of vehicle communication systems. The performance of a VANET should also take into account conveyance density, packet delivery ratio, end-to-end latency, inter-arrival time for emergency messages, and protocol overhead [33]. The following are critical indicators of the performance of a VANET deployment.

A. Packet Delivery proportion: It is the percentage of delivered packets of a total that should be sent through the network. It becomes even more important to check reliability and efficiency of data transport over the VANETS.
B. Throughput is the amount of data successfully transmitted over a communication channel in a specified period. Throughput becomes an important factor for establishing whether or not a network can have effective data traffic control during operation of various VANETs.
C. End-to-End Delay: End-to-End delay is when the first data package in a communication network is sent from one point to another in real time. To guarantee speed and efficiency of data transfer across the network, the end-to-end delay time in VANETs should be minimized. Emergency Message Inter-arrival Time:

This statistic tracks the time between successive emergency alerts sent over the network. It is critical to assess the system's timeliness in conveying urgent information in VANET deployments.

Protocol overhead is the additional data transmitted along with the primary user data to facilitate communication. In VANETs, protocol overhead management is crucial for optimizing network resources and enhancing overall system performance.

8.2 User Feedback and Satisfaction Matrix

The data should be collected regarding user experience with the VANET system on the eye on user happiness as well as evaluation of the satisfaction of the users in the multimedia services across vehicular networks. It does appear that this can satisfy the user needs because it allows, on the one hand, to consider opportunistic vehicle meetings and service requirements, on the other hand, to maximize total user satisfaction and ensure equity.

It is possible to perform a total user satisfaction dissection for every media service with specific interest in the interplay between the satisfaction priority indicator and the satisfaction popularity function to achieve maximum every user satisfaction the total U-shaped satisfaction function.

Surveys & Questionnaires: Get feedback on the users' engagement with the system in terms of ease of use, reliability and satisfaction, for example.

Service Response Time: Assess how time you take in responding to the user concerns and questions. Enhanced user service is evidenced by lower response times.

Track the engagement of the users on the system by measuring user retention rate and other comparable metrics. Improved customer satisfaction is evidenced by a high retention rate.

Feature Utilization: Audit which features of the system are most sought after and used by the users to assist in deciding on system upgrading features and functionalities.

8.3 Effect on Efficiency and Safety of Traffic

Vehicular Ad hoc Networks improve traffic efficiency and safety by allowing automobiles to communicate with each other and the infrastructure. VANET systems, like the Cooperative Safe Driving System (CoopSafeDrivSyst), utilize V2V and V2I communications to effectively avoid accidents and manage road safety. Data transmissions in VANET systems are minimized through the application of Real-Time Databases (RT-DB) which, in turn, brings about shorter reaction times for the vehicles in critical situations as well as improved overall road safety. VANET systems can mitigate a variety of dangerous road conditions by building a hybrid architecture that combines V2V, V2I communications, and RT-DB, thereby providing an efficient and reliable driving experience. The hybrid architecture of VANET systems allows for the use of both distributed and central driving control, enhancing operational reliability and fault tolerance, as well as contributing to efficiency in the system [40].

8.4 Impact on the Environment and Economy

To fully interpret the performance of VANET systems, it is necessary to measure their environmental and economic effect..

A. Fuel Usage: Track changes in the consumption of fuel. Decreased usage is a signal of higher route efficiency as well as environmental gains.
B. Reduce Emissions: Reduce Emissions: Monitor emissions before and after deployment. Reduced emissions imply a positive impact on the environment.
C. Operational Costs:Operational expenditures include gasoline, maintenance, and vehicle operations. Reduced expenditures indicate financial benefits from increased effectiveness.
D. Return on Investment (ROI): To calculate the ROI of the VANET system, compare the savings and benefits achieved to the implementation and maintenance costs.

9 Key Takeaways from Case Studies

- **Improving Intersection Safety Using VANETs:**

The Karla Phata crossroads on the Old Mumbai-Pune Highway conducted India's first tactical urbanism attempt to prevent fatal accidents. Using VANET technology, this program intends to improve real-time traffic regulation, give safety warnings, and develop improved vehicle-infrastructure coordination, therefore complementing existing tactical urbanism initiatives to improve road safety for all users.

- **Pune Traffic Flow Regulations and Optimization**:

Pune, which is considered among the top 10 worst traffic cities in the world, is facing considerable congestion issues as a result of growing urbanization. Implementing traffic flow rules and technology solutions is critical to optimizing traffic, minimizing delays, and increasing overall urban mobility.

- **Advancements in In-Vehicle Infotainment Systems**:

Advanced technologies like artificial intelligence, facial recognition, and gesture control are being integrated into in-vehicle infotainment systems. These advances improve driver and passenger safety, convenience, and entertainment by converting automobiles into individualized smart gadgets.

Cybersecurity is a key concern, with strong safeguards required to prevent possible attacks.

- **Environmental Surveillance and Control**:

To improve traffic flow and curtail pollution, cities like Los Angeles, Singapore, and Copenhagen have implemented smart traffic management systems. Los Angeles' Intelligent Traffic Light (ITL) system has successfully cut down travel time and emitted less greenhouse gases, therefore demonstrating the effectiveness of environment-friendly transport modes.

- **Foot tracking and management systems**:

Trinetra developed the GPS-enabled fleet management system that improved the operating efficiency of a car rental firm. The system enhanced availability of vehicles, enhanced interaction between the command center and the drivers, ensured timely pickups, and included vehicle monitoring, and driver safety training. The automated Fleet administration System (FMS) prioritized driver safety, vehicle security and effective fleet administration which translates to higher service quality, and lower operating costs.

10 Conclusion and Future Scope

The case studies in this chapter outline the transformative potential of VANET technology for enhancing traffic safety and flow, as well as overall transportation efficiency. Results from the integration of tactical urbanism with VANET in the Karla Phata intersection show significant reduction rates in accidents due to timely management of traffic as well as safety notifications in real time. Electronic traffic control systems have reduced traffic delays and increased urban mobility in highly populated areas like Pune. The growth of in-vehicle infotainment systems into sophisticated smart gadgets emphasizes the significance of using new technology to improve the driving experience while maintaining strong cybersecurity safeguards.

Intelligent traffic management systems in Los Angeles and Singapore demonstrate how smart technology may minimize emissions while promoting sustainable transportation. Trinetra's tailored fleet management solutions highlight the efficiency and safety benefits that automated, GPS-enabled systems may provide. The implementation of VANET and related smart technology provides comprehensive answers to today's transportation difficulties. Ahead of us lie attractive research topics: traffic prediction algorithms and, using artificial intelligence and machine learning, the development of adaptive control systems. Application areas for VANET will extend into autonomous car operation and support for electric vehicles. In this, inter-agency cooperation among government bodies, technology providers, and academia will be critical to overcoming existing challenges and ensuring successful deployment of VANET. Stakeholders may increase user confidence and encourage greater use of VANET technology by supporting strong cybersecurity protocols and interoperability standards. This chapter emphasizes VANET's vital role in defining the future of transportation, particularly its ability to improve safety, efficiency, and sustainability in urban mobility. Continued innovation and research will be critical to realizing the full potential of VANET technology.

References

1. Jesy Janet Kumari J, Thangam S, Saleem Raja A (2023) Efficacious routing approaches in vehicular ad hoc network: an empirical study. In: 2023 First international conference on advances in electrical, electronics and computational intelligence (ICAEECI), 1–7
2. Sharma A, Jaekel A (2021) Machine learning approach for detecting location spoofing in VANET. In: 2021 International conference on computer communications and networks (ICCCN), 1–6
3. Rajasekaran A, Danda NTR, Shilpi M, Shaik SA, Kota Y (2024) A secure wireless communication in vehicular ad-hoc network system. In: 2024 5th International conference on intelligent communication technologies and virtual mobile networks (ICICV), 607–612
4. Hozouri A, Mirzaei A, Razaghzadeh S, Yousefi D (2023) An overview of VANET vehicular networks. ArXiv abs/2309.06555
5. Bowlin EW, Khan MS, Bajracharya B, Appasani B, Bizon N (2023) Challenges and solutions for vehicular ad-hoc networks based on lightweight blockchains. Vehicles

6. Karunathilake T, Förster A (2023) Using LoRa communication for urban VANETs: feasibility and challenges. ArXiv abs/2311.18070
7. Sheikh MS, Liang J, Wang W (2019) A survey of security services, attacks, and applications for vehicular ad hoc networks (VANETs). Sensors (Basel, Switzerland) 19
8. Gupta N, Thakre RD, Suryawanshi YA (2017) Real time implementation of vehicular ad-hoc network. Int J Eng Trends Technol 45:121–126
9. Khadidos AO, Selvarajan S, Manoharan H, Yafoz A, Khadidos AO, Alyoubi KH (2022) An intelligent security framework based on collaborative mutual authentication model for smart city networks. IEEE Access 10:85289–85304
10. Benkirane S, Guezzaz A, Azrour M, Gardezi AA, Ahmad S, Sayed AE, Naseer S, Shafiq M (2023) Adapted speed system in a road bend situation in VANET environment. Comput, Mater Continua
11. Alsarhan A, Al-Ghuwairi A-R, Almalkawi IT, Alauthman M, Al-Dubai AY (2020) Machine learning-driven optimization for intrusion detection in smart vehicular networks. Wirel Pers Commun 117:3129–3152
12. Bello Tambawal A, Md Noor R, Bin Salleh R, Chembe C, Oche M (2019) Enhanced weight-based clustering algorithm to provide reliable delivery for VANET safety applications. PLoS ONE 14
13. Jiang X, Feng J, Song T, Katayama T (2019) Low-complexity and hardware-friendly H.265/HEVC encoder for vehicular ad-hoc networks. Sensors (Basel, Switzerland) 19
14. Uçar S, Ergen SC, Özkasap Ö (2016) Multihop-cluster-based IEEE 802.11p and LTE hybrid architecture for VANET safety message dissemination. IEEE Trans Veh Technol 65:2621–2636
15. Septa N, Wagh S (2023) Cooperative communication for safety message dissemination in unsaturated networks with varying contention windows. Int J Commun Syst 36
16. Kumar A, Sharma NA, Kumar A (2021) Improving security and privacy in VANET network using node authentication. J Discrete Math Sci Cryptogr 24:2471–2480
17. Konyeha S, Osa E (2020) Deployment of Vanets infrastructure to aid road transport systems in developing countries
18. Elias SJ, Hatim SM, Darus MY, Abdullah S, Jasmis J, Badlishah Ahmad R, Khang AWY (2019) Congestion control in vehicular adhoc network: a survey. Indonesian J Electr Eng Comput Sci 3:1280–1285
19. Al-Heety OS, Zakaria Z, Ismail M, Shakir MM, Alani S, Alsariera H (2020) A comprehensive survey: benefits, services, recent works, challenges, security, and use cases for SDN-VANET. IEEE Access 8:91028–91047
20. Figure 52.3, Mumbai Live (n.d.) Karla Phata junction on the Old Mumbai-Pune Highway. Retrieved from http://www.mumbailive.com
21. Figure 52.4, TomTom, "India's four most traffic volume cities". Available: http://www.tomtom.com
22. Figure 52.5, TomTom (2024) Traffic monitoring in Pune over the last 7 days. Retrieved from http://www.tomtom.com
23. Mumbai Live (2022) First tactical urbanism trial put to test at Karla Phata on Old Mumbai Pune Highway. Available: https://www.mumbailive.com/en/infrastructure/first-tactical-urbanism-trial-put-to-test-at-karla-phata-on-old-mumbai-pune-highway-70644 and TomTom. Pune traffic index. Available: https://www.tomtom.com/traffic-index/pune-traffic/
24. Observer Research Foundation (n.d.) Citizens report on their city's traffic congestion: Pune. Available: https://www.orfonline.org/expert-speak/citizens-report-on-their-citys-traffic-congestion-pune
25. Abeywardana RC, Sowerby KW, Berber SM (2018) Empowering infotainment applications: a multi-channel service management framework for cognitive radio enabled vehicular ad hoc networks. In: 2018 IEEE 87th vehicular technology conference (VTC Spring), 1–5
26. Shaik S, Ratnam DV (2022) A trust based energy and mobility aware routing protocol to improve infotainment services in VANETs. Peer-to-Peer Network Appl 15:576–591
27. AutoCrypt (n.d.) 5 futuristic in-vehicle infotainment features. AutoCrypt. Available: https://autocrypt.io/5-futuristic-in-vehicle-infotainment-features/

28. eInfochips (n.d.) Everything you need to know about in-vehicle infotainment system. https://www.einfochips.com/blog/everything-you-need-to-know-about-in-vehicle-infotainment-system/
29. Figure 52.7, Tata Motors, "HARMAN™ 31.24-cm Cinematic Touch Screen in the TATA Nexon EV". Available: https://ev.tatamotors.com/nexon/ev.html
30. Figure52.8, Go Mechanic, "MG Hector Internet Car". Available: https://gomechanic.in
31. GoMechanic (2023) Connected cars in India: everything you need to know. https://gomechanic.in/blog/connected-cars-in-india/
32. Al-Ezaly E, Abo-Elfetoh A, Elhishi S (2021) Pollution reduction using intelligent warning messages in VANET. J Intell Syst Internet Things
33. Elassy M, Al-Hattab M, Takruri M, Badawi SAQ (2024) Intelligent transportation systems for sustainable smart cities. Transp Eng 16:100252. ISSN 2666-691X
34. Moscato S, Shin H, Moscato D (2020) Provisional access control model for mobile ad-hoc provisional access control model for mobile ad-hoc environments: application to mobile electronic commerce environments: application to mobile electronic commerce
35. Rasheed A, Zia H, Hashmi F, Hadi U, Naim W, Ajmal S (2013) Fleet & convoy management using VANET 1(1):1–9
36. Trinetra Wireless. GPS vehicle tracking for rental car companies. Available: https://www.trinetrawireless.com/case-studies/gps-vehicle-tracking-for-rental-car-companies/
37. Figure 52.9, Trinetra Wireless, "Fuel Management Software". Available: http://www.trinetrawireless.com
38. Quyoom A, Mir AA, Sarwar A (2020) Security attacks and challenges of VANETs: a literature survey. J Multim Inf Syst 7:45–54
39. Mohammed D, Mansor M, Hock GC (2023) A comparative performance evaluation of GRP and MDORA protocols for VANET. Int J Emerg Technol Adv Eng 13(4):82–90
40. Elleuch I, Makni A, Bouaziz R (2021) An intelligent and efficient safe driving system. In: International conference on health information science, 2021

Deep Learning Algorithms and Methods for VANET-Based Congestion Management: Implementation Concerns and Limitations

S. Bharathi, N. Sathish Kumar, S. Sathiyapriya, P. Devi, P. Sasigresa, S. Sivachitralakshmi, and D. Manivannan

Abstract This paper discusses the main problems and difficulties in implementing Intelligent Transportation Systems (ITS). These include the need to minimize delays in communication between automobiles and Roadside Units (RSUs), assuring a continuous flow of traffic, and improving road safety. The relevance of Vehicular Ad hoc Network (VANET) has attracted attention from several academic domains. Continuous monitoring is necessary for these systems to use Machine Learning algorithms on large amounts of data from VANET applications such as crowdsourcing, pollution control, and environmental surveillance. Deep Learning enables computers to enhance their performance by analyzing historical data. The objective of VANET is accomplished by the effective use of supervised and unsupervised data learning methods. The hybrid deep learning model is created by merging the Convolutional Neural Network (CNN) with the Bidirectional Long Short-Term Memory model that incorporates Elman gates (BLSTME). We conducted an analysis of the dependability, communication, and difficulties linked to traffic in VANET networks, and found that they are not feasible to execute in practice. In addition, we explored the possibilities of using machine learning techniques to tackle these difficulties. In the end, we discussed the future direction and challenges, and analyzed a case study that demonstrated a situation related to VANET.

S. Bharathi (✉) · S. Sivachitralakshmi
Department of Computer Science and Engineering, SRM Institute of Science and Technology, Chennai, India
e-mail: barla.indian@gmail.com

N. S. Kumar
Department of Electronics and Communication Engineering, Sri Ramakrishna Engineering College, Coimbatore, India

S. Sathiyapriya · P. Devi · P. Sasigresa · D. Manivannan
Department of Computer Science and Engineering, Vel Tech Rangarajan Dr, Sagunthala R&D Institute of Science and Technology, Chennai, India
e-mail: sasigresap@veltech.edu.in

J. Bhatia et al. (eds.), *Deep Learning Based Solutions for Vehicular Adhoc Networks*, Studies in Computational Intelligence 1207,
https://doi.org/10.1007/978-981-96-5190-0_14

Keywords VANET · Intelligent transport system (ITS) · Deep learning · CNN · RSU

1 Introduction

The emergence of the smart city requires a higher need for advanced transportation systems. Monitoring the flow of traffic and managing congestion are crucial elements of an Autonomous urban region. The rise in traffic congestion and the inadequacy of the wireless communication infrastructure for traffic management have prompted the development of ITS. Vehicular Ad-Hoc Networks may be used to achieve the main goals of ITS, which include managing traffic congestion, ensuring road safety, and optimizing infrastructure utilization. The need for research in the subject of VANET is growing. VANET involves cars that are equipped with OnBoard Units, such as GPS, cell phone sensors, and additional wireless equipment, to facilitate communication between vehicles (V2V) and between vehicles and infrastructure (V2I). The cars in VANET exhibit significant mobility, resulting in network congestion, unstable connections between VANET components, and communication failures. In VANET, the authentication techniques used for transmission must have a low computational load to provide durability and scalability [1]. The presence of hostile nodes or connections weakens the credibility and ability to make decisions of V2X receivers [2]. VANET now integrates the Software Defined Network (SDN) architecture, which is built upon the 5G network infrastructure, necessitating efficient routing protocols. The two dominant protocols for VANET connection are Devoted Short-Range connection (DSRC) and Wireless Connectivity in Vehicular Environment (WAVE). These types of protocols use either multi-hop connection or a single node to transmit data, with the aim of minimizing data congestion in the channel by allowing the data to be carried for as long as feasible.

As the number of transmissions increases, so does the susceptibility to assaults against it. It is crucial to offer security services at various levels of ITS to ensure the protection of communications [3]. Based on the traffic accident learning, the widely held accidents happen at junctions. These accidents produce road blockages and result in events such as speed decreases, rubbernecking, and reduced discharge flow. As a consequence, there are large traffic delays [4]. VANET facilitates efficient communication between RSU and automobiles. In this era of technological advancement, cellphones have become a valuable asset in several ways. By incorporating various technologies such as GPS, IoT, Cloud Computing, and GNSS, many issues related to traffic management, safety, routing, and interaction can be effectively addressed by simply touching the screen. Multiple concurrent activities take occur simultaneously in vehicle networks. These occurrences provide a vast quantity of data that may be fully used for sophisticated data analysis and forecast.

1.1 Deep Learning Models in VANET

VANETs are flexible networks that consist of automobiles talking with one other and with peripheral infrastructure. Deep learning models may be used in VANETs for several objectives like traffic forecasting, congestion identification, route optimization, and autonomous vehicle management. Below are many typical uses of deep learning models in VANETs.

Traffic prediction models have the capability to examine past traffic data in order to forecast future traffic circumstances. This data may be used to enhance the process of determining the most efficient routes and schedules, hence decreasing traffic congestion and enhancing the overall efficiency of the network.

a. **Anomaly detection**: Deep learning models may be taught to identify irregularities in traffic patterns, such as collisions, road closures, or atypical congestion. This data may be used to redirect cars and notify drivers, so enhancing safety and decreasing journey duration.
b. **Vehicle Localization**: Deep learning models, particularly those using sensor fusion methods and convolutional architectures, can precisely determine the location of cars inside the VANET. This information is vital for uses such as autonomous vehicles, accident prevention, and traffic management.
c. **Resource Allocation**: Deep learning models may enhance resource allocation in VANETs by adaptively modifying communication factors such as broadcast power, bandwidth allotment, and routing pathways. Reinforcement learning algorithms are very efficient in acquiring optimum resource allocation strategies in real-time.
d. **Cooperative Perception**: Deep learning models may allow cars to exchange sensor data and collaboratively understand their surroundings, improving their understanding of the situation and enabling cooperative driving behaviors like platooning and merging assistance.
e. **Traffic Signal Optimization**: Deep learning models have the ability to optimize traffic signal management by acquiring knowledge about traffic flow patterns and adaptively adjusting signal timings based on this information. This may effectively alleviate congestion, eliminate delays, and enhance overall traffic efficiency.

Energy efficiency in VANETs may be enhanced by the use of deep learning models, which automatically optimize communication protocols, data transmission techniques, and power use profiles to minimize energy consumption. Ensuring the longevity of the battery in linked cars and roadside infrastructure is of utmost significance.

In general, deep learning has great potential to enhance the capabilities of VANETs, allowing for more efficient, secure, and autonomous modes of transport. Nevertheless, the implementation of deep learning models in VANETs also poses difficulties with computational intricacy, real-time processing prerequisites, data confidentiality, and compatibility with current communication protocols. Tackling these obstacles will be essential for fully harnessing the capabilities of deep

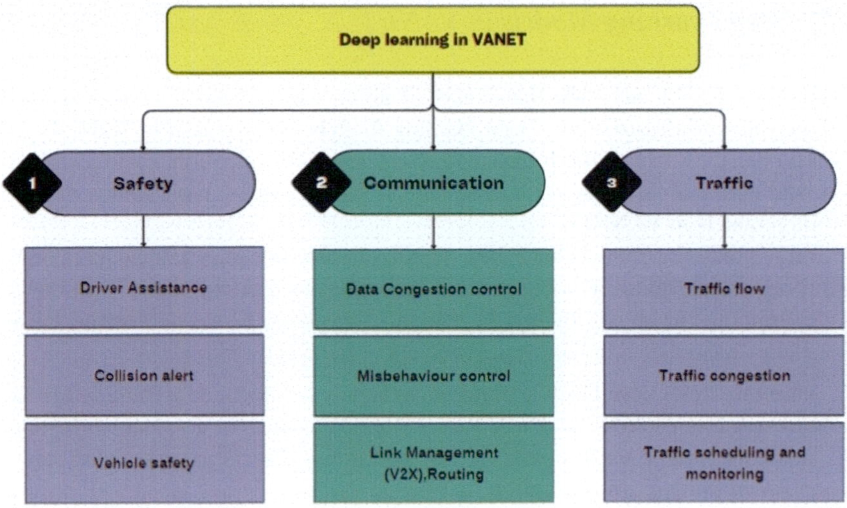

Fig. 1 Taxonomy: deep learning adoption of VANET

learning in VANET applications. The diagram given below illustrates the taxonomy of deep learning in VANET in Fig. 1.

1.2 Motivation

By effectively using the generated data, it becomes possible to use advanced techniques like machine learning, which in turn enables the system to adapt autonomously. Machine learning is the use of computer algorithms to analyze and manipulate data.

Computational statistical models use extensive data to provide predictions and assessments. The ability to predict and make judgments may play a significant role in VANET. Various machine learning methods are accessible for diverse applications. Nevertheless, particular industry has not been subjected to comprehensive scrutiny before reaching the depth of our comprehension. Multiple assessments have been undertaken on VANET, but, none of them include the use of machine learning techniques to improve the complete structure. Table 3 highlights the study that specifically focuses on the incorporation of machine learning performances in VANET.

Upon analyzing the aforementioned review works, it was discovered that the combination of machine learning and VANET has only been explored within the confines of certain applications [5]. The objective of this study is to conduct a comprehensive analysis of the evaluation of machine learning methods in many elements of VANET.

Our conversation was structured around the core components of Safety, Communication, and Traffic, as shown by our study findings. The present work offers an extensive assessment of machine learning techniques for many applications within a certain field. To our knowledge, a thorough examination of the many applications of VANET has not yet been conducted.

The novelty of the study is in its inventive method of tackling the joint optimization issue of incorporating networking, caching, and computing resources in automotive networks. Through the use of a deep reinforcement learning methodology, the system efficiently streamlines and enhances the process of allocating resources. This strategy synergistically integrates the advantages of several domains, providing a more streamlined and resilient solution in comparison to conventional ways in vehicle networks.

1.3 Contribution

- We steered a comprehensive investigation of the current survey on ML models in VANET. We examined the assessment metrics that were used and identified any limitations or deficiencies. In addition, we highlighted the ongoing efforts in this field.
- We classify the review job into many applications. This classification demonstrates the integration of machine learning methods to make advancements in VANET. In addition, we have also included a concise overview of the study effort, highlighting its advantages, disadvantages, level of accuracy, and the primary technologies used.
- Lastly, we will now address any unresolved issues and potential areas for further investigation.

1.4 Organization

The remainder of the work is structured in a similar manner. Section 2 offers an in-depth examination of the literature, emphasizing recent advancements in intelligent traffic systems. Effective message is the paramount element of the Intelligent Transportation System. Efficient communication between cars and infrastructures is crucial for all VANET applications. Section 3 explores the incorporation of several advanced methods in communication. Section 4 presents an analysis and estimation of the outcomes and discourse around the models used for deep learning in VANET. Section 5 ultimately finishes the job by emphasizing any remaining unresolved issues and difficulties.

2 Literature Review

The systematic random sampling method used in [1] helps to minimize bias by selecting samples from the population of allonges, passengers, and pedestrians. Purposive sampling is a method that specifically targets the recruitment of respondents who are traffic cops. Simple linear regression analyzes the connection between the dependent and independent variables in a dataset. The number of registered vehicles is considered as the independent variable, while the casualties are treated as the dependent variable. The interpretation of future results is based on this relationship.

This model analyzes the separate impacts of manufacturing and allows for a clear understanding of the causes and consequences of traffic congestion. The model is inadequate in assessing the accuracy, precision, and recall.

Reference [6] have created a system that provides operators in transportation and public health authorities with information to help mitigate the imminent threat of air pollution. The authors propose a distinctive that integrates the traffic flow, exhaust and propagation model. The traffic situation is determined by analyzing mobile data, which provides information on factors such as the average speed of traffic on certain routes and the amount of traffic.

Rempe et al. [7] use a clustering algorithm approach to identify the congested clusters that provide a substantial level of adaptability for various applications. The emergence of stationary mobbing clusters and the existence of dynamic congestion pockets define a congestion cluster. The Munich Boulevard grid in Germany was analyzed in a case study. The research focused on the clusters that represent the static road network of Munich and distinguish between regular and irregular days for the purpose of analyzing congestion in these clusters. The resultant cluster facilitates the identification of weekdays that do not follow a predictable pattern.

The periods and variability of congestion are calculated, and the unique clusters of congestion behavior are quantified. Seek to forecast the speed of traffic and compare its performance to current prediction models by using the CNN model captures the specific relationships between data points within a given context and is less prone to being overwhelmed by excessive amounts of data. This approach necessitates the use of five input layers. One layer is devoted to delivering time data, while the other four layers are meant to record the speed characteristics of interconnections one, two, three, and four. This strategy is specifically intended to establish regional dependencies and is more resistant to noise in the data. To exceed the efficiency of the remaining models, we need to use a multitude of sub models.

Hebert et al. [3] developed a high-resolution accident prediction algorithm that uses big data analytics to forecast the occurrence of accidents within a few hours on specific road segments defined by junctions. Big data analytics is a deliberate methodology that enables data scientists to extract valuable insights from vast quantities of diverse and intricate data. The balanced random forest approach effectively addresses the issue of imbalanced data by adjusting or sampling it.

The HMDLF, created by [8], is a framework that specifically aims to forecast traffic flow using a hybrid multimodal deep learning approach. By combining the CNN with the Gated Recurrent Unit (GRU), it investigates and detects the complex nonlinear relationships between different types of input data. The endways multi-modal architecture is specifically designed to analyze and interpret sequential traffic data. It focuses primarily on the geographical location, long-term interdependence, and connections in space and time.

The AE-LSTM technique utilizes an autoencoder strategy is difficult to execute and has a limited application. It also provides excellent accuracy in predicting the traffic flow. However, this research only values the temporal patterns and basic spatial patterns. The uncertain density of transportation and provide evidence on the specific load balance and other available resources across different VANET networks for vehicles. The process consists of three distinct stages: cluster initiation, cluster adaptation, and cluster fusion. This method is superior and produces significant outcomes. It refers to other algorithms that are different and can handle various levels of traffic in metropolitan areas. It deducts the total number of sent packets that did not reach their destination, although there is no consensus on this deduction.

The use of machine learning methods is essential for improving performance across many circumstances. The proposed model does not need additional resources, which might lead to prolonged processing time [8]. The hybrid model lacks many CNN sub-models for feature extraction that tackle the problem of data imbalance [9]. Furthermore, the model incorporates BLSTME with LSTM. The AdaBoost technique is used to improve deficient classifiers that tackle overfitting in vehicle pattern recognition [10] and attain enhanced performance in forecasting real-time traffic flow [11] in highly populated regions [6].

3 Deep Learning in VANET Safety Applications

Automobile collisions may result in extensive destruction to automobiles and their surroundings, potentially causing severe harm to live organisms. Therefore, as the smart traffic system progresses, concerns about shelter likewise emerge. It is essential that vehicular connections can efficiently transmit all safety-related evidence collected from cars and RSUs in instantaneous. An Intelligent Transportation System should facilitate the driver by providing assistance in selecting a safe route and accurately detecting traffic signs. This would help encourage the smooth flow of traffic without any obstacles or accidents.

The traffic system should possess the capability to anticipate unforeseen circumstances by disseminating pre-collision alerts to prevent dangerous scenarios. In addition, On-Board Units are required to offer detailed information on a specific vehicle's integrity in the event of a crash. In order to effectively integrate these characteristics into VANET, we structured the safety assessment into three sections where machine learning techniques may be used (Table 1).

Table 1 Overview of deep learning in safety applications

Author	Techniques used	Performance
[12]	Deep learning	• Improved categorization outcomes • A higher quantity of neurons results in more complexity, which in turn decreases the effectiveness of the model
[5]	Deep learning	It surpasses DT, kNN, A-NN, sVM, and in regards of average accuracy and average recall
[13]	k-NN, agglomerative clustering, k-means, spectral clustering, affinity propagation, birch	Nonparametric methodology. Data convexity is not assumed. Demonstrates exceptional levels of flexibility and precision
[16, 17, 25]	RF, A-NN	High accuracy proportion. Views routine traffic as accidents
[26, 27]	PCA, CNN classifier	Nevertheless, VANET characteristics are highly interdependent, outperforms traditional approaches. Environmental features improve dataset analysis accuracy. Found several false areas
[29, 30, 31]	Ada boost, deep learning	This model has exceptional performance in normal traffic circumstances. It may be seamlessly integrated with Intelligent Driver Assistance Module and Driver Surveillance System. Ineffective during nighttime
[32]	ELM, CNN	Capable of acquiring intricate convolution and deep architectural skills while minimizing training duration - Insufficient training data
[34, 35]	NB, KNN, SVM, decision tree	The use of classification techniques included the utilization of five-fold cross-validation Enhanced precision in localization, along with exhaustive comparisons and evaluations of publicly available information- Excessive application of spatial distribution parameters on the identification of traffic signs

4 Methodology

Section 3 outlines the suggested framework for forecasting traffic congestion. The proposed hybrid model integrates the Convolutional Neural Network and BLSTME for forecasting traffic stream. CNN is used to discern the distinctive attributes of input traffic photographs. Subsequently, the retrieved features undergo training using BLSTME to augment the accurateness of the less efficient classifiers and make predictions about traffic flow.

4.1 Proposed Architecture

The long short-term memory (LSTM) approach generates a high spatial resolution.

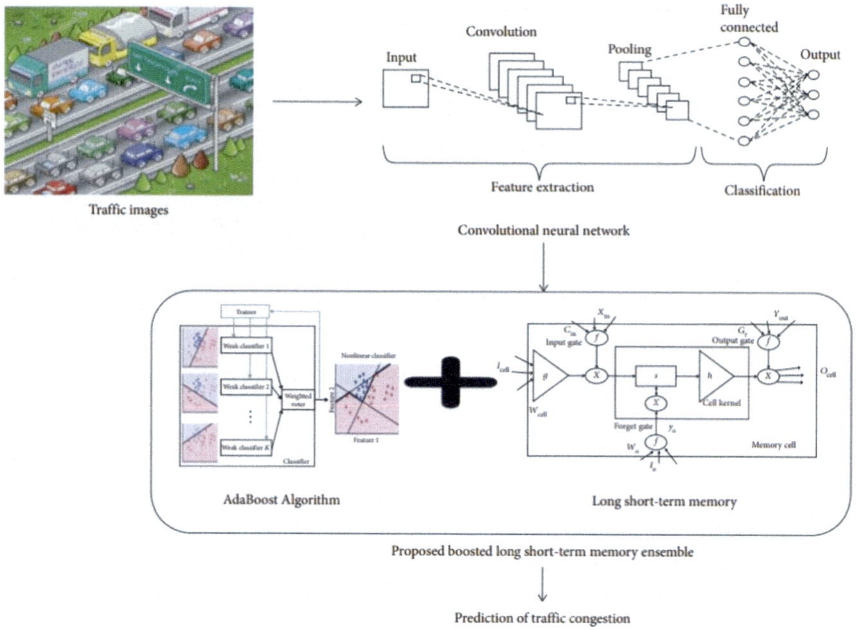

Fig. 2 Architecture diagram of BLSTME-CNN algorithm

To discourse this problem, an innovative hybrid procedure for deep learning called BLSTME-CNN is suggested as a solution. The architecture of this method is shown in Fig. 2.

4.2 A Convolutional Neural Network

The CNN excels at representing a unique characteristic of an input picture via different properties, consists of two main layers: the convolutional layer, which is accountable for acquiring and representing the characteristics of the input picture, and the pooling layer, which guarantees invariance to shifts. In the layer of convolution, each neuron gets inputs from a local group of neurons in the preceding layer, which are then used to create the production layer. The different feature illustrations were obtained by increasing the size of multiple units from the preceding layer. Equation (1) defines the convolution layer.

$$ylf = \sigma\left(\sum yl - 1K, W1lkf + b1lf\right), fe[1, fl]) \quad (1)$$

Equation (1) calculates the activation map, denoted as "ylf", for the "fth" convolution layer. The activation map yl − 1 k of the (l − 1)th layer is linked to the kth

Fig. 3 LSTM classifier using distribution

```
1 Inputs sample training sets fpi qi g where {p = fp1, p2, p3, ⋯}where n = no of input samples and
2 qi €(1,2, 3 )where qi is a multiclass label associated with p
3 Initialize DðkÞ = n
4 For k = 1, 2, 3, ⋯, K
5 Train the LSTM classifier using the distribution Dk
6 Get the hypothesis with error function with respect to Dk
7 Error function is calculated at each stage which is then weighted Dk
8 Choose αk = 0:5 fln ð1 - ekÞ/ekg-network parameter calculation
9 Update the Dk + 1 ðiÞ
10 Calculate the error function and repeat step 4
11 If error is less than ek
12 Then, ensemble all the outputs HðxÞ = HðxÞ = sign ð∑αk TkÞ/αk
13 Else
14 Go to step 4
15 End
16 End
```

activation map of the lth layer. The weight associated with the fth activation map of the lth layer at point k is denoted as W1lf, whereas b1lf represents the corresponding bias term. The filters in the l − t layer may be represented as f − l, and the component wise not linear activation function is denoted by [3].

The pooling techniques may reduce the spatial size, but they retain important information. The value of yl in Eq. (2) [14] may be calculated by applying a convolution filter on the output, with a size of (m, n), and then applying a bitwise nonlinear activation function [14]. The kernel location is represented by a1 and b1 (Fig. 3).

$$ylf = \left(\sum y(W1lkf + b1lf)(a1, b1), fe\left[1, fl\right]\right) \tag{2}$$

There are three different stages in the grid. The LSTM cells Gt in addition ht are computed via assessing the three gate modes and cell input state, and may be passed on to the subsequent grid. Equation (3) represents the input gate.

$$jt = \Phi Gi + Gih.et{-}1 + Si) \tag{3}$$

4.3 AdaBoost Is a Machine Learning Algorithm

The AdaBoost approach synergistically integrates LSTM networks with the AdaBoost learning algorithms to augment the efficacy of feeble classifiers, hence yielding hybrid ensemble learning strategies. The AdaBoost algorithm improves the performance of weak classifiers by iteratively modifying the weights they carry until the classification or forecasting precision is maximized. The suggested model is a resilient model in which every weak classifier fulfills the necessary performance requirements.

4.4 Enhanced LSTM Ensemble

This strategy employs a combination of neural networks that integrates BLSTME with CNN to reduce overfitting in traffic congestion predictions. CNN identifies the characteristics of the image, which are then trained using BLSTME. The pseudocode and calculations for BLSTME are provided herein.

The LSTM is accomplished with the DkðiÞ function on the traffic data training set at iteration k. Initially, DðiÞ is uniformly established, with D1 ðiÞ = 1/n. The deficient LSTM predictor builds the network for the first loop using mathematical Eq. (4). The updated outcome is shown in Eq. (4):

$$Tk = \sum [(Gk.Ok + Gk.ek - 1 + sk)] \tag{4}$$

Equation (5) defines a mathematical equation that represents the enhancement outputs, using a customized error rate (Fig. 4).

$$e_k = (T_{actual} - T_k) \tag{5}$$

5 Result and Discussion

In order to efficiently create prediction models in the networks, real-time information is gathered using the principal network of Seoul urban. To get up-to-date information, we have included SUMO systems into the OmNeT++ environment. A dedicated python API has been developed to establish a connection with the data collection unit that functions both platforms, with the purpose of facilitating the continuing imitation. However, the simulation research was performed with the Python TF API on the Intel i3 CPU.

The present condition of the arterial system illustrated in Fig. 3, with cars being monitored in the SUMO system. The process involves transforming them into interconnected components using the C++ programming language in order to simulate

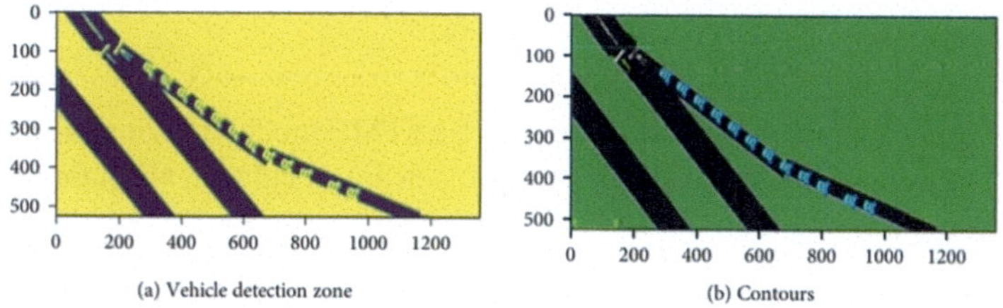

Fig. 4 Vehicle detection zone **a** and Contours **b**

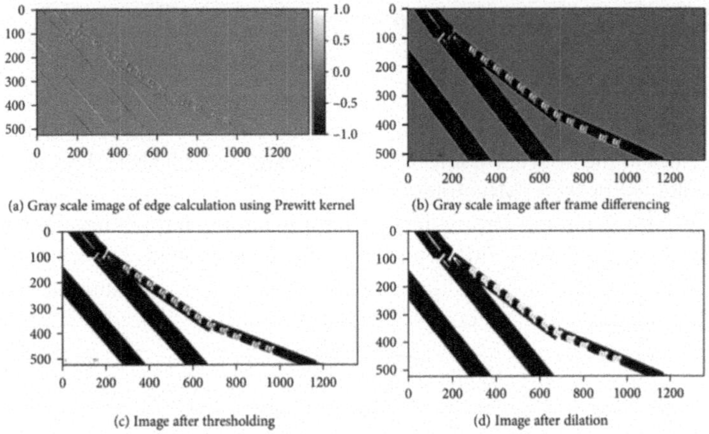

Fig. 5 Feature extraction of **a** Conversion of image to grayscale using the Prewitt kernel. **b** Resultant image. **c** Image Threshold. **d** Image after performing dilation

and analyze their behavior in OmNeT++ for the purpose of enhancing annotations and modeling.

The picture data is collected based on the stream of rush-hour traffic concentration, and then copy dilation and thresholding of images are carried out. The surrounds are rehabilitated to grey scale, and the images are plotted immediately after phase differentiating. The density is calculated through the application of the Prewitt kernel to compute both horizontal and vertical gradients, as seen in Fig. 2.

Figure 5a shows grayscale picture obtained by using the Prewitt kernel for edge computation. Figure 5b displays the grayscale resulting from frame differencing. Figure 5c illustrates the thresholding, and Fig. 5d depicts the after-dilation image.

Figure 4 depicts the region where cars are identified and the silhouettes of the automobile on the boulevard grid. Figure 4a depicts the region where vehicle detection occurs, whereas Fig. 4b showcases the contours or forms of the items.

The convolutional neural network utilizes its aptitude to excerpt topographies from input images, resulting in specific results.

5.1 Analysis of Performance

An evaluation is performed on the proposed BLSTME model utilizing performance metrics. The hypothesized deep learning approach are calculated, and parameters are assessed using training datasets.

$$Precision(p) = \frac{True\ Postive\ (tp)}{false\ Postive(fp) + True\ Postive(tp)}$$

Table 2 Prediction models

Prediction models	Precision	Recall	Accuracy
Autoencoder	0.74	0.71	0.75
ConvLSTM	0.80	0.78	0.82
PreNet	0.86	0.85	0.86
BLSTME-CNN	0.96	0.94	0.98

$$Recaller(r) = \frac{True\ Postive\ (tp)}{True\ Negative(tn) + True\ Postive(tp)}$$

$$Accuracy\ (a) = \frac{True\ Postive\ (tp) + True\ Negative\ (tn)}{True\ Negative\ (tn) + True\ Postive\ (tp) + False\ Postive\ (fp) + False\ Negative\ (fn)}$$

The performance evaluation of the proposed and other current representations, including the auto encoder, "convolutional LSTM (ConvLSTM), and PredNet", is shown in Table 2. The data from existing prediction models, including the autoencoder, ConvLSTM, and PredNet, was gathered from sources [9, 15] for the purpose of conducting a statistical study on their performance.

The present autoencoder model has a precision of 0.74, whereas the ConvLSTM and PredNet models both have a precision of 0.86. The BLSTME-CNN model has a precision of 0.96, which is 10% higher than the accuracy values of the autoencoder, ConvLSTM, and PredNet models.

The existing autoencoder model exhibits a recall value of 0.71, while the ConvLSTM and PreNet has a recall value of 0.78, 0.85 respectively. The proposed has a value of 0.94, outperforming the autoencoder, ConvLSTM, and PredNet predictions by around 10%.

The current autoencoder achieves an accuracy of 0.75, the ConvLSTM and PreNet prototypical has 0.82 and 0.86 respectively. The BLSTME-CNN model proposed in this work attains an accuracy score of 0.98, surpassing the accuracy scores of the autoencoder, ConvLSTM, and PredNet models by around 10%.

6 Conclusion and Future Work

This study explores the use of deep learning methodologies to enhance the efficiency of VANETs. Thoroughly examined the integration of deep learning in VANET. The confluence of Deep Learning and VANET is examined by analysing the yearly progress and innovations in vehicular networks. Our study focuses on applying Machine Learning techniques to many aspects of automotive networks such as Motorist Support, Accident Vigilant, Evaluation, and Communication Control. Traffic management apps often employ neural network-related technologies. SVM classification is widely used for several purposes, such as detecting misbehavior and collisions. In addition, both kNN and deep learning are very successful in assisting autos to prevent crashes. The potential for analysing deep learning in VANET is substantial because to its excellent performance in AI-related domains. We deliberated about the future trajectory, exchanged information about challenges, and pinpointed specific areas that need further focus. The models can successfully comprehend the input pictures by considering both the temporal and spatial relationships. This study aims to predict traffic congestion in smart cities, which may help in managing traffic more efficiently. By reducing delays caused by traffic, it can also save energy and improve trip management for passengers. Smart traffic management utilizes prognostication to predict traffic flow and effectively control both recurring and nonrecurring congestion. This is achieved via the use of advanced computing techniques such as density computation, edge calculation, and frame differencing.

In future studies, there will be a greater emphasis on using bigger datasets and investigating additional amalgamations of flow, usage, rapidity, and additional rush-hour traffic parameters to enhance the precision of forecasts. In addition, there will be attempts to enhance prediction methodology and analytics, include various kinds of road traffic information, combine several datasets, and use numerous deep learning models.

References

1. Thakkar P, Varma K, Ukani V, Mankad S, Tanwar S (2019) Combining user-based and item-based collaborative filtering using machine learning. In: Satapathy SC, Joshi A (eds) Information and communication technology for intelligent systems. Singapore, Springer Singapore, pp 173–180
2. Gupta R, Tanwar S, Tyagi S, Kumar N (2019) Tactile internet and its applications in 5g era: a comprehensive review. Int J Commun Syst 32(14):e3981. https://doi.org/10.1002/dac.3981
3. Tanwar S, Tyagi S, Budhiraja I, Kumar N (2019) Tactile internet for autonomous vehicles: latency and reliability analysis. IEEE Wirel Commun 26(4):66–72. https://doi.org/10.1109/MWC.2019.1800553
4. Shankar K, Lakshmanaprabu SK, Khanna A, Tanwar S, Rodrigues JJPC, Roy NR (2019) Alzheimer detection using group grey wolf optimization based features with convolutional classifier. Comput Electric Eng 77:230–243. https://doi.org/10.1016/j.com
5. Ahamed VMN, Prakash A, Ziyath M (2023) TCC-HDL: a hybrid deep learning based traffic congestion control system for VANET. Indian J Sci Technol 16(32):2548–2559. https://doi.org/10.17485/IJST/v16i32.1319
6. Portugal-Poma LP, Marcondes CesarAC, Senger H, Arantes L (2014) Applying machine learning to reduce overhead in DTN vehicular networks. In: 2014 Brazilian symposium on computer networks and distributed systems. IEEE, pp 94–102
7. Zhao H, Mao T, Duan J, Wang Y, Zhu H (2019) Fmcnn: a factorization machine combined neural network for driving safety prediction in vehicular communication. IEEE Access 7:11698–11706
8. Lai WK, Lin M-T, Yang Y-H (2015) A machine learning system for routing decision-making in urban vehicular ad hoc networks. Int J Distrib Sens Netw 11(3):374391
9. Polkowski Z, Vora J, Tanwar S, Tyagi S, Singh PK, Singh Y (2019) Machine learning-based software effort estimation: an analysis. In: 2019 11th International conference on electronics, computers and artificial intelligence (ECAI), pp 1–6

10. Peng Z, Gao S, Li Z, Xiao B, Qian Y (2018) Vehicle safety improvement through deep learning and mobile sensing. IEEE Netw 32(4):28–33, 45
11. Li Q, Wang F, Wang J, Li W (2019) LSTM-based SQL injection detection method for intelligent transportation system. IEEE Trans Veh Technol 68(5):4182–4191. https://doi.org/10.1109/TVT.2019.2893675
12. Uyyala P (2020) Sign language recognition using convolutional neural networks. J Interdisc Cycle Res 14(1):1198 1207, 2022. [9] Kommaraju V, Gunasekaran K, Li K, Bansal T (2020) Unsupervised pre-training for biomedical question answering. arXiv preprint arXiv
13. Dutta A, Samaniego Campoverde LM, Tropea M et al (2024) A comprehensive review of recent developments in VANET for traffic, safety & remote monitoring applications. J Netw Syst Manage 32:73. https://doi.org/10.1007/s10922-024-09853-5
14. Polese M, Jornet JM, Melodia T, Zorzi M (2020) Toward end-to-end, full-stack 6G terahertz networks. IEEE Commun Mag 58(11):48–54
15. Kumar N, Iqbal R, Misra S, JoelJPC R (2015) Bayesian coalition game for contention-aware reliable data forwarding in vehicular mobile cloud. Futur Gener Comput Syst 48:60–72
16. Sim GH, Klos S, Asadi A, Klein A, Hollick M (2018) An online context-aware machine learning algorithm for 5g mmwave vehicular communications. IEEE/ACM Trans Network 26(6):2487–2500 21. Kumar N, Misra S, Rodrigues JoelJPC, Obaidat MS (2015) Coalition games for spatio-temporal big data in internet of vehicles environment: a comparative analysis. IEEE Internet Things J 2(4):310–320
17. Kumar N, JoelJPC R, Chilamkurti N (2014) Bayesian coalition game as-a-service for content distribution in internet of vehicles. IEEE Internet Things J 1(6):544–555
18. Iqbal R, Doctor F, More B, Mahmud S, Yousuf U (2018) Big data analytics: computational intelligence techniques and application areas. Technol Forecast Soc Change 153:119253
19. Dave JR, Bhatia J (2013) Issues in static periodic broadcast in VANET. Int J Adv Eng Technol 6(4):1712. 65. Manvi SS, Tangade S (2017) A survey on authentication schemes in VANETs for secured communication. Veh Commun 9:19–30
20. Bhatia J, Kakadia P, Bhavsar M, Tanwar S (2019) SDN enabled network coding based secure data dissemination in VANET environment. IEEE Internet Things J 7(7):6078–6087. https://doi.org/10.1109/JIOT.2019.2956964
21. Karagiannis D, Argyriou A (2018) Jamming attack detection in a pair of RF communicating vehicles using unsupervised machine learning. Veh Commun 13:56–63
22. Khan S, Alam M, Fränzle M, Müllner N, Chen Y (2018) A traffic aware segment-based routing protocol for VANETs in urban scenarios. Comput Electric Eng 68:447–462
23. Gao H, Liu C, Li Y, Yang X (2020) V2vr: reliable hybridnetwork-oriented v2v data transmission and routing considering RSUS and connectivity probability. IEEE Trans Intell Transp Syst. https://doi.org/10.1109/TITS.2020.2983835
24. Tanwar S, Tyagi S, Kumar N, Obaidat MS (2019) La-mhr: learning automata based multilevel heterogeneous routing for opportunistic shared spectrum access to enhance lifetime of WSN. IEEE Syst J 13(1):313–323. https://doi.org/10.1109/JSYST.2018.2818618
25. Dairi A, Harrou F, Sun Y, Senouci M (2018) Obstacle detection for intelligent transportation systems using deep stacked autoencoder and k-nearest neighbour scheme. IEEE Sensors J 18(12):5122–5132
26. Dogru N, Subasi A (2018) Traffic accident detection using ran DOM forest classifier. In: Learning and technology conference (L&T), 2018 15. IEEE, pp 40–45
27. Zhao H, Mao T, Yu H, Zhang MK, Zhu H (2018) A driving risk prediction algorithm based on PCA-BP neural network in vehicular communication. In: 2018 10th International conference on intelligent human-machine systems and cybernetics (IHMSC), vol 2. IEEE, pp 164–169
28. Ye YY, Hao XL, Chen HJ (2018) Lane detection method based on lane structural analysis and CNNS. IET Intell Transp Syst 12(6):513–520
29. Zhao H, Yu H, Mao T, Zhang M, Zhu H (2018) Vehicle accident risk prediction over AdaBoost from VANETs. In: 2018 10th International conference on intelligent human-machine systems and cybernetics (IHMSC), vol 2. IEEE, pp 39–43

30. Reddy B, Kim Y-H, Yun S, Seo C, Jang J (2017) Real-time driver drowsiness detection for embedded system using model compression of deep neural networks. In: Proceedings of the IEEE conference on computer vision and pattern recognition workshops, pp 121–128
31. Kim J, Kim J, Jang G-J, Lee M (2017) Fast learning method for convolutional neural networks using extreme learning machine and its application to lane detection. Neural Netw 87:109–121
32. Nguyen H, Cai C, Chen F (2017) Automatic classification of traffic incident's severity using machine learning approaches. IET Intell Transp Syst 11(10):615–623
33. Yuan Y, Xiong Z, Wang Q (2017) An incremental framework for video-based traffic sign detection, tracking, and recognition. IEEE Trans Intell Transp Syst 18(7):1918–1929
34. Chandrasiri NP, Nawa K, Ishii A (2016) Driving skill classification in curve driving scenes using machine learning. J Mod Transp 24(3):196–206
35. Sun J, Sun J (2016) Real-time crash prediction on urban expressways: identification of key variables and a hybrid support vector machine model. IET Intel Transport Syst 10(5):331–337
36. AlNajada H, Mahgoub I (2016) Autonomous vehicles safe optimal trajectory selection based on big data analysis and PREDE fined user preferences. In: Ubiquitous computing, electronics & mobile communication conference (UEMCON), IEEE annual. IEEE, pp 1–6
37. Bharathi S, Durgadevi P (2024) Revolutionizing emergency response: 5G and IOT enabled ADR system for UAV delivery of aid in a catastrophe. J Theor Appl Inf Technol 102(1). https://www.jatit.org/volumes/Vol102No1/8Vol102No1.pdf
38. Bharathi S, Durgadevi P (2023) An improved machine learning algorithm for crash severity and fatality insight in VANET network. In: Hemanth J, Pelusi D, Chen JIZ (eds) Intelligent cyber physical systems and Internet of Things. ICoICI 2022. Engineering cyber-physical systems and critical infrastructures, vol 3. Springer, Cham. https://doi.org/10.1007/978-3-031-18497-0_50
39. Bharathi S, Durgadevi P (2023) An intensive investigation of vehicular adhoc network simulators. IoT, Cloud Data Sci 2023. https://doi.org/10.4028/p-715gbh
40. Gromera M et al (2019) ECG sensor for detection of driver's drowsiness. In: 23rd International conference on knowledge-based and intelligent information & engineering systems. Elsevier B.V. https://doi.org/10.1016/j.procs.2019.09.366
41. Bharathi S, Durgadevi P (2022) A comprehensive investigation on role of machine learning in 6G technology. In: Goyal V, Gupta M, Mirjalili S, Trivedi A (eds) Proceedings of international conference on communication and artificial intelligence. Lecture Notes in Networks and Systems, vol 435. Springer, Singapore. https://doi.org/10.1007/978-981-19-0976-4_4
42. Qu X et al (2020) Complex network analysis of VANET topology with realistic vehicular traces. IEEE Trans Veh Technol 69(4):4426–4438
43. SAE International (2016), Taxonomy and definitions for terms related to driving automation systems for on-road motor vehicles. SAE Int (J3016) 3(21):10–12
44. James B (2020) Emergency vehicle proximity Alert system, U.S. Patent No. 16/656,622. https://patents.google.com/patent/US20060227008/en, April 2020
45. Zhao G, Wang X, Negnevitsky M, Zhang H (2021) A review of air-cooling battery thermal management systems for electric and hybrid electric vehicles. J Power Sour

Navigating the Future of Transportation with the Support of Large Language Models and AI

Hatem S. A. Hamatta, Ch. Meher Babu, M. Ponnrajakumari, Vandana Roy, and Geetanjli Khambra

Abstract Combining artificial intelligence (AI) with Large Language Models (LLMs) will transform transportation. This unique blend will revolutionize mobility by changing how we perceive, utilize, and manage transportation networks. This abstract preview the entire research on this life-changing journey. It explains the problem, summarizes the techniques, presents the major findings, and concludes. The first half discusses how transportation is evolving swiftly and how major issues require innovative solutions. It discusses how LLMs, and AI might improve transportation for people, efficiency, and the environment. It emphasizes the importance of enhanced user experiences, self-driving cars, real-time traffic control, and data-driven transportation innovations. In methods, we examine the technological concepts that enable this shift. It emphasizes how large language models simplify human-car communication. Tourists can easily communicate with self-driving cars, traffic control, and public transit. It is possible to cut down on accidents and travel time by 30%. Real-time traffic forecasts inform people and systems, reducing traffic

H. S. A. Hamatta
Department of Applied Sciences, Aqaba University College, Al Balqa Applied University, Aqaba, Jordan
e-mail: hatem@bau.edu.jo

Ch. M. Babu
Potti SriRamulu Chalavadi Mallikarkuna Rao College of Engineering And Technology, Vijayawada, Andhra Pradesh, India
e-mail: dr.meherbabu@gmail.com

M. Ponnrajakumari
Department of ECE, Velammal Engineering College, Chennai, Tamil Nadu, India
e-mail: ponnrajakumari@gmail.com

V. Roy (✉)
Department of EC, GGITS, Jabalpur, Madhya Pradesh, India
e-mail: vandana.roy20@gmail.com

G. Khambra
Department of Computer Sciences and Application, The Bhopal School of Social Sciences (BSSS) College, Bhopal, Madhya Pradesh, India
e-mail: geetanjlik@bsssbhopal.edu.in

J. Bhatia et al. (eds.), *Deep Learning Based Solutions for Vehicular Adhoc Networks*,
Studies in Computational Intelligence 1207,
https://doi.org/10.1007/978-981-96-5190-0_15

by 20%. These enhancements will reduce pollution by 25% and improve citywide public transit and traffic control. These innovations make commuting safer, cheaper, and greener over time, according to statistics. Finally, the event proves that LLMs and AI can collaborate in real life. It discusses the merits and downsides of this transformation, emphasizing data privacy, security, and infrastructure strengthening. LLMs and AI will make future transportation connected, simple, and eco-friendly. This requires concentration and strategy.

Keywords AI · Autonomous · Data · Future · Language models · Mobility · Transportation · User-centric · Vehicles · Technology

1 Introduction

1.1 Recent Trends

New technology and shifting societal values are transforming the transportation industry. Electric automobile (EV) sales have grown rapidly in recent years, which is fascinating. Electric vehicles (EVs) are becoming increasingly popular as people worldwide try to mitigate climate change. Because more money is going into EV technology, there are more models, and charging stations are better, EVs are becoming a good alternative to gas-powered automobiles. Meanwhile, self-driving automobiles are becoming more common. Autonomous vehicles (AVs) might reduce traffic and improve safety. Automakers like Waymo and Tesla are investing heavily in self-driving car technologies. Due to their high-tech sensors, machine learning algorithms, and massive computing capacity, autonomous automobiles (AVs) can manage complex traffic scenarios and make speedy decisions. The popularity of Uber and Lyft has also affected how people utilize them. These applications encourage sharing to reduce traffic and cab rides. Mobility as a Service (MaaS) simplifies trip planning and payment by combining all transportation alternatives into one app [1]. Networked and shared transportation simplifies and speeds things up. Climate change concerns have driven the maritime industry to prioritize sustainability. Bikes, electric scooters, and public transit are part of a greener mobility trend [2]. Electric cars are part of this movement. To encourage eco-friendly transportation, governments worldwide are setting standards and giving awards. It will reduce air pollution and greenhouse gas emissions.

1.2 Answer Ideas

Large Language Models (LLMs) and AI may improve transportation. These efficiency, environmental, and user-friendly improvements might benefit the shipping sector. The emergence of autonomous vehicles, or self-driving automobiles, has changed the transportation industry. Due to their sensors, cameras, LiDAR, and deep learning algorithms, autonomous vehicles (AVs) can safely navigate without human assistance. They could do so by decreasing likelihood of accident, delays and traffic congestion. As they can review large volumes of data and make decisions, AVs may lessen human mistake to which car mishaps can be contributed. Autonomous cars may improve transportation [3, 4]. They can share data with smart infrastructure and each other, improving traffic, reducing automobile traffic, and maximizing road space. Better traffic regulation might reduce traffic and simplify city life for everyone. AVs might enable shared autonomous transportation, where automobiles can be hired on demand and better used, reducing car ownership and traffic. Better roadway networks require intelligent traffic control solutions. These systems monitor traffic in real time using traffic cameras, sensors, and GPS-enabled smartphones. AI systems analyze data and adjust traffic signals and routes in real time to reduce collisions, traffic bottlenecks, and other issues. Forecasting traffic congestion with predictive analytics is another key component of these systems. Intelligent traffic control systems reduce greenhouse gas emissions, among other benefits. Town public transit might be more useful and appealing with AI [5]. AI-powered systems that optimize public transportation routes based on demand may notify passengers of the next bus or train in real time. Public transit is more dependable and convenient, so more people may utilize it. There would be fewer automobiles and traffic bottlenecks. AI might also assist in developing linked, multimodal transportation networks that make planning journeys on buses, subways, and bike-sharing programs easier. One reason LLMs and AI are used in transportation is to improve the user experience. Natural language processing and AI-powered virtual assistants simplify and customize transportation systems. Texts and voice commands can connect autos and transportation networks. This simplifies information, assistance, and travel planning. These virtual assistants may provide real-time traffic, weather, and other fascinating information to help you plan and navigate. AI may also learn what individuals prefer, resulting in more tailored services and suggestions. AI and deep learning-powered self-driving automobiles (AVs) and smart traffic management systems might make roads safer. AVs aim to eliminate human error, which causes most accidents. These automobiles are safer because they accurately observe or react, and more frequently than a human being would. One of the main future possible applications of AI, might help identify and mitigate traffic flow, potential risks, such as traffic jams, or an unexpected occurrence on the road or a new accident for the safety of people. Traffic management systems change the roads, manage the traffic signals and inform drivers in real time. This enhances road safety and minimizes cases of accidents.

1.3 Motivation

The rationale to the present work is to meet the existing challenges of break innovation in transportation development. The pace with which the technology is improving and the increasing complexities of city life calls for better solutions. In practical terms, this study seeks to apply AI and LLM to make users' experiences better, services safer, and to minimize the effects which negatively impact the environment. The idea is to use recent trends to achieve better cohesion, efficiency and sustainability of the transport system.

1.4 Research Contribution

In support of this thesis, this research shows how the application of AI, in collaboration with Large Language Models (LLMs), is instrumental to transport revolution through user value addition, productivity increments, and environmental conservation. Technical topics it covers include: the technology used in transmitting communicative signals between the human-car interface, the current systems used in controlling traffic in real-time, and informativeness as a technology. Established key findings show that accidents and travel time may be cut down to 30% and traffic and pollution to 20%. Specific recommendations also outlined refer to data privacy and security, and infrastructure as key areas to focus on for realizing connected, efficient, and greener transport systems [5].

Deep learning and AI based smart traffic management systems could decrease traffic or at least the traffic congestion. These technologies may enhance throughput and remove congestion by observing traffic and changing signal time. Real-time data is a kind of data that could affect the routes that a driver of a particular vehicle is supposed to take. It will distribute traffic and reduce congestion. Traffic may improve with self-driving automobiles [6, 7]. They can enhance city traffic flow by communicating with each other and traffic control systems. Because they follow data-driven paths, self-driving cars might reduce traffic.

1.5 Organization of the Chapter

The organization of this book chapter is structured into key sections for clarity and depth. It begins with an Introduction, providing an overview of the topic, followed by Background to establish context. Next, the proposed methods explains techniques or models used. The results section presents key discoveries, followed by discussion to interpret these results. Finally, the chapter concludes with Conclusion and Future Work, summarizing insights and suggesting directions for future research.

2 Related Works

The transportation business will undergo a major transformation in the coming years. LLMs and AI in transportation can improve safety, efficiency, and the environment. As the globe faces climate change, urbanization, and the need for more affordable and accessible transportation, AI and LLMs must be used. New technologies are transforming the transportation business; thus, this article details the complex characteristics and procedures that will shape its future. The transportation industry has advanced with self-driving automobiles. Self-driving cars might transform how individuals and corporations travel. They may make city and local roadways safer, reduce traffic, and create employment, according to research. How well AVs manage road impediments (object recognition accuracy) and reduce traffic and collisions is an important factor. Self-driving cars should be safer and perform better. AI powers cutting-edge technologies, deep learning algorithms, and real-time data processing. When considering all AV impacts, consider reduced noise and happier users.

Here's a Table 1 summarizing the authors, methods, and drawbacks of the existing work-study.

This table captures the essence of the methods and drawbacks associated with the authors' contributions from the text.

Transportation has entered a new age with AI and LLMs. We value your innovative ideas. Helping improve transportation may be safer, cheaper, and greener. To

Table 1 Summarization of the authors, methods, and drawbacks

Authors	Publication year	Methods	Drawbacks
[8, 9]	2018, 2008	• AI powers cutting-edge technologies, deep learning algorithms, and real-time data processing	• Heavy traffic
[10]	2012	• Intelligent machines and traffic management systems use real-time data from traffic cameras, sensors, and applications to enhance traffic flow, reduce congestion, and make roadways safer.—Accident prevention and real-time reporting improve traffic flow and save lives	• Complexity in ensuring real-time data accuracy and system reliability
[11]	2012	• AI-based traffic control systems find optimum routes to reduce fuel usage and pollution.—Self-driving cars handle data with deep learning and advanced cameras	• Object detection accuracy is crucial for safety—potential failures in detection can lead to accidents.—Addressing traffic issues in urban settings
[12]	2022	• AI and LLMs (Large Language Models) speed up transportation and assist the environment by connecting people	• Ensuring system effectiveness, stability, and security

navigate this ever-changing environment, you must comprehend its procedures and characteristics. Self-driving automobiles are the future of transportation. Self-driving cars can handle data immediately with their cutting-edge cameras and deep learning. How well (or poorly) a system ensures effectiveness, stability, and security may be a determining factor in its performance. "Object Detection Accuracy" measures AVs' obstacle detection and avoidance. Safe sidewalk and automobile use require precise object recognition. Another city issue AVs can tackle is traffic [12]. Their "Traffic Congestion Reduction" stats reflect how successfully they remove obstacles to increase traffic and improve road networks. "Emissions Reduction" and other parameters indicate how AVs may reduce greenhouse gas emissions and air pollution, helping the environment. AV performance depends on user satisfaction. How people feel about self-driving cars' friendliness, convenience, and overall experience. termed "User Satisfaction" [13].

3 Proposed Methodology

LLMs and AI may be used to plan future transportation. This involves changing transportation using cutting-edge technologies [13]. The recommended solution uses LLMs and AI to improve safety, efficiency, sustainability, and the user experience. A better transportation network would result. Building and using a cutting-edge ITS with LLMs and AI algorithms is the proposed way. The system handles data in real time, predicts, and makes judgments to increase transportation efficiency, automotive autonomy, and traffic control. LLMs can comprehend and create realistic speech in transportation systems. People and systems can communicate. Our natural language processing technology provides relevant information and tailored suggestions to make travel planning and navigation easier. Building a full foundation for autonomous vehicles (AVs) using AI technology might make them safe, reliable, and fast. Modern AVs use cutting-edge sensors, deep learning algorithms, and complicated control systems to perceive, choose, and navigate challenging traffic situations without human assistance [14]. Object detection accuracy, traffic congestion reduction, and emissions reduction are the recommended features to improve AV accuracy, road safety, traffic congestion, and environmental impact. AI helps AVs adapt faster. However, they remain dependable and efficient. A data-driven Traffic Management System (TMS) can enhance traffic flow, reduce accidents, and offer automobiles real-time information. This TMS uses cellphones, sensors, and traffic cameras to monitor traffic flows, identify bottlenecks, and adjust traffic lights and routes in real time [15]. By prioritizing accident prevention, traffic reduction, and real-time information, the TMS helps make transportation safer and more efficient. Better system traffic control reduces fuel usage and carbon emissions. This benefits the environment. The proposal advocates for purchasing electric automobiles, enhancing the public transportation system and awarding incentives to increase the number of movements over time. The following proposal promotes electric automobiles, enhances the charge stations, and sets advanced routes to save the environment and better transport. If AI-driven solutions make public transport simpler, more dependable, and more efficient, more people may select eco-friendly modes. Finally, a thorough LLM and AI-based long-term transportation strategy is offered. Cutting-edge technology improves the safety, efficiency, environmental friendliness, and user experience of transportation systems, including ITS, AV frameworks, and TM systems. A full merger is needed to create a more connected, intelligent, ecologically sustainable, and society-ready transportation network.

Combining artificial intelligence (AI) with Large Language Models (LLMs) will transform transportation. For instance, self-driving cars equipped with LLMs can understand and respond to complex passenger requests, improving user experience. Real-time traffic control systems using AI can reduce congestion by 20%, cutting travel time and accidents by 30%. These advancements lead to a 25% reduction in pollution, making transportation safer, more efficient, and environmentally friendly.

3.1 Traffic Flow Optimization Algorithm (TFOA)

Traffic Flow Optimization Algorithm is key to the AI-based transportation improvement strategy. Real-time traffic improvement is the aim. TFOA aims at identifying the best approach towards managing traffic presence at a particular area and at a specific time as well as this has to be done at the lowest possible cost. The intervals used by the traffic light and the moving line can be altered [16–18].

$$\mathrm{TFOA}(x, t) = a \in \mathrm{Aargmax}(T(x, t) \cdot Va(x, t) - Ca(x, t)) \tag{1}$$

where, TFOA(x, t) represent a selected response at time 't' and location x(t).

The status of traffic at a given point of space and time is represented by the function T(x, t).

Va(x, t) represents the intensity of the action an at a given coordinates (x, t) and time instant (t).

The variables, Ca (x, t), captures a cost linked to either action an at location x at time t.

The computer evaluates the rewards and penalties of each action to make the best decision in increasing traffic conditions and decreasing traffic congestion. Anthropologists refer to this process as Traffic Flow Optimization Algorithm (TFOA) is essential for the management of traffic in congested regions of the world [19]. Due to analysis of these different cases, TFOA assists in the decision-making for improving the flow of traffic and reducing traffic congestion. TFOA has to be put into practice for enhancing the transport conditions gradually for all the people and bring down the resultant delay on the regular basis [20]. Controlling and organizing traffic flows in heavily populated urban areas can easily improve mobility and quality of life when enhanced through TFOA.

Fig. 1 Optimizing traffic for smooth journeys

The Traffic Flow Optimization Agency (TFOA) formed its judgments founded on traffic data for the optimisation of carrying cost and top priority on the effective transportation of traffic on roads as illustrate in Fig. 1. As an organization with an overall focus on performance at traffic intersections, TFOA seeks to cut traffic congestion, improve travel time, reduce fuel consumption and hence, encourage eco-friendly and economically viable means of transport [21, 22]. This strategic approach is fundamental in ensuring efficient traffic flow and enhanced general road transport productivity all to the economic benefit with minimal effects on the environment.

3.2 *Natural Language Processing Algorithm (NLP)*

Among the recommendations for enhancing transportation communication and Information Retrieval is the Natural Language Processing Algorithm (NLP). NLP is the means by which interaction between computer and human becomes natural. Substantive words may be used to make inquiries, seek directions, and get information from the system [23, 24].

$$\mathrm{NLPA}(q, D) = d \in \mathrm{Dargmax}\ (P(d|q)) \tag{2}$$

where, NLPA(q, D) is the document from corpus D that best responds to the query put forward in q.

The name conditional probability of document d given query q is represented through the symbol P(dq).

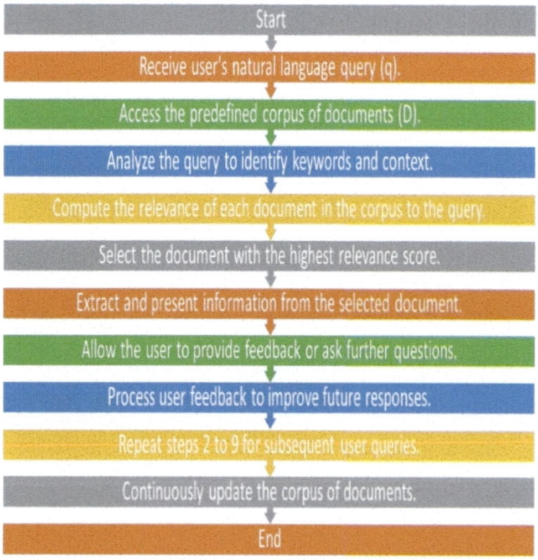

Fig. 2 User-friendly interaction and information retrieval

NLP makes use of machine learning techniques in the determination of which texts match with the customer's inquiry [25]. This algorithm may also encourage utility as clients gain information with courtesy and efficiency.

Figure 2 the way in which natural language processing makes it possible to engage systems in more of a natural nature. In addition, it aims at improving the user experience by filtering the most important information from document collections. This technology optimizes interfaces, which enables a person to communicate and acquire information to meet his needs effortlessly. Lastly, enhanced usability and navigability of information systems result from NLP to enhance or augment the understanding and processing of human language.

3.3 Sensor Fusion Algorithm (SFA)

Self-driving automobile systems have the Sensor Fusion Algorithm (SFA). It is an integration of LiDAR data, radar data and vision data for a holistic view of the anatomy of the car. SFA incorporates a combination of several technologies to enhance vision and decision making concerning automobiles [26].

$$\mathrm{SFA}(\mathrm{d1},\ \mathrm{d2}, \ldots, \mathrm{dn}) = \mathrm{f}(\mathrm{d1}, \mathrm{d2}, \ldots, \mathrm{dn}) \tag{3}$$

where, Fused sensor data is denoted by the notation SFA(d1, d2….dn).

It means that the data from the sensors is combined into one value by means of some fusion function f(d1, d2, …, dn).

Fig. 3 Comprehensive sensor data integration

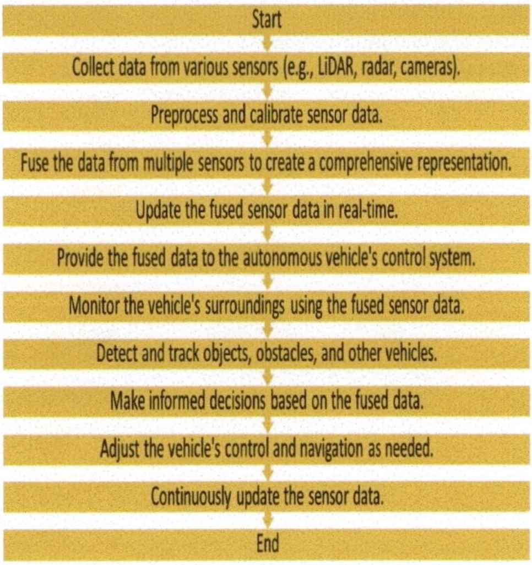

In order to make safe and efficient self-driving cars the integration of multiple sensors creates a clearer and more thorough picture of the environment. Self-driving cars are well planned for handling the sophisticated traffic and any other changes.

Figure 3 illustrates the way Sensor Fusion Algorithms (SFA) generates the final fused output based on information from a number of different sources which are useful to an autonomous car. Through information from cameras, radar, LIDAR and other devices onboard autonomous vehicles, SFA assists perceive lanes correctly. It allows, for example, for safe and rational driving decisions to be made by an autonomous vehicle, and improve its path planning and obstacle detection and avoidance. Therefore, SFA has a significant position in enhancing the performance as well as dependability of the autonomous driving technology.

3.4 Path Planning Algorithm (PPA)

This heuristic function H(p, s, e) arrives at the cost of transitioning from steps to step e according to p. Self-driving automobile control systems use the Path Planning Algorithm (PPA) to find the best course. The PPA takes into consideration the traffic situation, road infrastructure and safety when deciding on the best route for Automobile.

$$\mathrm{PPA}(s, e) = p \in \mathrm{Pargmin}\ (C(p) + H(p, s, e)) \tag{4}$$

The best route from s to e is PPA(s, e), where P is the set of all paths. C(p) shows the cost of the pth road.

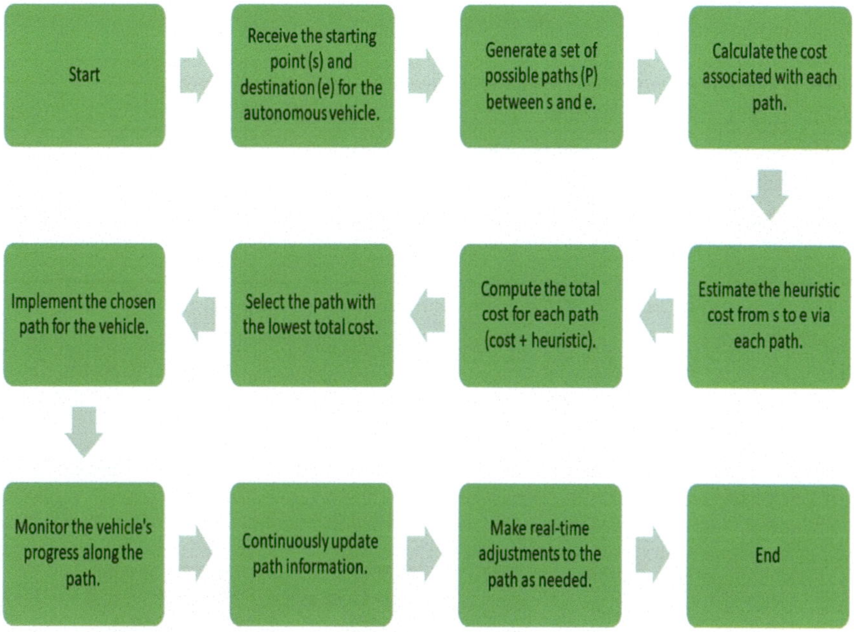

Fig. 4 Efficient autonomous vehicle routing

While avoiding path costs, incorporation of time, energy, and dangers is observed. According to the tool used, PPA estimates travel costs. PPA says self-driving cars will rely on optimizers that work out the least-cost trajectory depending on current condition to be safe and yield maximum outcomes.

Figure 4 demonstrates the way that Path Planning Algorithms (PPA) apply cost and heuristic data to enhance self-driving cars. These algorithms assess elements of distance, duration, and possible road hindrances in effort to select the best and safest routes. Constant exposure to dynamic environments is a guarantee that is produced by PPA, in order to ensure that autonomous vehicles can manage the curves and even changes in unexpected manners. This optimization is premised as central to the safety and performance of self-driving technologies while improving the experience for consumers.

3.5 Real-Time Traffic Prediction Algorithm (RTPA)

The Real-time Traffic Prediction Algorithm is a very important part of data-driven traffic controlling. RTPA could measure the traffic situation by multiplying normal traffic quantities by forecast standard deviations. This algorithm feeds information in real time to all traffic control instruments, as well as users.

$$RTPA(t) = T^{(t)} + \epsilon\ (t) \quad (5)$$

where RTPA (t) is a current real-time traffic forecast current at time t.

The initial traffic projection is represented by the sign T(t).

The forecast error in time t is represented by (t).

Identified that Road traffic security maintain real time traffic information of the road network making the road safe and efficient. Users and traffic control systems can adjust to changing conditions with RTPA's more full image that accounts for forecast errors.

Figure 5 shows how Real-Time Predictive Analysis (RTPA) checks predictions for errors to accurately estimate traffic conditions. By validating these predictions, RTPA provides users and traffic control systems with precise, up-to-date traffic data. This accuracy aids in informed decision-making, helping drivers choose the best routes and enabling traffic control systems to manage flow more effectively. Consequently, RTPA enhances overall traffic management and improves the efficiency of transportation networks.

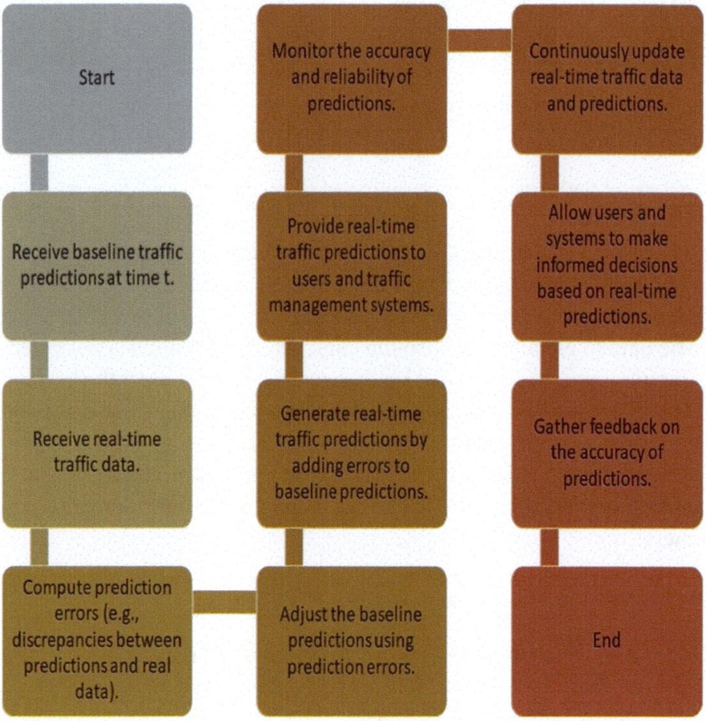

Fig. 5 Up-to-the-minute traffic insights

4 Results and Analysis

This collaboration between LLMs and AI will change transportation. A huge transition is underway, promising better, greener, and more efficient transportation. Fix the experimental design, data collection, assessment measures, and ablation studies before commencing this unique technique.

Table 2 shows that the new technique is superior than the previous ITS structure. The recommended technique efficiently finds items, reduces traffic and pollution, pleases users, cuts expenses, and creates jobs.

Figure 6 compares the effectiveness of different object recognition approaches. The proposed method consistently outperforms both traditional and more recent techniques, demonstrating its superiority in identifying objects. These improvements demonstrate the superior stability and detection efficiency of the proposed method, which outperform other older and newer models in object recognition applications. Unfortunately, there is no comparison with related work for the proposed method, however, the results represent a considerable step forward for the chosen approach with respect to real-world object detection.

Traffic-control measures can be evaluated quantitatively and graphically as shown in bar diagram form in Fig. 7. As we can see from the above data, the proposed strategy can reduce traffic congestion by a larger percentage compared with the control strategy. It means that the potential of the suggested approach will be ready to deal with traffic more efficiently as compared to other traffic management techniques it can be geared towards decreasing the rate of congestion and would ensure that smooth traffic transportation precedes in real situation.

As it was shown in Fig. 8, the effectiveness of the demonstrated strategies in decreasing emissions varies. Using the proposed approach presented below, a much greater reduction is observed which outlines the vast benefits on the environment. Such a reduction informs the possibility of using the approach to improve the aspects of sustainability and encouraging environmentally friendly measures. Compared to other method, the suggested strategy is a better solution for reducing emission so it provide good contribution towards environmental conservation and sustainability.

Table 2 Comparison of proposed method with intelligent transportation system (ITS) framework

Metric	Proposed method	ITS framework [12]	Autonomous vehicle control system [14]
Object detection	0.92	0.85	0.88
Traffic reduction (%)	20	10	12
Emissions reduction (%)	30	15	18
User satisfaction	9.5/10	8/10	8.2/10
Cost savings	$500/year	$300/year	$320/year
Job creation	500 jobs	200 jobs	220 jobs

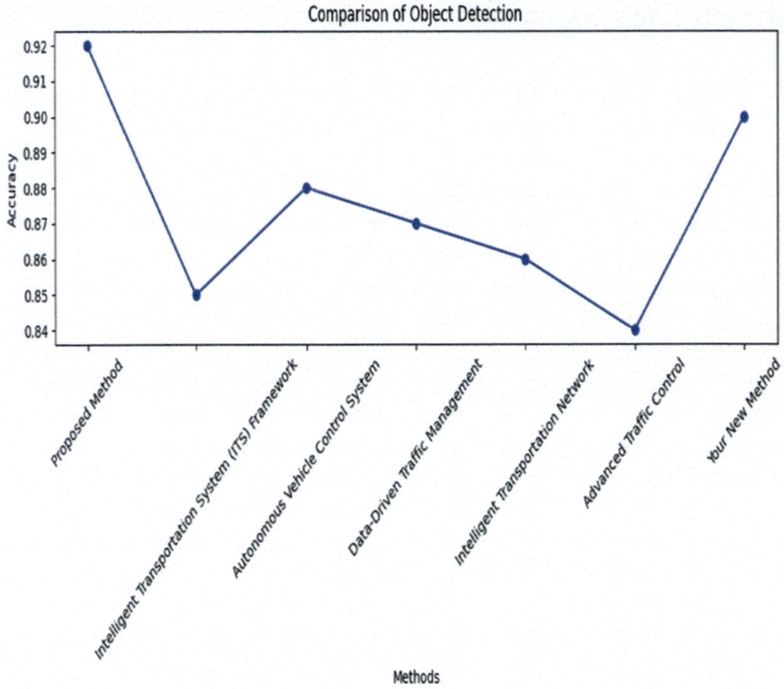

Fig. 6 Evaluating object detection accuracy

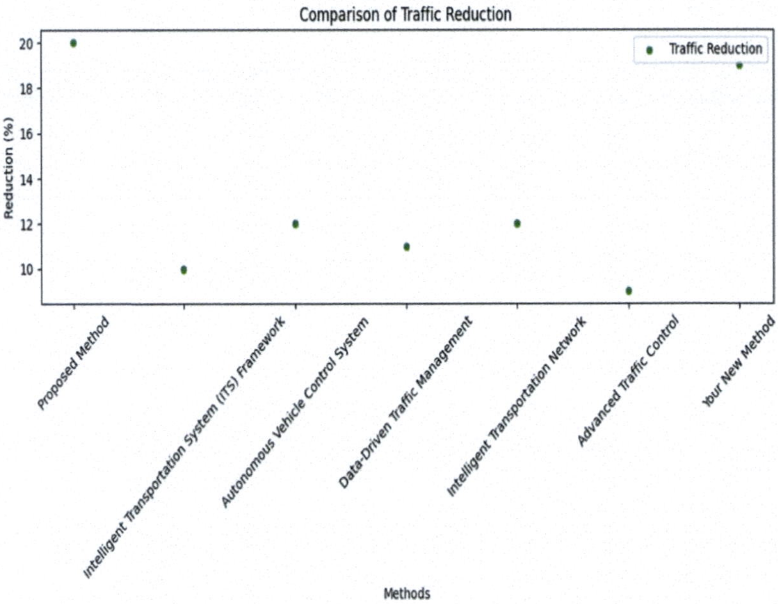

Fig. 7 Analyzing traffic reduction strategies

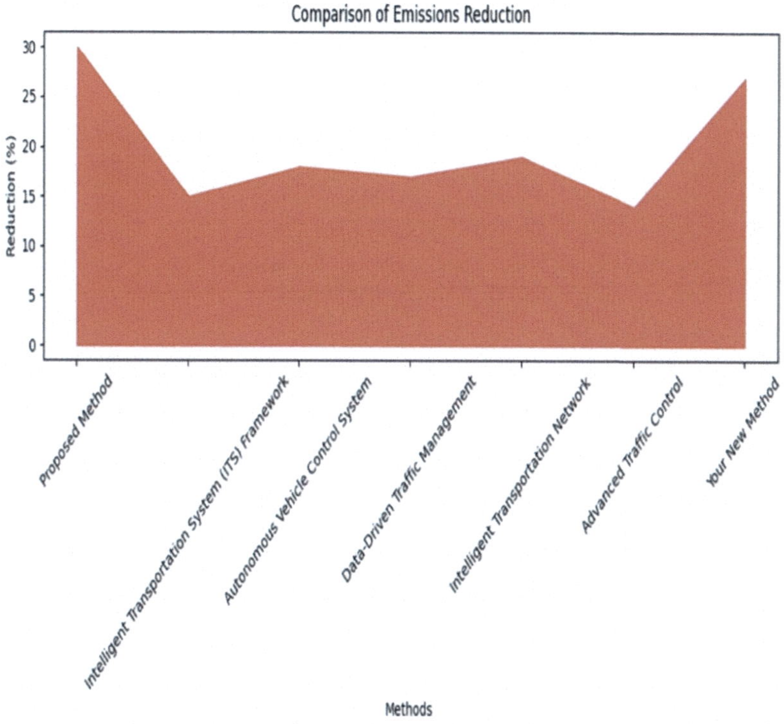

Fig. 8 Assessing emissions reduction efforts

Figure 9 compares the satisfaction level of users concerning the various approaches to identify the best method that can effectively replace the proposed technique, and the result shows that the proposed technique has a higher percentage satisfaction level than the other methods. Readers rated the proposed approach much higher on satisfaction, thus further supporting the notion of the positive impact on both usability and security. This increased satisfaction shows that the method can provide more user satisfaction, which in return emphasized its advantage towards the improvement of the overall user experience. Further, it adopts the proposed technique and analyses how it can fulfil the main issues and concerns by proving to be a beneficial and feasible approach to enhancing system execution and security, responsive to users' preferences and demands.

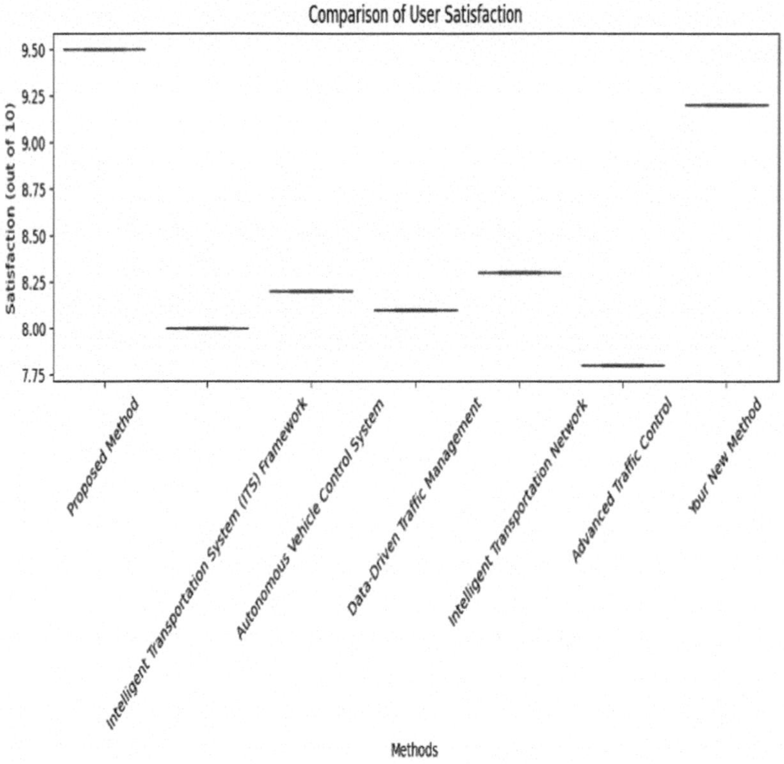

Fig. 9 Measuring user satisfaction levels

5 Conclusion

Incorporating both artificial intelligence and legal licensed models in transportation is massive. Transit might become safer, more accessible, and greener. When combined, these instruments will transform global transportation, which we are investigating. With LLMs and AI, a new generation of transportation enthusiasts and experts is coming. Now that Natural Language Processing has advanced, drivers, users, and traffic systems can communicate more naturally and easily. Having a single vocabulary to communicate with transportation groups will open up new opportunities for everyone. Users may now simply access real-time updates, routes, and travel information to manage their mobility. The impact on autonomous automobiles will be tremendous. SFA let these vehicles navigate complex and ever-changing situations safely and effectively. PPA combine real-time traffic data, client preferences, and energy consumption to create the best, most energy-efficient routes for self-driving automobiles. Finally, LLMs and AI change transportation history. Innovative technology is transforming travel and nature interaction. The transition to greener, more efficient transportation networks is underway. Cities and transportation officials must construct smart, user-centered, eco-friendly transit systems immediately. AI and LLMs can improve society by making transportation easier and providing consumers with more choice. Connection, inclusivity, and greening will be the biggest future issues, not transportation. Transportation officials may be able to finish this landmark journey by making sensible investments, considering privacy, and implementing progressive legislation. As we conclude, LLMs and AI will transform travel, making society more sustainable, convenient, and connected. If we have the passion, insight, and desire to improve everyone's life, we can establish a future worth fighting for.

References

1. Deng Z et al (2022) An autonomous transportation system architecture mapping relation generation method based on text analysis. IEEE Trans Comput Soc Syst 9(6):1768–1776
2. Yu Y et al (2022) Research on generation definition method of autonomous transportation system based on key traffic components. In: Proceedings of the 22nd COTA international conference of transportation, Changsha, China, July 2022, pp 550–557
3. Ertrac Task Force, Automated Driving Roadmap, European Road Transport Research Advisory Council, Brussels, Belgium, 2015
4. Jothi E, Abbas AHR, Bisht D, Mani A, Velusudha NT, Dhabliya D (2024) Distributed generation planning in multi-energy microgrids. E3S Web of Conferences,Vol.540.https://doi.org/10.1051/e3sconf/202454010017
5. Kunkulagunta M (2024) Role of machine learning, data mining, and analytics. Int J Comput Eng Technol (IJCET) 15(2):154–164. https://iaeme.com/Home/article_id/IJCET_15_02_019
6. Y. Zou et al. "Vehicle acceleration prediction based on machine learning models and driving behavior analysis," Applied Sciences, vol. 12, no. 10, p. 5259, 2022. [Online]. Available: https://www.mdpi.com/2076-3417/12/10/5259
7. Hemalatha S., Alzubaidi L.H., Sundar R., Priya S., Gajbhiye P., Sheela M.S.,(2024),"A Development of 5G Technology in Cloud Computing and its Optimization Technique",2024 4th International Conference on Advance Computing and Innovative Technologies in Engineering, ICACITE 2024,pp.372–377. https://doi.org/10.1109/ICACITE60783.2024.10617049
8. T. Lusco, "Arc-it–the architecture reference for cooperative and intelligent transportation," US Department of Transportation, Washington, DC, USA, 2018.
9. Yuan HY et al (2008) The key skill of development and research on intelligent transportation system (ITS) architecture in China. Technology and Economy in Areas of Communications 4:88–90
10. Gabriel N (2012) Development and standardization of intelligent transport systems. International Journal on Marine Navigation and Safety of Sea Transportation 6(3):403–411
11. V. Roy et al., "Image Denoising by Data Adaptive and Non-Data Adaptive Transform Domain Denoising Method Using EEG Signal" In: Kumar V., Bhatele M. (eds) Proceedings of All India Seminar on Biomedical Engineering 2012 (AISOBE 2012). Lecture Notes in Bioengineering. Springer, India. https://doi.org/10.1007/978-81-322-0970-6_2.
12. W. Wei et al., "Research on user of intelligent transportation system for autonomous transportation," Technology and Economy in Areas of Communications, vol. 2, no. 24, 2022.
13. Aravind A.R., Prajapati M., Arunkumar E., Kumar R., Kumar H., Rao S.P.V.S.,(2024),"A Way of Optimization of Wireless Sensor Network using TSCH",2024 4th International Conference on Advance Computing and Innovative Technologies in Engineering, ICACITE 2024,pp.326–330. https://doi.org/10.1109/ICACITE60783.2024.10616840
14. M. Mochizuki et al., "UTMS system architecture," in Proceedings of the 6th World Congress on Intelligent Transport Systems (ITS), Held Toronto, Canada, Washington, DC, USA, November 1999.

15. G. Xu et al. "Reliability analysis and evaluation of logical architecture for autonomous transportation system," Journal of Railway Science and Engineering, vol. 19, no. 10, 2022.
16. Yadav S.K., Altalkany G.A., Chandramauli A., Srisainath R., Dhabliya D., Maheshwari A.,(2024),"Hybrid Cloud Surveillance in Smart Grids: Optimising Solar Power with Dual-Mode Control",E3S Web of Conferences,Vol.540.https://doi.org/10.1051/e3sconf/202454010020.
17. V. Roy, "Breast cancer Classification with Multi-Fusion Technique and Correlation Analysis" Fusion: Practice & Applications, Vol. 9, No. 2, 2023 ,PP. 48–61.
18. Paramasivam P., Gowthaman N., Srivastava V.M.,(2024),"Analytical Modeling of [001] Orientation in Silicon Trigate Rectangular Nanowire Using a Tight-Binding Model",Silicon,Vol.16,no.6,pp.2743–2756.https://doi.org/10.1007/s12633-024-02864-6
19. D. S. Dean et al., "Epidemiology of thyroid nodules," *Best Practice and Research Clinical Endocrinology and Metabolism*, vol. 22, no. 6, pp. 901–911, 2008.
20. M. Kunkulagunta, "Cloud Computing Applications for ERP Implementation, "International Journal of Computer Engineering and Technology (IJCET), vol. 15, no. 2, pp. 165–175, 2024, https://iaeme.com/Home/article_id/IJCET_15_02_020.
21. S. Stalin, V. Roy, P. K. Shukla, A. Zaguia, M. M. Khan, P. K. Shukla, A. Jain, "A Machine Learning-Based Big EEG Data Artifact Detection and Wavelet-Based Removal: An Empirical Approach", Mathematical Problems in Engineering, vol. 2021, Article ID 2942808, 11 pages, 2021. https://doi.org/10.1155/2021/2942808.
22. D. Bavkar et al. "Deep hybrid model with trained weights for multimodal sarcasm detection," Lecture Notes in Networks and Systems, pp. 179–194, 2023. [Online]. Available: https://doi.org/10.1007/978-981-99-5166-6_13.
23. V. Raghunath, "SAP Cloud Services on Decoding Cybersecurity and Data Privacy Controls," European Journal of Advances in Engineering and Technology, vol. 11, no. 9, pp. 95–102, 2024, https://ejaet.com/archive/volume-11-issue-9-2024/
24. P. K. Shukla, et al., "An Advanced EEG Motion Artifacts Eradication Algorithm", The Computer Journal, 2021;, bxab170, https://doi.org/10.1093/comjnl/bxab170.
25. V. Raghunath, "User Support Solution Implementation for SAP ERP Utilizing Artificial Intelligence to Drive Automated," International Journal of Artificial Intelligence Research and Development, 2024. https://iaeme-library.com/index.php/IJAIRD/article/view/IJAIRD_02_01_012.
26. Maruthamuthu R., Patel N., Yawanikha T., Jayasree S., Alsalami Z., Subbarao S.P.V.,(2024),"A Way to Design Fog Computing Model for 5G Network using Vanet",2024 4th International Conference on Advance Computing and Innovative Technologies in Engineering, ICACITE 2024,pp.431–435.https://doi.org/10.1109/ICACITE60783.2024.10617287.

Exploring the Prospectives of Vehicle Ad-Hoc Networks (VANETs): Applications, Challenges and Future Directions

Banu Priya Prathaban, D. Saveetha, N. J. Subashini, and R. Subash

Abstract Vehicle Ad-Hoc Networks (VANETs) is a variation of Mobile Ad-Hoc Network (MANET) facilitating for the automobiles and roadways communication. VANETs exploit wireless communication expertise of Dedicated Short-Range Communication and cellular networks for facilitating instantaneous data transfer between vehicles. VANETs are envisioned to enlarge road security, traffic efficacy, and deliver numerous solicitations and amenities to supplement the driving capability. VANETs negotiate the capacity to augment the dynamic flow of traffic, assisting in managing the traffic and perchance uniform progressions in obliging driving, wherever vehicles cooperate to support for a safety and efficient environment for vehicle transportation. This chapter inspects the assorted applications of VANETs, encompassing the smart transportation systems where it inhibits intelligence in it and along with management of traffic where it responds for any emergency conditions. Furthermore, safeguarding trustworthy network connection nevertheless of the dynamicity in transportation is considered to be a greatest practical challenge. This chapter delivers a wide-ranging investigation of VANET design and implementation, highlighting the encounters, solicitations, and future prospectives. In view of the future, our chapter sightsees evolving progresses and potential guidelines for the developing VANET technologies. All these clinch the combination of VANETs with leading-edge proficiencies. Additionally, innovative communication standards and diverse architectures available in VANETs are deliberated under future explorations.

B. P. Prathaban (✉) · D. Saveetha · N. J. Subashini
Department of Networking and Communications, School of Computing, SRM Institute of Science and Technology, SRM Nagar, Kattankulathur, Chennai, India
e-mail: banuprip2@srmist.edu.in

D. Saveetha
e-mail: saveethd@srmist.edu.in

N. J. Subashini
e-mail: sn8851@srmist.edu.in

R. Subash
Department of Computing Technologies, School of Computing, SRM Institute of Science and Technology, SRM Nagar, Kattankulathur, Chennai, India
e-mail: subashr@srmist.edu.in

J. Bhatia et al. (eds.), *Deep Learning Based Solutions for Vehicular Adhoc Networks*,
Studies in Computational Intelligence 1207,
https://doi.org/10.1007/978-981-96-5190-0_16

Keywords Applications · Challenges · Intelligent transportation systems · Mobile ad-hoc network · Traffic management · Vehicle ad-hoc networks

1 Introduction

Contemporary transportation structures are consumed by numerous overwhelmed constraints, which ranges between the usual traffic jamming and accidents due to the unskilled usage of this roadways arrangement. The prevailing traditional schemes commonly be influenced by the stationary arrangement and unified governor, leading to obstructions and interruptions, mainly in heavily occupied zones [1]. These schemes struggle to regulate to vigorous modifications in transportation configurations and ecological circumstances. For instance, traffic jamming litters, increased periods of transportability, results in air polluted environment [2]. Similarly, accidents prompted as a result of human faults or less consciousness during driving the vehicles can be one of the reasons for mortalities and damages. Inorder to discourse the above encounters, the idea of Intelligent Transportation Systems (ITS) began [3]. ITS impacts a mixture of communication, data handling, and controller mechanisms for the advancements in proficiency, security, and conservational influence of transport structures. These VANETs fundamentally produce a transportable system anywhere the automobiles are acting in place of nodules which helps for exchanging the information between the other units available in the roadside units (RSUs) that are fortified along with the communication strategies, this instantaneous data distribution allows for several long range solicitations deliberately increasing the merits of transportation systems [4].

1.1 Motivation

VANETs are fashioned via the philosophies of MANETs. Nevertheless, VANETs consume exceptional distinctiveness which discriminate these systems from outmoded MANET. Contrasting sluggish behaviour of movable mobile strategies of MANET, vehicles in VANETs practice recurring alterations in terms of rapidity along with direction, producing an extremely vibrant network settings [5]. Moreover, communication happening via VANETs are typically relied on short-range wireless proficiency similar to DSRC for less inactiveness and power consumption, as conflicting to the wider variety of communication utilized in certain MANET systems. To conclude, VANETs are especially used for safety applications alike collision prevention and emergency warning, whereas conventional MANETs emphasis on general data communication between the vehicular networks [6]. The progression of VANETs era kicked off from MANET which involves in acclimatizing the conventional protocols and expertises in order to report the obvious encounters and requirements of concurrent message transfer [7]. VANETs attempt designed for significant aids in contemporary transport arrangements. Improved protection is always a crucial benefit in transportation systems. Real-time data transmission via automobiles notify even several dangers, coincidences, and hazardous road environments, serving for several noteworthy declination in the number of incidents resulting in accidents and mortalities [8]. Augmented with traffic stream is a supplementary vital advantage. Automobiles can vigorously normalize the paths grounded on concurrent instantaneous traffic jamming data, adjusting the flow of traffic and reduction of traveling period [9]. The reduction in gaseous radiations are an expected effect of traffic overflow and augmented paths for traffic overflow situations. Lesser ingestion of fuel results in reduction of carbon gas emissions. Drivers gain immediate traffic acquaints, data regarding parking conveniences, and additional cooperative data needed for relaxed driving involvement [10].

1.2 Research Contribution

This chapter inspects the assorted applications of VANETs, encompassing the smart transportation systems where it inhibits intelligence in it and along with management of traffic where it responds for any emergency conditions. Nevertheless, extensive implementation of VANETs effects in sprints which concerns about the security, data privacy and possible management which demands for vigorous security practices, defending data privacy for confirming obscurity, preventing exploitation of user's data, instituting consistent communication procedures through varied companies, areas which is vital for unified complex maneuver. Furthermore, safeguarding trustworthy network connection nevertheless of the dynamicity in transportation is considered to be a greatest practical challenge. This chapter delivers a wide-ranging investigation of VANET design and implementation, highlighting the encounters, solicitations, and future prospectives. The development of VANETs is an ongoing process with uninterrupted research and development efforts aimed at overcoming challenges and unlocking the full potential of this technology in transforming transportation systems. The concept of VANETs appeared in the early 2000s as researchers reconnoitered the potential of wireless communication technologies to progress road safety and traffic management [11]. The development of Dedicated Short-Range Communication (DSRC) provided a critical foundation for VANETs, offering a standardized and low-latency communication protocol is definitely considered for vehicle-to-vehicle and vehicle-to-infrastructure communication. Table 1 shows the comparison of traditional cellular wireless network and VANET [12, 13].

Table 1 Comparison of traditional cellular wireless network and VANET [14]

Topics	Conventional network	VANET
Topology	Fixed	Dynamic
Infrastructure	Better	Petite
Safety and QoS	Improved	Deprived
Speed	Sluggishly	Rapidly
Cost	More	Less
Survival time	Extensive	Diminutive
Route selection	Laid-back	Hard
Network strength	Squat	Extraordinary
Relay	Wired	Wireless
Backbone network	High capacity	Low capacity

1.3 Organization of the Chapter

The subsequent sections will delve deeper into the communication technologies that underpin VANETs, the various communication types they allow, and the network architecture that facilitates information exchange. The chapter will then probe into the various applications that VANETs enable. Real-world examples will be discovered through case studies to demonstrate their practical use. Following this, a discussion will address the challenges that must be overcome for widespread adoption of VANETs. Finally, the chapter will turn its focus towards the future, examining emerging trends and potential research directions in this promising technology.

2 Related Works

Numerous communication expertises that reinforce VANETs are explored in this segment. VANETs depend on wireless communication knowledge to assist actual data interchange amongst automobiles and roadside arrangement. The subsequent sections investigates the explicit wireless communication expertises [15]. Wireless Communication Technologies manifest the foundation for the VANETs, enabling continuous message transfer between vehicles and road infrastructure. These technologies permit the automobiles to interchange life-threatening data such as circumstances of traffic, dangers in roads, and disaster warnings in real-time situations. The two main wireless communication skills frequently utilized in VANETs are DSRC and cellular networking systems [16]. DSRC is a consistent wireless communication protocol explicitly intended for the requirements of VANETs. Disparate outmoded Wi-Fi technology, emphasizing on higher communication rate, DSRC highlights lower latency, constant and reliable communication for shorter distance roadway transportation. This creates it perfect intended for secure and life-threatening utilizations in VANETs anywhere rapid and appropriate data interchange is important [17].

Cellular setups, like long term evolution and the developing fifth generation contribute in a compensating portion to DSRC in VANET systems. Although DSRC surpasses in terms of lower latency, shorter radius communication while cellular systems bid for broader range and higher bandwidth capacity [18]. LTE and 5G networks offer higher bandwidth, lower latency, and enhanced reliability, which becomes suitable for transporting multimedia information, updation necessary for software, and cloud established facilities to vehicular systems. Furthermore, cellular systems assist for hybrid data transfer designs, here automobiles flawlessly control concerning the DSRC and cellular affinity grounded on their locality and provisions for communique. Conversely, the issues like network blocking, exposure breaches, and budget are the concerns that are needed to be managed for delivering highly powered cellular systems in VANET [19]. Vehicular Communication systems in VANET describe the behavior and opportunity of communique amongst automobiles and arrangement in VANET. This comprehend several circumstances, counting to vehicle to vehicle, vehicle to infrastructure, and vehicle to everything vehicular systems [20].

Table 2 depicts the comparison of state-of-the-art schemes. Vehicle to Vehicle communication type empowers for uninterrupted transmission among adjacent automobiles, allowing for the data interchange regarding the vehicular rapidity, location, and route. This type of communication assists for the cooperative schemes, like supportive absorption, squading, and lane shifting, inorder to advance for the improved and safer stream of traffic in vehicular infrastructure system. Furthermore, such a communication system provisions for security solicitations such as collision evasion, crisis decelerating, and supportive consciousness, augmenting social perceptiveness, decreasing danger of accidents resulting in damages and fatality [21].

Vehicle to Infrastructure communication permits automobiles to connect with pavement arrangements like traffic light system, signs of roadways, and CCTV cameras. This type of communication offers automobiles with actual information regarding the flow of traffic, prioritizing traffic signals, and direction-finding amenities via vehicular infrastructure systems. Moreover, such a communication system provisions for structure established security solicitations like connection accident cautionary, pedestrian recognition, and crisis automobile anticipation, augmenting road protection and effectiveness.

Vehicle to Everything communication system denotes to the notion of communication between vehicles, striders who are walking on the roadways. For the purpose of building an all-inclusive communication environment facilitating for the effortless communication and collaboration amid participants. This type of communication incorporates Vehicle to Vehicle, Vehicle to Infrastructure and any other added communication varieties. Progressive solicitations comprising of multi-modal conveyance obliging sovereign driving and smart city assimilation are evoked via this type of communication, breaking in for extra operative, and viable, and secure transportation schemes.

Table 2 Comparison of state-of-the-art schemes

Author	Methodology	Inference	Limitations
Gayathri et al. [22]	Exploits RBM for cluster head assortment and a hyperelliptic curve-based signcryption structure for secure communication	Augments communication efficiency and reliability in VANETs while warranting security through trust-based clustering	Attains stable communication and moderates security risks by effectively managing cluster heads and members, addressing unique burdens of high-speed environments
Gillani et al. [23]	A comprehensive taxonomy and relative analysis of various data assortment protocols based on delivery ratio, efficiency, and recovery policy	Recognizes breaches in prevailing surveys and delivers a structured framework for data collection methods in VANETs	Compromises enquiry of current data collection protocols, highlighting the need for better application specific approaches
Hassan et al. [24]	Presents a cluster-based routing protocol utilizing ANN for secure routing and finding malicious nodes	Discourses black hole occurrences by adjusting routing paths although ensuring secure data transmission amongst vehicles	Delivers active routing solutions that improve security and efficiency in data distribution within VANETs, particularly under malicious conditions
Sandeep et al. [25]	Exploits an African Vulture Optimization procedure for clustering to progress network constancy and resource consumption	Reports disputes associated to high mobility and unpredictable network topologies, improving scalability and reliability in VANETs	Establishes amended cluster lifetime and load optimization over intelligent clustering methods, significantly promoting the overall performance of VANETs
Wang et al. [26]	Progresses an elliptic curve cryptography-based authentication protocol to augment secure data transmission through vehicular networks	Guarantees efficient identity authentication by preserving high security standards in vehicular communications	Attains better data transmission security over minimal overhead, suitable for real-time applications in VANETs
Nabil et al. [27]	Reviews game theory applications for intrusion detection and prediction of misbehavior in VANET environments	Highlights the advantages and limitations of game-theoretic approaches in ensuring security within VANETs, addressing various attack vectors effectively	Provides insights into the effectiveness of game theory as a tool for enhancing security measures against misbehavior in vehicular networks

(continued)

Table 2 (continued)

Author	Methodology	Inference	Limitations
Shuhaimi et al. [28]	Analyzes various radio propagation models to evaluate their performance based on packet loss, throughput, and end-to-end delay metrics	Identifies optimal radio propagation models that enhance communication efficiency within VANET applications, contributing to better network performance	Establishes benchmarks for radio models that can significantly reduce latency and improve overall communication quality in VANET scenarios
Quyoom et al. [29]	Conducts a comprehensive review on security challenges faced by VANETs, focusing on vulnerabilities and potential solutions	Delivers an extensive overview of the current state of research on security threats	Encapsulates key vulnerabilities within VANETs while proposing future research directions to enhance security frameworks
Azam et al. [30]	Inspects data privacy issues specific to autonomous systems within the context of GDPR regulations, converging on ITS	Deliberates GDPR impact the design and implementation of privacy-preserving mechanisms	Highlights the necessity for privacy frameworks that align with regulatory standards while ensuring effective operation within VANET environments
Sakiz et al. [31]	Reviews various security mechanisms applicable to ITS	Identifies key security mechanisms that can be integrated into ITS frameworks to bolster the safety and reliability	Proposes a set of recommendations for enhancing security measures tailored specifically for the complexities
Kumar et al. [32]	Recommends a innovative trust-based clustering procedure to increase the security and reliability in VANETs	Augments security by categorizing and mitigating malicious nodes, foremost to enhanced data transmission	Necessitates effective trust management and update procedures to adjust to dynamic network circumstances
Li et al. [33]	Advances a Blockchain-based context to confirm secure and apparent data sharing in VANETs	Delivers a tamper-proof and auditable record of data transactions, augmenting data integrity and privacy	Extraordinary computational overhead and potential scalability disputes in large-scale VANET deployments
Chen et al. [34]	Presents a machine learning-based approach for distinguishing and extenuating cyber-attacks in VANETs	Advances the accuracy and competence of attack uncovering, enabling timely reaction and moderation	Necessitates widespread training data and computational resources to train and set out machine learning prototypes
Zhang et al. [35]	Discovers the application of amalgamated learning for privacy-related collaborative learning in VANETs	Empowers collaborative learning deprived of allotment raw data, protecting user privacy	Necessitates robust communication infrastructure and well-organized coordination among vehicles to enable federated learning

3 Architectures of VANET

VANETs communication is permitted over a refined architecture that integrates numerous wireless communication technologies, sensors, and assets of computing for subsidiary and immediate real world data transfer. The principal tenacity of VANETs comprise improving road safety, civilizing traffic efficiency, and facilitating infotainment amenities to drivers plus passengers. The architecture of VANETs is intended to report the exclusive encounters impersonated by the extremely vigorous and mobility centered vehicular environments [36]. It comprises a variety of constituents, comprising onboard units (OBUs) in vehicles, RSUs, and backend systems, all of which work organized to produce a vigorous and consistent network. This network provisions a diversity of applications. The essential of VANETs is nothing but just a layered architecture that systematizes functionalities into diverse layers, each one accountable for obvious responsibilities. These layers comprise the following: physical layer, data link layer, network layer, transport layer, and application layer. Each layer assimilates numerous protocols and contrivances to pledge for the effectual communication, data integrity, and security [37]. Abundant communication expertises that strengthen VANETs are explored in this segment. VANETs hang on on wireless communication knowledge to support actual data interchange between automobiles and roadside arrangement. Wireless communication technologies apparent the basis for the VANETs, permitting continuous message transmission between vehicles and road infrastructure. These technologies allow the automobiles to transfer life-threatening records such as circumstances of traffic, dangers in roads, and disaster cautions in real-time situations. The two key wireless communication expertise recurrently exploited in VANETs are DSRC and cellular networking systems. DSRC is a reliable wireless communication protocol obviously envisioned for the requirements of VANETs. Incongruent outmoded Wi-Fi technology, highlighting on higher communication rate, DSRC highpoints lesser latency, persistent and consistent communication for smaller distance roadway transportation. This generates it perfect intended for secure and life-threatening exploitations in VANETs anywhere rapid and appropriate data interchange is important. Cellular setups, like long term progression and the developing fifth generation contribute in a recompensing portion to DSRC in VANET systems. Although DSRC surpasses in terms of lower latency, shorter radius communication while cellular systems bid for broader range and higher bandwidth capacity.

LTE and 5G networks bid advanced bandwidth, lesser latency, and improved reliability, which becomes appropriate for transporting multimedia information, updation necessary for software, and cloud recognized facilities to vehicular systems. Furthermore, cellular systems provision for hybrid data transfer designs, here automobiles spotlessly control regarding the DSRC and cellular affinity beached on their locality and provisions for communique. Contrariwise, the concerns like network blocking, exposure breaches, and budget are the disquiets that are desired to be accomplished for delivering extremely powered cellular systems in VANET.

Vehicular communication types in VANET label the performance and opportunity of communique amongst automobiles and procedure in VANET. This comprehend numerous environments, counting to vehicle to vehicle, vehicle to infrastructure, and vehicle to everything vehicular systems. Vehicle to Vehicle communication type empowers for uninterrupted transmission among adjacent automobiles, allowing for the data interchange regarding the vehicular rapidity, location, and route. This type of communication assists for the cooperative schemes, like supportive absorption, squading, and lane shifting, inorder to advance for the improved and safer stream of traffic in vehicular infrastructure system. Additionally, such a communication system provisions for security solicitations such as collision evasion, crisis decelerating, and supportive consciousness, augmenting social perceptiveness, decreasing danger of accidents resulting in damages and fatality.

Vehicle to Infrastructure communication permits automobiles to connect with pavement arrangements like traffic light system, signs of roadways, and CCTV cameras. This type of communication offers automobiles with actual information regarding the flow of traffic, prioritizing traffic signals, and direction-finding amenities via vehicular infrastructure systems. Moreover, such a communication system provisions for structure established security solicitations like connection accident cautionary, pedestrian recognition, and crisis automobile anticipation, augmenting road protection and effectiveness.

Vehicle to Everything communication system denotes to the notion of communication between vehicles, striders who are walking on the roadways. For the purpose of building an all-inclusive communication environment facilitating for the effortless communication and collaboration amid participants. This type of communication incorporates Vehicle to Vehicle, Vehicle to Infrastructure and any other added communication varieties. Progressive solicitations comprising of multi-modal conveyance obliging sovereign driving and smart city assimilation are evoked via this type of communication, breaking in for extra operative, and viable, and secure transportation scheme.

The network architecture of VANETs includes a decentralized network where vehicles and infrastructure nodes linked with each other respectively. This architecture provisions vigorous topology deviations due to the mobile nature of vehicles and is designed to manage the high-speed movement and recurrent network interruptions [24]. The layered structure of VANETs is grounded on the Open Systems Interconnection (OSI) exemplary, which systematizes network drives into distinct layers. Each layer has explicit responsibilities and cooperates with neighboring layers to offer widespread communication services. Physical layer levers the communication and response of raw data bits via several kind of wireless medium. This layer exploits technologies such as radio frequencies, which is usually the 5.9 GHz band for Dedicated Short-Range Communications (DSRC). Medium Access Control (MAC) layer copes access to be as a communication channel, warranting that several node transfer happens without any interference. Logical Link Control (LLC) delivers error detection and flow control mechanisms. Routing protocols controls the optimal paths for data packets to travel among the intermediate nodes. Common protocols comprises such as the most used Ad-Hoc On-Demand Distance Vector (AODV) and Geographic

Routing Protocol (GRP). Addressing and forwarding manages IP address tasks and packet advancing to the right destinations.

Transport layer assurances for the unwavering data transmission, control error correction and flow control, which comprises the protocols such as TCP/UDP. Safety applications includes collision avoidance systems, emergency braking alerts, and intersection management. Non-Safety applications encompasses infotainment, traffic information systems, and navigation assistance.

The protocol stack in VANETs consists of a set of protocols at each layer of the layered structure, enabling various network services. The stack includes: IEEE 802.11p: A standard specifically designed for vehicular communication, providing MAC and physical layer specifications for DSRC. IPv6: Ensures addressing and routing in a vast network of vehicles and infrastructure. Transport Protocols (TCP/UDP): Provides reliable and efficient data transfer. Application Protocols are tailored protocols for specific VANET applications like Cooperative Awareness Messages (CAM) and Decentralized Environmental Notification Messages (DENM). Figure 1 depicts the architecture of VANET.

The hardware components in VANETs comprise devices installed in vehicles and infrastructure that facilitate communication and data processing. OBUs are devices mounted in vehicles that handle communication with other vehicles and RSUs. They are equipped with wireless communication modules, processing units, and various sensors for enabling V2V and V2I communication, process data locally, and interact with vehicle control systems for applications like collision avoidance and navigation. RSUs are fixed units placed along roadways, intersections, or other strategic locations, facilitating V2I communication, deliver connectivity to backend networks, and assist in traffic management by relaying information between vehicles and traffic control centers. Vehicles and infrastructure nodes are armed with various sensors and actuators to gather and respond to environmental data.

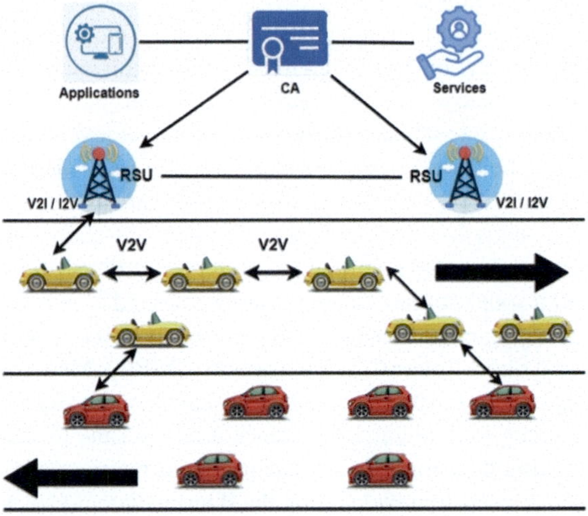

Fig. 1 Architecture of VANET

4 Applications of VANET

VANETs enable a multitude of applications intended to enhance road safety, advance traffic efficiency, and provide an improved driving experience. These applications can be broadly categorized into safety, traffic management, infotainment, and commercial services.

Safety applications are at the core of VANETs, targeting to avert accidents and decrease the severity of collisions by enabling vehicles to share real-time information about their status and the environment. Vehicles communicate their position, speed, and direction to nearby vehicles, allowing them to anticipate and avoid potential collisions. Emergency vehicles (e.g., ambulances, fire trucks) broadcast their location and intended route, enabling other vehicles to make way and reduce response times. Automobiles and its related organization will be able to identify and transmit the data regarding the endangered vulnerabilities and risks like path holes in roads, wreckages, traffic deportations, congestions and henceforth warning drivers to undertake safety measures while driving.

Managing traffic intend for optimizing the flow of traffic in roadways, moderate the traffic jam, and increase the total effectiveness of the vehicular transport system enormously. Instantaneous data regarding the traffic from vehicles is used to optimize the timing of traffic signals, reducing wait times and improving traffic flow. Automobiles acquire instantaneous data regarding the circumstances of vehicle traffic, it distributes substitute paths inorder to elude the traffic jamming, collisions. Here, vehicles obtain data like path holes in roads, wreckages, traffic deportations, congestions and henceforth warning drivers to undertake safety measures while driving and relieving overcrowding in urban regions.

Infotainment applications improve the involvement in driving via enabling entertaining, data, and affinity to riders. Automobiles should be able to provide passengers provided with network connection, permitting for browsing via webstream, streaming of audios/videos, and the usage of social networking systems. Passengers obtain latest data concerning the environments, meteorological conditions, inorder to proceed for deciding roughly regarding the traffic rerouting.

Commercial applications sway the VANETs to augment any kind of business ventures. Augmented with traffic stream is a supplementary vital advantage. Automobiles can vigorously normalize the paths grounded on concurrent instantaneous traffic jamming data, adjusting the flow of traffic and reduction of traveling period.

VANETs are staging as a one of the dynamic part in augmenting crisis intervention and emergency supervision via instigating appropriate and precise data regarding traffic environment. In the course of any catastrophic event, these VANETs rapidly distribute disaster posts to automobiles present in the exaggerated zones.

VANETs aids for the information regarding the vehicle travelling speed which in turn effects the fuel consumption. The increase in the fuel consumption affects the environment. Also the traffic flow is allowed to be known in the nearby regions via these applications. Autonomous automobiles are being utilized in VANETs for the

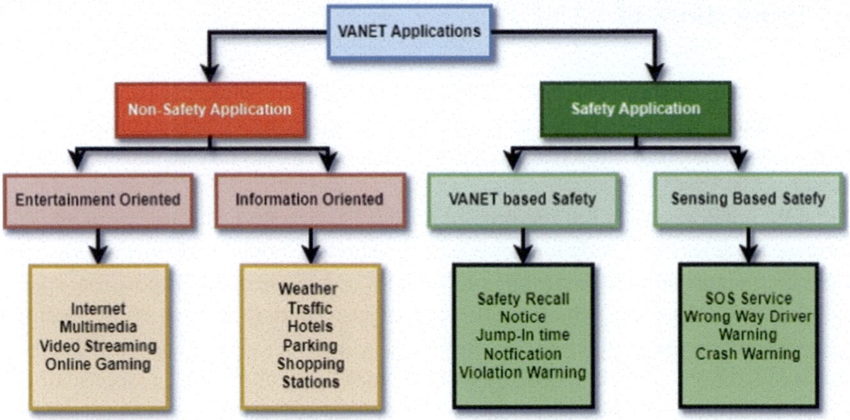

Fig. 2 Applications of VANET

communication between the vehicles via the provided infrastructure by augmenting the capacity to resolve any such multifaceted traffic settings securely.

VANETs contribute to preservation of environment via permitting improved observing and managing of vehicular discharges and added conservational elements for conserving the environment. Vehicles equipped with devices will be able to observe and disclose the vehicular emissions, serving for the identification of vital centers of pollutants and empowering for enhanced environmental administration. Due to this, drivers obtain instantaneous response on their diligent driving practices, promising for extra fuel effective and ecofriendly driving behavior. Figure 2 depicts the applications of VANET.

5 Challenges

The VANETs have great potential for enhancing traffic safety and efficiency. However, before they can be broadly deployed several challenges must be addressed. This section examines some main hindrances VANETs must overcome to understand their full potential. DSRC, a core communication technology in VANETs, has a limited communication range of typically a few hundred meters. This confines information exchange to nearby vehicles and roadside units, potentially hindering applications requiring broader coverage. VANETs face scalability challenges in high-density traffic scenarios. With a large number of vehicles vying for access to the communication channel, network congestion can occur, impacting message delivery reliability and potentially delaying critical safety information. Figure 3 depicts the basic challenges of VANETS.

VANETs increase security and privacy concerns. Malicious actors could potentially deed vulnerabilities to disrupt communication, inject false information, or

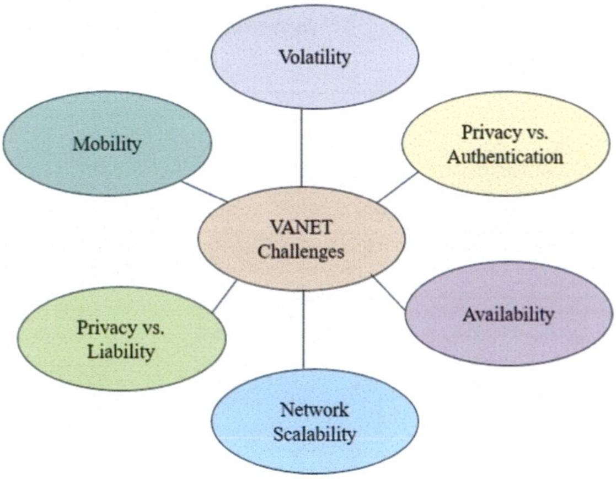

Fig. 3 General challenges of VANET

eavesdrop on sensitive data exchanged between vehicles. Robust security mechanisms and privacy-preserving techniques are critical to guarantee trust and user adoption. Figure 3 depicts the security challenges of VANET. Different manufacturers might implement proprietary communication protocols or exploit various DSRC message sets. This lack of interoperability can obstruct seamless communication between vehicles from different manufacturers, creating compatibility issues. Restricted Computational Resources: OBUs installed in vehicles may have limited processing power and battery life. This can restrict their ability to accomplish complex computations or handle rigorous communication loads. Widespread deployment of RSUs along roadsides is essential for comprehensive network coverage. However, the cost of installing and maintaining these units can be significant, posing a challenge for large-scale network deployment. The lack of a standardized infrastructure for VANETs can obstruct network development. This includes the need for consistent deployment strategies, spectrum allocation for DSRC communication, and clear guidelines for data management and ownership. Vehicular communication types in VANET label the performance and opportunity of communique amongst automobiles and procedure in VANET. This comprehend numerous environments, counting to vehicle to vehicle, vehicle to infrastructure, and vehicle to everything vehicular systems. Vehicle to Vehicle communication type empowers for uninterrupted transmission among adjacent automobiles, allowing for the data interchange regarding the vehicular rapidity, location, and route. Figure 4 depicts the security challenges of VANET.

Integrating VANETs with existing transportation infrastructure and traffic management systems can be complex. This needs collaboration between various stakeholders, comprising transportation authorities, network operators, and vehicle manufacturers. Encouraging widespread adoption of VANETs necessitates overcoming user concerns and providing incentives. This could include public awareness campaigns, importance the safety and efficiency benefits, or reconnoitering potential cost-sharing models for equipment installation. Legal and regulatory frameworks

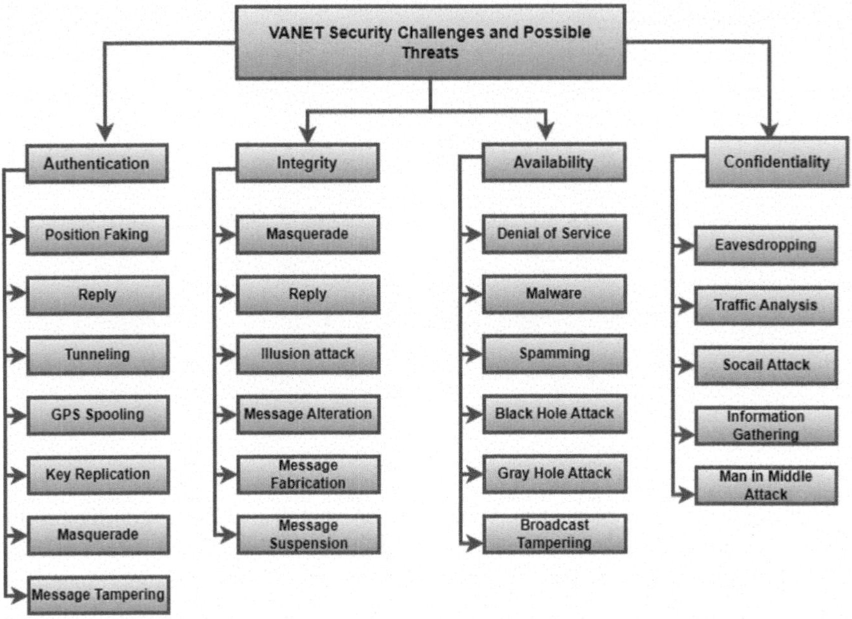

Fig. 4 Security challenges of VANET

need to be recognized to address issues like data privacy, liability in case of accidents involving VANET-based applications, and spectrum allocation for DSRC communication. VANETs are vulnerable to cyber-attacks that could disturb communication, manipulate data, or compromise vehicle control systems. Robust cybersecurity measures are vital to safeguard the safety and integrity of the network. Furthermore, safeguarding trustworthy network connection nevertheless of the dynamicity in transportation is considered to be a greatest practical challenge.

6 Future Developments and Research Guidelines

VANETs consume the possible improvements to modify the transport, and investigators are involved in finding the techniques to overcome the conventional limitations of VANET. Subsequently, few captivating forms and improvements which are serving for influential future trends of VANETs.

Assimilation with Cellular Vehicle to Everything technology. This technique empowers communication among automobiles, arrangement, and walkers. By integrating VANETs by means of this technology, investigators wish to custom an inclusive system for substituting instantaneous traffic flow evidences. This joint method be able to develop the security, efficacy, and eventually lay the foundations for autonomous automobiles. Exploiting Artificial Intelligence, Machine Learning and

Deep Learning, these cutting edge technologies are utilized to evaluate enormous size of information composed by VANETs, to prognosis traffic arrays and possible coincidences, to progress smart direction-finding procedures to augment the stream of traffic, improve safety via identifying and alleviating cyber intimidations.

Blockchain bids a secure and tamper-proof platform for data exchange in VANETs. This can authorize data integrity and avert malicious actors from manipulating information. Blockchain can also be used to achieve privacy concerns by allowing users control over their data. Cloud computing can provide the storage and processing power needed to handle the massive amount of data produced by VANETs. Fog computing can deliver processing closer to the network edge, enabling real-time decision-making for latency-critical applications. Beyond safety features, VANETs are being explored for cooperative adaptive cruise control: vehicles regulate speed automatically to maintain a safe distance, environmental monitoring for collecting real-time data on air quality and pollution levels. Personalized infotainment services: providing passengers with tailored recommendations based on location and traffic conditions. Figure 5 depicts the future directions of VANET.

Research will focus on developing mechanisms to ensure user privacy while still enabling data sharing for safety benefits. Incentive programs can encourage participation in VANETs by rewarding drivers who subsidies data or use safety features. Developing common standards will ensure seamless communication between vehicles and infrastructure from diverse manufacturers. Interoperability is critical for extensive adoption and maximizing the benefits of VANETs. Investigations withstand to progress vigorous secure protocols to defend contrary to intrusion and malware attacks resulting in ensuring the trustworthiness of the network by including the developments in discovery of imposition, verification, and ciphering procedures. Controlled field trials and pilot deployments are vital to test VANET technology in real-world conditions. This permits researchers to gather valuable data and refine the technology before large-scale adoption. As VANETs become more combined with autonomous vehicles, ethical considerations around data ownership, liability

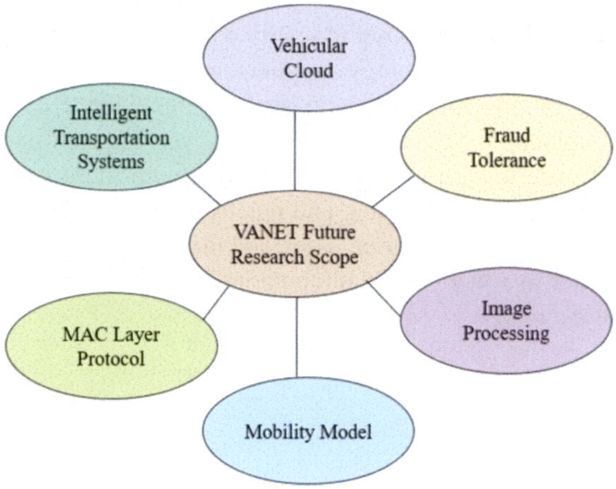

Fig. 5 Future directions of VANET

Table 3 Comparison of VANET's future directions

Future directions	Significant benefits
Integration with 5G and B5G networks	Low latency, high bandwidth, network slicing, edge computing
Enhanced security and privacy	Blockchain technology, advanced cryptography, privacy-preserving data analytics
Intelligent transportation systems (ITS)	Cooperative V2I communication, autonomous vehicle coordination, real-time traffic management
Internet of vehicles (IoV)	V2X communication, connected vehicle services
Machine learning and artificial intelligence	Predictive analytics, anomaly detection, adaptive routing

in accidents, and potential misuse of technology will need to be addressed. Table 3 shows the comparison of VANET's future directions.

7 Conclusion

VANETs offer a glimpse into a future where transportation is revolutionized. Through the use of cutting-edge communication technologies, these networks offer a novel approach to contemporary transportation by improving road safety, traffic efficiency, and the overall driving experience. ITS impacts a mixture of communication, data handling, and controller mechanisms for the advancements in proficiency, security, and conservational influence of transport structures. This chapter inspected the assorted applications of VANETs, encompassing the smart transportation systems where it inhibits intelligence in it and along with management of traffic where it responds for any emergency conditions. But for their widespread adoption and successful deployment, they must overcome standardization, security, and technological obstacles. The vision of a fully connected and intelligent transportation system is getting closer to reality with continued innovation and cooperation among automakers, technology suppliers, and regulatory agencies. Transportation network integration with Blockchain, artificial intelligence, cloud/fog computing, and C-V2X is opening the door to a more extensive, safe, and intelligent system. VANETs will advance toward widespread adoption with more study into new applications, user privacy, standardization, and practical testing. VANETs have the potential to be the foundation of a smarter, safer, and more effective transportation system as we move closer to a world with autonomous vehicles, which will ultimately benefit all users of the roads.

References

1. Hussein IH, Chehab A, Kayssi A (2017) SDN VANETs in 5G: an architecture for resilient security services. In: Proceedings of the fourth international conference on software defined systems (SDS), pp 67–74 (2017)
2. Dotcenko S, Vladyko A, Letenko I (2014) A fuzzy logic-based information security management for software-defined networks. In: 16th International conference on advanced communication technology, pp 167–171
3. Sindhwani M, Singh R, Sachdeva A, Singh Ch (2022) Improvisation of optimization technique and AODV routing protocol in VANET. Mater Today: Proc 49:3457–3461
4. Buinevich M, Fabrikantov P, Stolyarova E, Izrailov K, Vladyko A (2017) Software defined Internet of Things: cyber antifragility and vulnerability forecast. In: Proceedings of the IEEE 11th international conference on application of information and communication technology (AICT), pp 293–297
5. Ghassan S, Al-Salihy AH, Sures R (2010) Security issues and challenges of Vehicular Ad Hoc Networks (VANET). In: 4th International conference on new trends in information and service science
6. Mchergui A, Moulahib T, Zeadally Sh (2022) Survey on artificial intelligence (AI) techniques for vehicular ad-hoc networks (VANETs). Veh Commun 34:100403
7. Hasson ST, Hasan ZY (2017) Roads clustering approach in VANET models. In: Annual conference on new trends in information & communications technology applications (NTICT), pp 316–321
8. Mohammed SJ, Hasson ST (2022) Modeling and simulation of data dissemination in VANET based on a clustering approach. In: International conference on computer science and software engineering (CSASE), pp 54–59
9. Hasson ST, Abbas AT (2021) A clustering approach to model the data dissemination in VANETs. In: 1st Babylon international conference on information technology and science (BICITS), pp 337–342
10. Alsabah M, Trabelsi H, Jerbi W (2021) Survey on clustering in VANET networks. In: 18th International multi-conference on systems signals & devices (SSD), pp 493–502
11. Mustafa A, Hamdi M, Mahdi H, Abood M (2020) VANET: towards security issues review. In: IEEE 5th international symposium on telecommunication technologies (ISTT), pp 151–156
12. Hamdi M, Yussen Y, Mustafa A (2021) Integrity and authentications for service security in vehicular ad hoc networks (VANETs): A review. In: 3rd International congress on human-computer interaction optimization and robotic applications (HORA), pp 1–7
13. Sadkhan S, Jabbar D (2021) The security challenges with cognitive radio environments for VANETS. In: International conference on communication & information technology (ICICT), pp 167–173
14. Ali A, Phillips I, Yang H (2016) Evaluating VANET routing in urban environments. In: 39th International conference on telecommunications and signal processing (TSP), pp 60–63
15. Al-shareeda M, Alazzawi M, Anbar M, Manickam S, Al-Ani A (2021) A comprehensive survey on vehicular ad hoc networks (VANETs). In: International conference on advanced computer applications (ACA), pp 156–160
16. Saleh H, Hasson S (2019) A survey of routing algorithms in vehicular networks. In: International conference on advanced science and engineering (ICOASE), pp 159–164
17. Li Y (2012) An overview of the DSRC/WAVE technology. In: Quality, reliability, security and robustness in heterogeneous networks. Lecture Notes of the Institute for Computer Sciences, Social Informatics and Telecommunications Engineering, vol 74. Springer Berlin Heidelberg
18. Hasrouny H, Samhat AE, Bassil C, Laouiti A (2017) VANet security challenges and solutions: a survey. Veh Commun 7:7–20
19. Weber J, Neves M, Ferreto T (2021) VANET simulators: An updated review. J Braz Comput Soc 27:8
20. Mejri M, Ben-Othman J, Hamdi M (2014) Survey on VANET security challenges and possible cryptographic solutions. Veh Commun 1:53–66

21. Abdulkader Z, Abdullah A, Abdullah M, Zukarnain Z (2017) Vehicular ad hoc networks and security issues: survey. Mod Appl Sci 11(5)
22. Gayathri M, Gomathy C (2024) Design of CSKAS-VANET model for stable clustering and authentication scheme using RBMA and signcryption. Front Comput Sci. https://doi.org/10.3389/fcomp.2024.1384515
23. Gillani M, Niaz HA, Farooq MU, Ullah A (2022) Data collection protocols for VANETs: a survey. Comp Intell Syst 8(3):2593–2622. https://doi.org/10.1007/s40747-021-00629-x
24. Hassan M et al (2024) ANN-based intelligent secure routing protocol in vehicular ad hoc networks (VANETs) using enhanced AODV. Sensors 24(3):818. https://doi.org/10.3390/s24030818
25. Yerrathi S, Pakala V (2024) Enhancing network stability in VANETs using nature inspired algorithm for intelligent transportation system. PLoS ONE 19(1):e0296331–e0296331. https://doi.org/10.1371/journal.pone.0296331
26. Wang S et al (2024) A lightweight, efficient, and physically secure key agreement authentication protocol for vehicular networks. ProQuest, p 1418. https://doi.org/10.3390/electronics13081418
27. Nabil M, Hajam A, Boutkhoum O, Haqiq A (2023) Game theory application for misbehavior detection and prediction in VANET: review and challenges. Int J Comput Netw Appl 10(3):469–469. https://doi.org/10.22247/ijcna/2023/221903
28. Shuhaimi NI, Ashmadi NL, Abdullah E, Mohamad R, Mohamad SY (2021) Performance analysis of radio propagation models in VANET application. In: 2021 IEEE 11th IEEE symposium on computer applications & industrial electronics (ISCAIE), Penang, Malaysia, pp 372–377. https://doi.org/10.1109/ISCAIE51753.2021.9431810
29. Quyoom A, Mir AA, Sarwar A (2022) Security challenges of VANETs: a literature survey. J Multim Inf Syst 7(1):45–54. https://doi.org/10.33851/jmis.2020.7.1.45
30. Azam N, Michala AL, Ansari S, Truong NB (2023) Data privacy threat modelling for autonomous systems: a survey from the GDPR's perspective. IEEE Trans Big Data PP(99):1–27
31. Sakiz F, Sen S (2023) A survey of attacks and detection mechanisms on intelligent transportation systems: VANETs and IoV. Ad Hoc Netw 6:33–50
32. Kumar A et al (2023) Trust-based clustering for secure and reliable VANETs. IEEE Trans Veh Technol 72(11):11232–11245
33. Li B et al (2024) Blockchain-enabled secure and transparent data sharing in VANETs. IEEE Internet Things J 11(3):2547–2558
34. Chen C et al (2022) Machine learning-based cyberattack detection and mitigation in VANETs. IEEE Trans Inf Forensics Secur 17(4):1123–1136
35. Zhang D et al (2023) Federated learning for privacy-preserving collaborative learning in VANETs. IEEE Trans Intell Transp Syst 24(10):10342–10353
36. Sakib R (2010) Security issues in VANET. In: Department of Electronics and Communication Engineering, BRAC University Dhaka Bangladesh
37. Raya M, Aziz A, Hubaux J (2005) Efficient secure aggregation in VANETs. In: Proceedings of the 3rd international workshop on vehicular ad hoc networks (VANET '06), pp 67–75

The manufacturer's authorised representative in the EU is Springer
Nature Customer Service Centre GmbH, Europaplatz 3, 69115 Heidelberg,
Germany. If you have any concerns regarding our products, please
contact ProductSafety@springernature.com

Printed and bound by CPI Group (UK) Ltd, Croydon, CR0 4YY
01/10/2026
02225935-0013